Non-native Species and Their Role in the Environment

Non-native Species and Their Role in the Environment

The Need for a Broader Perspective

By

Radu Cornel Guiașu

BRILL

LEIDEN | BOSTON

Cover illustrations: All photographs taken by Doris Wong, except the photographs of the black locust tree, the gray squirrel, and the Norway maple, which were taken by Radu Guiașu, and the photograph of the mute swan, which was taken by Ileana Guiașu.

Library of Congress Cataloging-in-Publication Data

Names: Guiașu, Radu Cornel, author.
Title: Non-native species and their role in the environment : the need for a broader perspective / by Radu Cornel Guiașu.
Description: Boston : Brill, 2016. | Includes bibliographical references and index.
Identifiers: LCCN 2016020872 (print) | LCCN 2016028627 (ebook) | ISBN 9789004172661 (hardback : alk. paper) | ISBN 9789047426134 (E-book)
Subjects: LCSH: Introduced organisms--Environmental aspects.
Classification: LCC QH353 .G848 2016 (print) | LCC QH353 (ebook) | DDC 578.6/2--dc23
LC record available at https://lccn.loc.gov/2016020872

Typeface for the Latin, Greek, and Cyrillic scripts: "Brill". See and download: brill.com/brill-typeface.

ISBN 978-90-04-17266-1 (hardback)
ISBN 978-90-47-42613-4 (e-book)

This book is printed on acid-free paper and produced in a sustainable manner.

To my parents, Ileana and Silviu, and my partner Doris

Contents

Acknowledgements

First of all, I would like to thank my parents, Ileana and Silviu Guiașu, for always encouraging me to pursue my dreams and academic goals and providing the opportunities for all sorts of fascinating intellectual explorations and rewarding outdoor activities, including hiking through the woods, mountain climbing, and visits to natural history museums and zoological and botanical gardens, throughout my Romanian childhood. I have been fortunate to grow up in a household where so many wonderful books of all kinds were always readily available, and this stimulated my lifelong love of reading and writing.

Even though I pursued an academic career in a field very different from those of my parents – my father is a mathematician and my mother is a nuclear physicist – I always had their support and understanding. My mother, Ileana, also offered essential assistance with the library research for this book, and my father, Silviu, now a Professor Emeritus in the Mathematics and Statistics Department at York University, collaborated with me on several biodiversity studies in the last few years and assisted with the index and the two graphs found in the book.

My interest in the diversity of the living world started early, as I was collecting insects and plants soon after I learned how to walk – I have no idea if the many insects or plants I collected as a child were considered native or not, and I still do not think this was of particular importance. I received my first microscope while I was still in elementary school, and discovered the microscopic living world with fascination for myself, only about 300 years after Antoni van Leeuwenhoek had done so for the very first time.

My uncle, Dr. Mihai Bînă, was a surgeon with a passion for zoology, who showed me that it was possible for people to coexist harmoniously in an apartment with exotic fish, birds, and a dog, and gave me my first tropical aquarium fish, when I was ten years old. I have had aquaria in my home ever since and published my very first scientific studies in the field of ichthyology.

My long-time partner and fellow biologist, Doris Wong, showed unflinching support throughout this project, spent countless hours discussing many aspects of the work in progress with me, and asked many questions of the "How would you answer this criticism?" type, which led, hopefully, to a better book. Doris also drew one of the key illustrations in the book, took several of the photographs shown within these pages, and edited all the photographs prior to their inclusion in the manuscript.

Several professors have guided and enriched my university studies over the years. Among them, I would like to acknowledge, in particular, Michael

Boyer, Martin Lewis, and James Tait, for their assistance and advice during my undergraduate studies in biology at York University. My ideas about introduced species started to take shape during my graduate and postdoctoral studies in zoology, at the University of Toronto, in what was presumably Canada's largest department dedicated to biological studies, and one of the largest such departments in the world. My Ph.D. thesis supervisor, Professor David Dunham, allowed me the freedom to explore and pursue my ideas, while at the same time offering gentle guidance, sound advice, and many enjoyable conversations. Professor David Barr was partly responsible for inspiring my interest in the positive aspects of non-native species, as well as being very supportive of my research and career.

At the University of Toronto, I was fortunate to work with talented and resourceful research assistants and friends such as Julie Ray and Tuhin Giri, who have accompanied me in numerous field trips and helped me collect essential data.

My friend and colleague, Dr. Spencer Mukai, discussed aspects of the book with me and offered much needed encouragement.

Coping with the inevitable uncertainties of graduate and postdoctoral research and lengthy research projects can be difficult enough at the best of times, but the decision to take the long path from the beginning of the undergraduate studies in biology to an academic career was made easier by the support of my family and friends. Staying on this path eventually led to the completion of this work.

Some of the ideas and case studies discussed in a fair bit of detail in this book were first mentioned in a cover story I wrote in 2008 for the *Royal Ontario Museum Magazine*. The editor of the magazine, Lee-Anne Jack, provided good-natured support and valuable editorial suggestions prior to the publication of the article.

Maureen Zubowski, of the Royal Ontario Museum (ROM), generously made the voluminous ROM crayfish database available to me, and this was very helpful for my work on the distributions of the crayfishes of Ontario and the discussion of the concept of "native range".

I am thankful to the former chairs of my department, Michael Barutciski and Guillaume Bernardi, former Associate Principal Françoise Mougeon, and the former Principal of Glendon College, Kenneth McRoberts, for recommending and approving a temporary teaching load reduction to assist me with the completion of the book. Principal McRoberts, Professor Barutciski, and my colleagues in the Department of Multidisciplinary Studies, the Environmental and Health Studies Program, and the Biology Program provided encouragement and support, and contributed to the creation of a pleasant working environment, throughout my years at Glendon. The Glendon campus of York

University is located in a scenic wooded ravine, in the middle of Toronto, and these urban forests contain an interesting and constantly changing mix of both native and exotic species. During my frequent walks on campus, I was able to explore this vibrant novel ecosystem found just steps away from my university office, and I have developed some of the ideas found in this book during these explorations.

Many people familiar with this project and the research which led to it – including family members, friends, colleagues, and students – were kind enough to take the time to bring various relevant articles and examples to my attention, or send valuable reprints and photographs. The list of people who deserve my gratitude for this may be incomplete, but certainly should include, in alphabetical order: Matthew Chew, Amanda Colwill, Lynda Corkum, Mark Davis, Brenda Detfurth, Jane Duggan, Brie Edwards, Omid Fekri, Andrew Forbes, Matthew Greenlees, Ileana Guiașu, Silviu Guiașu, Brendon Larson, Adrian Leu, Jocelyn Martel, Paul Moore, Spencer Mukai, Nikolai Myhre, Matt Ronan, Richard Shine, Keith Somers, Michiel Thijssen, Jacklyn Volkhammer, Doris Wong, and Stan Yavno.

Former students Jane Duggan, Omid Fekri, Tia McNeil, and Matt Ronan conducted research projects related to introduced species under my supervision and presented co-authored papers based on this research at scientific meetings in the last few years, thereby contributing to current debates in invasion biology. My colleague Colin Coates was one of the organizers of the Toronto conference which included the session on invasive species mentioned in the last chapter of this book, and I am grateful to him for his encouragement and for introducing me to the field of Environmental history.

Professor Richard Shine, from the University of Sydney, was very kind to send me a photograph of the cane toad for inclusion in this book, and Professor Paul Moore, from Bowling Green State University in Ohio, generously made a photograph of the rusty crayfish available to me. I am grateful to Professor Emeritus Lynda Corkum, from the University of Windsor, for sending me photographs of the common carp and the tubenose goby, and to Dr. Stan Yavno, from Tel Aviv University, for the round goby pictures. Professor Andrew Forbes, from the University of Iowa, promptly sent a photograph of the apple maggot fly. My old friend Adrian Leu and his daughter Beatrice kindly agreed to send me photographs of a domesticated European rabbit – a valued family pet – taken at their home in England. The text was enhanced by photographs of several introduced species taken by Ileana Guiașu, Doris Wong, Matthew Greenlees of the University of Sydney, Lynda Corkum, Stan Yavno, Andrew Forbes, Paul Moore, and myself.

I am grateful to Patricia Zline, the Rights and Permissions Assistant for The Rowan & Littlefield Publishing Group, for the permission to use the book cover photograph of Bill Murray and the gopher from the movie *Caddyshack* for this work.

Professor Mark Davis – a leading expert in the field of invasion biology – and an anonymous reviewer provided detailed and very useful comments on the book manuscript during the peer-review process. I very much appreciated Professor Davis' positive comments and very helpful suggestions, which have allowed me to improve and update certain sections of the manuscript.

This book would not have been published without the wonderful support and expert advice received from Michiel Thijssen, the Acquisitions Editor for biology and history of science, medicine, scholarship, technology and the environment at Brill publishers, as well as the essential assistance received from Assistant Editors Wendel Scholma and Sabine Steenbeek during various phases in the completion of the manuscript. I wish to thank Michiel Thijssen, in particular, for his patience, good humour, and guidance during the years when the book was being written, and Wendel Scholma (Assistant Editor for biology and history of science) and Ellen Girmscheid (Team Leader CDE Books) for supervising the proceedings and offering timely and very helpful advice and practical help throughout the complex process of preparing the book for publication.

Introduction

> To protect the nature that is all around us, we must think long and hard about the nature we carry inside our heads.
>
> WILLIAM CRONON

> Thus science must begin with myths, and with the criticism of myths.
>
> KARL POPPER

As I was reading the paper one Sunday morning, a rather alarming front-page headline caught my eye: *Attack of the killer worms – foreign invaders threaten saplings and trilliums*. Upon seeing this, I immediately thought of the entertaining movie *Tremors*, which depicts the deadly threat posed by some enormous human-devouring worm-like creatures (of a previously unknown species) to an isolated small community in the United States. For a brief moment, I couldn't help wondering if my hiking days through the local woods may have to come to an end. However, reading further, I discovered that the article was actually referring to some inconspicuous, and quite peaceful, non-native earthworm species introduced into the Canadian province of Ontario hundreds of years ago whose only apparent crime is eating leaf litter. The article was somewhat vague about the exact nature of the threat, but appeared to suggest that these earthworms may have a negative impact on the soil. However, in the fourth paragraph, the author confessed that no formal studies on this had been conducted in Ontario. Mere "anecdotal reports" of the earthworm invasion were circulating (Scrivener, 2007).

Introduced earthworms – whose aeration of the soil and recycling of nutrients are generally good for plant life – are not the only non-native species to receive negative press coverage lately. In recent years, many of the general nature articles I read have mentioned one threat or another from an ever-growing number of "invasive" species said to be poised to take over our water, soil, and forests, destroying everything in their path. The simple fact that a species may have originated elsewhere – even if it arrived in North America centuries ago and has had no discernible, scientifically proven negative impact on its new surroundings – seems to be enough to generate this type of bad publicity. Even worse, often enough, the proposed solution for getting rid of these non-native species is a controversial control program. These programs, which can be gruesome, expensive and unnecessary, can sometimes create problems for native species and the environment as well.

 | DOI 10.1163/9789047426134_002

Such news articles have not arisen out of nowhere. They mirror a trend started by some ecologists and conservation biologists in the last few decades. This trend, which seems to have intensified in the last few years, is called invasion biology, a new branch of the field of conservation biology, or a subfield of ecology, depending on how one wishes to classify and connect various related biological disciplines. Through my teaching of conservation biology, ecology, and environmental science courses at York University's Glendon College and the field research I have conducted primarily in the area of freshwater ecology for the past two and a half decades or so, I have become very much aware of the growing popularity of invasion biology.

Rapid growth in the young field of conservation biology since the 1980s has generally been a positive development, overall. It has led to increased public awareness of human threats to the natural world, and enhanced efforts to minimize our negative impact and reverse some of the damage we have done to the environment and several of our fellow species. Many good and dedicated researchers are working hard, sometimes in difficult conditions, to learn more about a variety of species and ecosystems, and the valuable knowledge gained through such vital research is essential for protecting countless endangered animals and plants and preserving many vulnerable natural environments. In many ways, this essential work is just beginning, since we still know very little about most of the approximately 1.8 million living species that have been identified to date, and perhaps even less about the complex ways in which the vast majority of these species interact in, and contribute to, the ecosystems they are a part of.

The publication of the book *The ecology of invasions by animals and plants* in 1958, by British ecologist Charles Elton (1900–1991) – one of the founders of animal ecology – is widely acknowledged as the starting point for the modern field (or subfield, or subdiscipline, depending on which authors one relies on for the exact label) of invasion biology (Ricciardi and MacIsaac, 2008; Richardson and Pyšek, 2008; Richardson, 2011). As pointed out in the good reviews by Cadotte (2006) and Davis (2006 and 2009), non-native species are mentioned in the writings of several naturalists and biologists well before the publication of Elton's book. The first such written accounts can be traced as far back as the 1700s, and several prominent figures in the history of the life sciences wrote about introduced species during the 19th century and the first half of the 20th century. For example, Charles Darwin mentioned introduced species on several occasions, and specifically discussed "the naturalisation of plants through man's agency in foreign lands" in the chapter on natural selection, within his famous book *On the origin of species*, published in 1859. So, Elton was certainly not the first to think about non-native species, but he is, nevertheless, generally considered to be "the father" of invasion biology, and some of the ideas he incorporated in his rather slender 1958 volume on invasive species have remained very influential in this field, for better or worse (Davis et al., 2001; Ricciardi and MacIsaac, 2008; Richardson and Pyšek, 2008; Davis, 2009). In fact, on July 17th, 2012, I conducted an online search using Google Scholar and found that Elton's *The ecology of invasions by animals and plants* had been cited in 3,920 other publications – more than twice the number of citations received by the landmark *Animal ecology* book by the same author. A similar search conducted on January 19th, 2015 (also using Google Scholar) showed that the number of citations for Elton's book on invasive species had jumped to 5,066, indicating a sustained high interest in this work, as well as the general topic of invasion biology.

It should be noted, however, that the sharp rise in the number of publications related to invasion biology did not take place until the early 1990s, more than three decades after Elton's book on invasive species was published. In fact, the specialized flagship journal *Biological Invasions* did not appear until 1999. Thus, the impact of Elton's book on the work and research interests of other ecologists was clearly not immediate, and the possible reasons for this are still being debated (Ricciardi and MacIsaac, 2008; Davis, 2009; Simberloff, 2011). So, in reality, although a few publications referred to introduced species before, for all intents and purposes, the modern field of invasion biology is only about 25 years old, at most, since virtually all the studies conducted in this field were undertaken within the last two and a half decades. As this discipline gathers momentum, and new followers, I think it is important to keep these very recent intellectual and conceptual origins (and databases) in mind, since 25 years is a very short time indeed – a 25-year-old field is really still in the early stages of infancy – especially for making accurate and confident predictions about the long-term impact of species in their new environments, and following up such predictions with thorough, long-term conclusive field studies.

I have to admit that my own interest in invasion biology was not initially sparked by Elton's *The ecology of invasions by animals and plants*, since I did not read this book until a few years after I had already familiarized myself with many other, much more recent, specialized articles and books in this field. Thus, in my case, at least, the exposure to the literature on invasive species did not start with Elton's words, although later on I went back to Elton's book to get a sense of perspective about the early history of invasion biology. So, Elton may have started this field, at least according to some, but I suspect that for many present day ecologists, the interest in invasion biology probably did not begin with Elton, even though we all cite Elton's book in our publications.

My own personal recollections tend to match the results of the formal literature surveys of this field. For instance, in the early to mid-1980s, when I was an undergraduate student in biology at York University, in Toronto, my department did not offer, as I recall, any specific conservation biology courses, and I don't remember any discussions related to invasive species at all in any of my ecology classes (even though I stayed awake in all of them and my memory is still intact). This was neither an oversight on the part of the departmental leadership, nor an indication that people did not care as much about the well-being of ecosystems and species, back then. In those days, conservation biology was just beginning to emerge as a separate discipline, and the sharp division of the species of the world into native and non-native categories (or *our* species and the alien ones) was not yet in fashion. As a graduate student at the University of Toronto, I started to hear a bit more about introduced species, partly because the Zoology department (as it was known at the time) was very large, and some of the many ecologists teaching there did mention invasive species a few times, here and there. However, in my own personal experience as a biologist, the current emphasis on invasive species and invasion biology really started in earnest at some point in the 1990s, and this very young field seems to have been expanding greatly, and also experiencing some, perhaps inevitable, growing pains, ever since. This book is intended to, among other things, help the maturation process along, if possible. In my view, there is a genuine need for a broader, more inclusive, objective, analytical, fact-based, tolerant and open-minded perspective on introduced species, and, although different opinions and voices are being heard with increasing frequency these days in invasion biology, even more vigorous and intellectually rigorous debates on key issues in this field should be both healthy and necessary.

Although many of the assumptions and claims made by some invasion biologists will be analyzed and sometimes questioned or challenged in the pages of this book, my purpose is not to deny that invasive species can sometimes cause problems. We might as well get that out of the way right now. Indeed, sensationalism aside, some invasive species can, under certain circumstances, threaten human health, undermine economic interests, and sometimes add to the risks faced by vulnerable local species. For example, the recent arrival in Ontario of the notorious West Nile virus – which is known to infect and kill humans, as well as other mammals and many types of birds – is decidedly unwelcome, of course. Until this virus appeared in Ontario, a few years ago, mosquitoes used to be merely a nuisance in this part of the world. Now, each mosquito bite comes with a relatively small but nevertheless genuine risk of serious illness. Since I have been bitten by mosquitoes hundreds of times (a very conservative estimate), especially during my research in areas such as

Ontario's Algonquin Provincial Park, the West Nile virus has been added to my already lengthy list of fieldwork-related concerns, together with hungry black bears, powerful rapids and dangerous undertow in some of the rivers where we collected data and specimens, major storms, broken down vehicles, finding adequate funding, and so on.

The problems (real or perceived) caused by some exotic species have been documented in great detail in a large number of recent general and specialized scientific articles and books, and the literature focusing primarily, or exclusively, on the negative aspects associated with the introductions of species to areas outside their so-called native ranges is already voluminous and rapidly growing. Many of these primary studies and more general articles are cited in this book.

On the other hand, although this may not be advertised perhaps quite as much as the negative effects of some invasive species, there seems to be a general understanding by most ecologists that the vast majority of introduced species do not cause problems in their new surroundings. As Williamson (1996) put it: "But most successful, i.e. established, invaders are not pests, and most invaders are not successful." This is often briefly acknowledged, in some form, in conservation biology textbooks, at the beginning of a section or chapter on invasive species, but then a much more detailed description of the negative impacts of a relatively small minority of exotic species usually follows (this is the case, for example, in Primack, 2010).

In fact, most non-native species and populations do not survive for long in their new environments, because the conditions are unsuitable for them, and among the relatively few surviving exotics, many remain in relatively small areas and have a negligible impact. The Eurasian tree sparrow (*Passer montanus*), for example, was introduced to North America more than one hundred years ago, and this species is still restricted only to a small region near the city of St. Louis, in Missouri (Lomolino et al., 2010).

At the same time, it should be recognized that native species can also have negative impacts on each other. In fact, a major review study found that introduced species were less of a threat for endangered species in Canada than interactions among the native species themselves (Venter et al., 2006). This study will be discussed in more detail later in the book. Native species can cause their own share of problems for people, as well. In other words, human health problems and damage to some of our economic interests, as well as certain potential threats to vulnerable species, are not caused exclusively by non-native species, even though the emergence of invasion biology may sometimes obscure these facts. For example, as Davis et al. (2011) recently pointed out, the mountain pine beetle, *Dendroctonus ponderosae*, is a native species which is probably responsible for killing more trees than any other insects in North America. Therefore, this native North American species currently poses a greater threat to the forests of this continent than introduced species, such as the much maligned Asian long-horned beetle. This information is not meant to denigrate or demonize native species, of course, but simply to show that, at a particular point in time, at a given location, and under certain environmental conditions, some species will be more successful than others. Occasionally, the spectacular success of certain species may temporarily interfere (or be perceived to interfere) with human interests, and this may happen regardless of whether we consider such species to be native or introduced. Thus, when it comes to the ecological and economic impacts of various species, the distinction between native and non-native species may often be irrelevant, from a practical standpoint.

Furthermore, introduced species can also have positive effects on the environment and the other species they may interact with. Such positive news about exotic species are usually not mentioned, and little studied, if at all, by most invasion biologists, since the very raison d'être of their field, overall, is to document the negative impact of invasive species and create or enhance a general awareness of the threats posed by such presumably irredeemably nasty "invaders". The field of invasion biology tends to single out introduced species as the main, or even the only, culprits responsible for many current environmental problems, but the reality is often much more complex and far less clear.

Therefore, since there are already plenty of sources and authors presenting and emphasizing the negative side of exotic species, in this book I will be going in a different direction, in an attempt to broaden the discussion about this important topic, stimulate new debates, and perhaps add some balance to what has been so far a fairly one-sided discussion about the recent movements of species around the world, possibly with purposeful or inadvertent human assistance. So, this book is not meant to be a comprehensive exploration of everything currently known about introduced species. Such a task would be nearly impossible, in any case, for any one author, in any one book, and would probably require a huge, 1,000-page plus volume, or series of volumes, with contributions from numerous experts. The many specialized, as well as the more general, references cited in this book should offer additional valuable sources of information for readers interested in examining particular topics or points of view in more detail. The current book makes a particular case and reflects my own opinions with respect to certain aspects of invasion biology, as I perceive this field. The book is the result of several years of doing research in various fields of biology (including ecology, evolutionary biology, conservation biology, and animal behaviour), reading widely, and thinking carefully about the distributions, interactions, and impact of species in new environments. It is my sincere hope that the book will contribute to the current discussions in invasion biology, and other fields, about the way we treat, study, and perceive our fellow species, and about the types of assumptions underlying our research, and the nature of the goals we hope to achieve through our scientific work.

CHAPTER 1

The Troublesome Matter of Subjective Definitions

> The most erroneous stories are those we think we know best – and therefore never scrutinize or question.
>
> STEPHEN JAY GOULD

The young field of invasion biology is essentially based, first and foremost, on the shaky assumption that we have very clear-cut, comprehensive, rigorous, and logically consistent ways of differentiating native species from non-native ones, and on the notion that the definitions for terms such as native, introduced, and invasive species are universal and based on unambiguous and objective scientific criteria and facts. After reading some of the key articles and books published in the areas of conservation and invasion biology, members of the general public interested in such issues may be left with the impression that ecologists always know exactly how to separate native from introduced (or exotic) species, and that the validity of these terms is obvious and never in doubt. Such an impression is also created and reinforced by some rather alarmist and simplistic articles or news items appearing in, or distributed by, various types of media. These articles are often based on information provided by certain invasion biologists, eager to get their anti-introduced species message across, usually in very categorical and sometimes nearly apocalyptic terms.

The reality, however, is considerably more complex. To begin with, it is often difficult to ascertain whether a particular species is native to a given area or not. In many cases, we just don't know exactly where a species originated and precisely when or how it reached a certain region. It is virtually impossible to establish the exact boundaries of a so-called "native range". In some cases, even the native ranges of species widely regarded as invasive, such as the yellow crazy ant (*Anoplolepis gracilipes*), are completely unknown (Cooling and Hoffmann, 2015), which makes it impossible to recreate the entire dispersal history of such "invaders", and raises the possibility that these species may be considered non-native even in the areas where they had originated.

As an additional example, the water lettuce (*Pistia stratiotes*) is an aquatic plant listed as invasive in a number of jurisdictions. However, the exact part of the world where this species originated is not known. Apparently, there are a variety of expert opinions, or educated guesses, on this subject, and regions as different and far apart as the south-eastern United States, Africa, southern Asia, and South America have been proposed as the possible centers of

 | DOI 10.1163/9789047426134_003

origin for this plant. Interestingly, the water lettuce is now banned in states such as Louisiana, Florida, Mississippi, and Texas, even though this species may actually be native to this region (Booy et al., 2015; www.invadingspecies.com/invaders/plants-aquatic/water-lettuce/, 2015).

In other cases, the distinction between the native and non-native ranges of certain "invasive" or "alien" species appears to be quite puzzling. On the IUCN's (the World Conservation Union) 100 of the World's Worst Invasive Alien Species web site, which contains information gathered by the Invasive Species Specialist Group (ISSG), the common carp (*Cyprinus carpio*) is listed as "alien" in Romania, but "native" in neighbouring countries such as Bulgaria, Hungary, and Moldova. In addition, this fish species is also listed as "native" in the Danube River (www.issg.org/worst100_species.html, 2014 and 2016). This river separates southern Romania from northern Bulgaria, and thus represents most of the border between the two countries. How is it possible for the common carp to be "native" to Bulgaria and the Danube River, but "alien" in Romania? Why are these fish considered "native" only as long as they stay on the Bulgarian side of the Danube, and why do they suddenly become "alien" if they cross over to the Romanian side of the same river? Furthermore, on the same web site, we can find the following information about the native range of the common carp in the Global Invasive Species database: "Common carp originated in Europe in rivers around the Black Sea and the Aegean basin, especially the Danube" and "Prior to human influence the common carp was found in the Black, Caspian and Aral Sea drainages, east into Siberia and China and west as far as the Danube River." So, again, given this information, and the fact that Romania includes several major rivers within the Black Sea drainage, as well as a part of the Black Sea coastline, why is the carp listed as an "alien" species in this country, according to the ISSG database?

At the same time, the various definitions for the terms "native" and "non-native" as they are applied to biological species can be more or less arbitrary, and many different and partially contradictory definitions for these terms have been used in the relevant scientific literature. The variety, arbitrary nature, and inconsistent use of such essential definitions in this field have been noticed and discussed by several authors, including Brown (1997), Schwartz (1997), Kendle and Rose (2000), Shrader-Frechette (2001), Colautti and MacIsaac (2004), Falk-Petersen et al. (2006), Guiașu (2007a, 2008a, and 2009), and Humair et al. (2014).

For example, some authors consider introduced or exotic species to be those that have arrived in a particular area since the beginning of recorded human history, and native or indigenous species those that have either evolved in the region or moved there during prehistoric times (Myers and Bazely, 2003). Schlaepfer et al. (2011) defined non-native species as those "that occur outside of their historic range", suggesting – perhaps inadvertently – that information about the prehistoric ranges of species may not be relevant or important. Furthermore, this definition used by Schlaepfer et al. (2011) does not take into account the reason or reasons why a species may occur outside its so-called "historic range" – for example, the species may have expanded its range on its own, and many range expansions and contractions may go undetected for long periods of time. In fact, as we shall see later in the book, we often don't have much information about either the "historic" or the "prehistoric" ranges of various species. And we should expect the distributions of species to change all the time, of course.

In other studies, species are classified according to their presence or absence in a particular region at an arbitrary point in time. The year 1500, for example, is often used in Europe to assess the status of various plant species (Myers and Bazely, 2003; Pyšek et al., 2003). In his detailed analysis of various categories of so-called "man-accompanying" or synanthropic plants found in central Europe, Kornás (1990) introduced a rather complex system of classification for these species. Kornás (1990) made it clear that the synanthropic plants include both native and "alien" species. He then further subdivided the permanently established alien synanthropic plant species of central Europe into two main categories: the older immigrants (or archaeophytes), which have presumably arrived before the year 1500, and the so-called "newcomers" (or kenophytes, or neophytes), which may have arrived after 1500. The year 1500 is chosen as the boundary between these two categories of plants, because, according to Kornás (1990), this was the beginning of the Modern Era, when the first "New World weeds" appeared in Europe, as a result of the European discovery of America. One could argue that Columbus actually reached America for the first time in 1492, not 1500, and we may be left wondering why the year 1500 would be used as a demarcation line for plants expanding their ranges from one region of the so-called Old World to another, or, in general, for plants not known to have arrived in Europe from the New World, for example. We may also wonder if the term "newcomer" can really be applied to a species which has lived in Europe for about 500 years or so. In other words, for how long does a species have to exist at a certain location, before we can accept it as a "natural" part of the local ecosystems?

Such questions highlight, again, the rather subjective and arbitrary nature of all such boundaries and definitions. In his discussion, Kornás (1990) also stated that the beginning of the expansion of the ranges of alien species into central Europe was clearly connected to the arrival of agriculture for the first time in this region, thousands of years ago. This is another questionable boundary which will be analyzed further below.

Prehistoric Perspective

In a much cited discussion of the definition of the term "native" as it should presumably be applied to species in the UK and Ireland, Webb (1985) stated that native species in these countries are those which had either evolved in this region, or arrived before the start of the Neolithic period, or since the Neolithic in a way completely independent of human activities. The Neolithic, or the last part of the Stone Age, is a period normally associated with the beginning of agriculture, when people first started domesticating certain animals and plants (Renfrew and Bahn, 1997; Jurmain et al., 2004; Hart-Davis, 2007). Since the Neolithic period started at different times in different parts of the world, it would be difficult to apply Webb's definition universally to other regions of the planet. In fact, in some parts of the world, certain human populations maintained a largely hunting and gathering lifestyle until modern times, or even until the present day. Even within Europe and nearby regions, the Neolithic began at very different times in various locations. The earliest development of agriculture may have taken place in the Middle East, perhaps about 12,000 years ago, but the transition from a nomadic hunting and gathering lifestyle to a largely sedentary, agricultural way of life did not take place until about 7,500 years ago in central Europe, and approximately 6,500 years ago, or even more recently, in north-western Europe. In other words, the Neolithic started several thousands of years later in north-western Europe in comparison to the south-eastern regions of the same continent. Different dates for the start of the Neolithic are given for other regions, on different continents (Hart-Davis, 2007). Thus, generalizing Webb's definition to other parts of the world would mean that, for example, a species which reached the Middle East about 9,000 years ago would be considered as non-native (unless we would somehow know with certainty that the arrival of the species in that region had absolutely nothing to do with human activities, which may be very difficult, if not impossible, to determine), whereas if the same species arrived in central or western Europe 9,000 or 8,000 years ago, the species would be regarded as native (even if it might have been brought there by hunters and gatherers).

Brown (1997) considered the definition of the term "native", as it is currently applied to British trees and woodlands, to be "arbitrary and ecologically flawed", and questioned the very usefulness of such a definition. By assuming that the only "native" tree species in Britain are those that arrived there between the end of the last glacial period, about 11,000 years ago, and the start of the Neolithic human civilization in this region, perhaps 6,500 or 6,000 years ago, some invasion biologists are greatly restricting the period of time during which species could qualify as indigenous in this part of the world. Such severe restrictions leave out many species, including those belonging to the tree genera *Tsuga* and *Pterocarya*, which are now classified as "exotics", even though they had been present in Britain before the last Ice Age, during previous interglacial periods, but presumably did not return on their own to the area after the ice sheets retreated (Brown, 1997). The plant species *Rhododendron ponticum*, which is now considered as an "alien weed" in the UK, and is the target of eradication programs, may also have been native to the British Isles before the last glacial period (Webb, 1985; Kendle and Rose, 2000). As Kendle and Rose (2000), who did not endorse this opinion, pointed out, some botanists take the view that no new species can ever be added to the list of British native plants, since all newly discovered species are likely to have arrived with some sort of assistance from people. In my experience, this seems to be the view of certain invasion biologists working in other parts of the world as well. In other words, according to this perspective, we are essentially stuck with the native species we now have, and we can never legitimately gain any new native species. We can only lose them. So, presumably, we should "circle the wagons", stand on guard, adopt a perpetual siege mentality, and protect the currently recognized native species at all costs, while fiercely fighting by any necessary means against all other plants and animals that may come into our respective jurisdictions. After all, in this human-dominated world, how can we ever be sure that a species has expanded its range without our help? And if any species previously believed to be native are discovered to have been introduced, somehow, then we should turn on them as well.

However, it may be worth mentioning that for many (and probably most, if not all) locations around the world we simply do not have comprehensive lists of species which may have been present there before any signs of "significant" human activities. The first human colonizers in various regions were not professional botanists or zoologists with expertise in taxonomy, and they did not keep detailed records of all the species they first encountered upon reaching new lands, or lists of all the species they may have introduced into these regions. Thus, as Brown (1997) correctly pointed out, the assumption that we can now reliably separate the effects of human activities from so-called "natural" influences or factors when we are trying to assign native or exotic status to various species is at best naïve, and conservation actions based on this assumption can be dogmatic and counterproductive.

In his influential definition of native species, Webb (1985) assumed that people did not have a significant impact on the distributions of species before the start of agriculture. This assumption, however, is almost certainly wrong, and virtually impossible to test. Our species, *Homo sapiens*, is approximately 200,000 years old (the oldest *H. sapiens* fossils are 195,000 and 160,000 years old, respectively, and have been found in Ethiopia), and even in the regions of the world with the earliest evidence of agricultural traditions and activities, only the last 5% or so of our history as a separate species has been spent in a sedentary lifestyle based on agriculture. The very nature of the nomadic life of hunters and gatherers strongly suggests that people have been important agents for the dispersal of various species, either on purpose or inadvertently, well before the start of the Neolithic.

Using the beginning of recorded human history, or the end of the Neolithic, or any other period in human history, as the chronological boundary between "native" and "introduced" species at a particular location is obviously arbitrary and logically inconsistent. Prehistoric groups of hunters and gatherers have had a major impact on the environments they first colonized in a variety of ways – for instance, by dramatically changing the landscapes through deforestation or the use of fire, and by contributing to the extinction of many species of animals through intensive hunting. The arrival of the first prehistoric, or indigenous, human colonizers in various parts of the world has been associated with the extinctions of a multitude of species. Some of these species were undoubtedly keystone species, meaning that they had a crucial role in the survival of several other species in those ecosystems, and a big impact on the organization of the biological communities they were a vital part of.

For example, until the first arrivals of prehistoric humans on these continents, North America was home to several large herbivores, such as various species of mammoths, mastodons, camels, horses, and giant ground sloths, as well as large predators like sabre-tooth cats, lions, cheetahs, giant short-faced bears and dire wolves, while Australia had a variety of giant marsupial species including the herbivorous rhino-size diprotodonts, giant short-faced kangaroos, and carnivores such as the marsupial lion. All of these species, and many others, became extinct following the arrival of prehistoric human groups on these continents (Martin, 1984; Stuart, 1991; Lange, 2002; Hambler, 2004). In North America, this took place starting about 12,000 years ago, whereas in Australia these extinctions started after the arrival of the first Aboriginal people, more than 46,000 years ago. While there may have been other contributing factors as well, there is persuasive evidence that the activities of prehistoric human hunters and gatherers played a prominent role in the Pleistocene megafaunal extinction, which led to the disappearance of numerous large (weighing 44 kg or more) animals in many parts of the world (Koch and Barnosky, 2006).

More recently, the arrival of the first human colonizers led to the extinctions of many species of animals and some plants on various Pacific islands – some of the very last places on Earth to be reached by people. The extinction of all species of moa birds, mainly because of overhunting and extensive habitat destruction by the Maori people of New Zealand, and the disappearance of the giant endemic palm tree of Easter Island because of the activities of the first Polynesian colonizers are two well-known such examples (Kirch, 2002; Ponting, 2007). The moas were large, flightless birds related to modern-day ostriches and emus. Before the arrival of humans, New Zealand's land-based mammalian fauna consisted of only two species of bats. Thus, in the absence of large mammals, the moas diversified and became some of the dominant animals in the New Zealand Archipelago. There may have been anywhere between 10 and 30 species of moas in New Zealand (Trotter and McCulloch, 1984; Piper, 2009). The uncertainty about the exact number of species in this group of birds, despite the fact that they became extinct only a few hundred years ago, and despite the discovery and analysis of abundant moa remains, highlights the difficulty of estimating the exact composition and nature of past faunas and floras. The endemic palm trees of Easter Island were probably the dominant and most conspicuous element of the island's plant life, just before the arrival of the Polynesians. These may well have been the largest palm trees in the world at the time, and they seemed to have been closely related to the Chilean Palm (*Jubaea chilensis*), which is still present in forest fragments in western South America (Ponting, 2007; Chester, 2008).

The prehistoric extinction of the elephant species, and the other large herbivores such as ground sloths, in North America, due, to a large degree, to hunting by early groups of human hunters and gatherers, must have had a significant impact on the composition of plant life and the distributions of various types of ecosystems on this continent. Living elephant species, such as *Loxodonta africana*, are known to have a profound impact on some ecosystems in Africa, by contributing to the natural clearing of forest patches, and transforming wooded savanna ecosystems into grasslands. This transformation favours grassland species, which benefit from the presence and forest clearing activities of the elephants. In a game reserve in South Africa, where no elephants were present for about 100 years or so, three species of antelope have become locally extinct and other grassland species such as wildebeest are much more rare than they used to be (Renfrew and Bahn, 1997).

The moa birds of New Zealand were all herbivores, and their diet included the twigs, leaves, fruits, and seeds of local plants. It is likely that at least some of the larger moas were keystone species in New Zealand, and their disappearance had important effects on some of the plant species these birds were browsing on and had coevolved with (Trotter and McCulloch, 1984; Cooper et al., 1993; Barlow, 2000; Worthy and Holdaway, 2002; Guiașu, 2007b). The extinction of the moas undoubtedly had profound effects on other New Zealand species as well. For example, Haast's eagle (*Harpagornis moorei*) – once the world's largest eagle species – used to prey on the moas. Just like the moas, *H. moorei* was endemic to New Zealand. Hunting by humans and the disappearance of its moa prey led to the extinction of this impressive aerial predator as well (Piper, 2009; Lomolino et al., 2010).

Thus, by contributing significantly to the extinction of many megaherbivores, and some large predators, prehistoric humans had a dramatic effect on the animal and plant species compositions of most terrestrial ecosystems almost all over the world. These were major and irreversible human-generated changes in the biodiversity of the planet, and they all occurred before recorded history. Therefore, the assumption that humans only started having a significant impact on Earth's ecosystems and biodiversity in historic times, and particularly in modern times, is clearly flawed.

At the same time, as humans spread all around the world, from their original "native" range of East Africa, during the last 100,000 years or so, they surely must have had an important role in dispersing various species of plants, animals, fungi, bacteria, and so on, throughout the planet. Therefore, such human-assisted "invasions" by many species must have taken place, unrecorded, for tens of thousands of years, long before the existence of historical records of any kind, and very long before the rise of current concerns about invasive species. This calls into question the very validity and usefulness of the concept of a "native range" for virtually any given species. Furthermore, our species, *Homo sapiens*, has first and foremost ensured its own dispersal, far away from its initial area of origin, into almost every corner of the world's land areas. As a result, all modern humans can be considered as members of a highly invasive species, and our moral right to discriminate against other species that may have expanded their original ranges at least in part as a result of our activities is highly questionable. Some may argue that the fight against so-called invasive species is a necessary attempt to restore some sort of "balance" to various ecosystems and protect "native" species. However, it could also be argued that the current battle against non-native species is just the latest chapter in the long human war against the natural world, and the latest attempt to shape nature according to our changing whims and ideologies. Only this time, the battle is fought under the respectable banner of "conservation".

The titles of certain books about invasive species, such as *Nature out of place* (Van Driesche and Van Driesche, 2000) or *Life out of bounds* (Bright, 1998), suggest that some conservation biologists seem to think they know exactly where each species belongs, or should belong, and that it is their sacred duty to make sure the ranges of species never change. The implication appears to be that we have to somehow keep nature in its predetermined place, and not let it expand outside the artificial geographical and ecological boundaries within which invasion biology has placed and confined many species. This mindset often ignores not only the facts of historical biogeography, but also the knowledge derived from evolutionary biology. In other words, the view of the living world frequently offered by invasion biologists appears to be rather narrow, static, and relentlessly grim.

The misuse and misunderstanding of evolutionary biology terms, including the key concept of natural selection, by some of the proponents of the notion that native species are somehow "superior" to, or "better" than, non-native ones was discussed in some detail by the prominent evolutionary biologist Stephen Jay Gould (2002). At the same time, Davis et al. (2001) analyzed and lamented the isolation and dissociation of invasion biology even from other branches of ecology. So, in some ways, invasion biology stands apart, by isolating introduced species from the rest of the living world, and generally targeting such species for discrimination and control, simply because of their perceived status as perennial "outsiders" that (at least according to some of the workers in this field) can never belong in, or bring truly useful contributions to, their new environments.

Earthworms in Ontario

I started the book with a brief description of an alarmist newspaper article about the presence of supposedly invasive earthworms in Ontario. It is widely assumed that earthworms used to exist in North American northern forests before the last glaciation, but then disappeared during the last Ice Age, and did not return after the ice sheets retreated from these regions, about 10,000 years ago or so. Presumably, the earthworms currently found in Ontario have been reintroduced by European settlers, and are therefore considered to be "non-native" in this province (Schwert, 1979; Bohlen et al., 2004). However, the status of the earthworms in some northern parts of North America may not be quite as simple to determine as the previous statements may indicate. First of all, even Bohlen et al. (2004) – in an article which focuses on the "exotic earthworm invasion" – admit that native earthworm species actually do exist in the northern forests of North America, although the authors consider such species to be currently rare in these ecosystems. More importantly, Schwert (1979) found a fossil earthworm cocoon, most likely belonging to the species *Dendrodrilus rubidus*, in southern Ontario sediments which had been deposited shortly after the end of the last Ice Age. The author estimated that these sediments were older than 7,000 years, and possibly as old as about 10,000 years or so. The evidence presented indicates that the earthworm species *Dendrodrilus rubidus* is actually native to Ontario, which shows that not all local earthworm species currently found here were introduced by European settlers after all. *D. rubidus* is apparently common in southern Ontario, and is also found in parts of Europe, South America, Africa, Asia, and Australia. In other words, this is clearly a cosmopolitan species, which also arrived in Ontario on its own thousands of years ago, following the end of the last glaciation (Schwert, 1979).

Since earthworms were present in the soils of Ontario forests before, as well as shortly after, the last Ice Age, it is unlikely that too many local plant species evolved in the absence of earthworms. Bohlen et al. (2004) mentioned that native earthworms slowly recolonized areas they used to occupy before the last Ice Age. Slow or not, the steady advance northward of these previously "native" earthworms can be expected to continue during the current interglacial period. Furthermore, the addition of exotic earthworms to Ontario forests may be regarded as a way to restore ecosystem services native earthworms used to provide in this region before the last Ice Age. Surely, some plants can be expected to benefit from the return of the earthworms, and a variety of animals will undoubtedly feed on them as well.

Cryptogenic Species

Because of the uncertainty surrounding the terms native and non-native, in 1996, American biologist James Carlton proposed a third category – cryptogenic species. These are species whose native or exotic status simply cannot be determined based on the available evidence by using any of the standard criteria. Carlton (1996) stated that cryptogenic species may be "remarkably common" in many ecosystems. In his analysis, Carlton used the concept of cryptogenic species to emphasize the fact that we may be underestimating the number of introduced species in some ecosystems, because of the lack of knowledge and evidence about the regions where many species may have originated. However, in the absence of any reliable evidence, such an interpretation is really nothing more than pure speculation, of course, and we could just as well use the high number of cryptogenic species to make an equally hypothetical case that perhaps we are underestimating the number of native species in these ecosystems. These types of arguments and studies, and Carlton's introduction of the cryptogenic species concept, also highlight how little we really know about the geographic origin and possible subsequent range expansions or reductions of many species in various ecosystems. This great and pervasive uncertainty should perhaps give thoughtful invasion biologists a good reason to reexamine some of the key assumptions their field is based on.

In fact, in a very interesting study, which should serve as a cautionary tale for invasion biologists, van Leeuwen et al. (2008) found that at least six plant species assumed to be non-native in the Galápagos Archipelago are actually native to these islands after all. Prior to this study, one of these plant species, *Hibiscus diversifolius*, had been considered as an invasive species which could threaten the ecosystems of the islands. Even though *H. diversifolius* is currently quite rare on the island of Santa Cruz (one of the larger islands found in the center of the Galápagos Archipelago), the currently perceived range expansion of this species on the island was interpreted as evidence for its invasive status. Another of the plant species analyzed in this study, *Ranunculus flagelliformis*, was first discovered in the Galápagos Islands only in 1972, and was therefore considered to be a very recently introduced species in this region. As a result, and as the authors of the study indicate, both *H. diversifolius* and *R. flagelliformis* may well have become targets for eradication in the archipelago. The other four species referred to in the study were also considered "weedy" or introduced and possibly invasive in many Pacific islands, including the Galápagos ones. It was assumed that all of these six plant species had been introduced to the Galápagos region since the Europeans first visited the islands in 1535. This is believed to have been the very first human visit to these remote islands. However, the recent scientific analysis of the abundant fossil pollen deposits found in sediment sequences from the island of Santa Cruz revealed that all of these six plant species had been present in the Galápagos Archipelago for thousands of years before the first humans reached the islands. Therefore, the pollen analysis clearly identified all of these plants as native to the Galápagos Islands, and the authors of the study concluded that it was probable that these particular plant species are also native to other Pacific islands where they may have been incorrectly classified as introduced, or even invasive, species (van Leeuwen et al., 2008). The study also showed that both *H. diversifolius* and *R. flagelliformis* had been much more common and abundant on Santa Cruz in the past, than they are during the present time. So, it turns out that these species are actually rare natives possibly in need of protection, rather than invasives that may well have been targeted for destruction. And the current perceived range expansion of *H. diversifolius* may simply represent the return of this species to some of the habitats it used to occupy in the more distant past.

Fortunately, in this case, the conditions were right for the preservation of the fossilized pollen grains belonging to these plant species, and a group of innovative scientists took the time to carefully gather and analyze the relevant data. However, at many other locations, the environmental circumstances may not allow the preservation of such vital fossil evidence. Even in places where the evidence could still be found, there may not be any qualified scientists available to properly look for and interpret it. Thus, often, the assignment of the native or non-native, or even invasive, status to various species is made without any solid and direct scientific proof about the geographic origins of these species.

Remaining in the Galápagos Islands region for the next example, it is difficult to know if the presence of the coconut palm (*Cocos nucifera*) in this archipelago is due to an initial introduction by human colonists or not. The fruits of the coconut palm tree can survive in sea water for long periods of time, and can therefore be dispersed naturally by oceanic currents to many tropical beaches all over the world. Thus, this species certainly has the ability to colonize remote tropical islands on its own. On the other hand, it is also possible that human visitors or settlers may have brought this tree species to the Galápagos Islands, and some other locations where *C. nucifera* now grows (McMullen, 1999). Therefore, the coconut palm may be "native", introduced, or both, in many regions.

Sometimes, populations of introduced and native specimens of the same species may coexist at the same locations, complicating things even further. In Ontario, both native and introduced populations of the self-heal plant (*Prunella vulgaris*) can be found. As the common name indicates, this apparently edible plant has medicinal properties (Kershaw, 2002; Dickinson and Royer, 2014). The introduced plants are of Eurasian origin, and, in fact, I can remember seeing these wildflowers growing in meadows near the forests in the Carpathian Mountains during my childhood. Years later, I encountered the plants again, this time near an urban woodlot in Ontario – I have no idea if these populations were native or introduced, nor do I think there is any particular reason why this should matter much.

The northern pike (*Esox lucius*) is a commercially and ecologically important freshwater fish species widespread throughout much of the Northern Hemisphere. It was generally believed that this fish species was introduced by people to Ireland during medieval times. Apparently, this belief was based mainly on one study published in the 1950s. As a result of its non-native status, and since the pike is a top predator, control programs against this fish were undertaken in Ireland in order to protect the native brown trout (*Salmo trutta*). However, a new genetic study has revelead the presence of two main distinct lineages of pike in Ireland – one of these lineages is older and more widespread, and may, in fact, have reached Ireland without human help, while the other lineage was indeed probably introduced by people during the Middle Ages. According to the genetic data, the pike populations of Britain and Ireland were separated naturally between 4,000 and 3,500 years ago – presumably by the formation of the Irish Sea after the end of the last glaciation – and this led to the emergence of a genetically unique native Irish northern pike population. Then, about 1,000 years ago, additional pike specimens were added to the freshwaters of Ireland, this time as a result of human-caused introductions (Pedreschi et al., 2014). This information is controversial, because the designation of some pike populations as native would be expected to lead to changes in the management of this species. As a result, the debate has already started. In response to the study by Pedreschi et al. (2014), Ensing (2015) agreed that there are indeed two genetically distinct pike lineages in Ireland, but hypothesized that both of these lineages could have reached the island as a result of separate introductions by people, and, therefore, both lineages would likely be non-native. Pedreschi and Mariani (2015) found Ensing's argument unconvincing and maintained that the analysis of the data supports the native status of the older Irish pike population, which originated about 4,000 years ago. A good part of the debate focused on whether or not people had the know-how and the technology required for the possible long distance transport of live freshwater fishes by boat during the Neolithic – something that seems difficult to establish one way or another with much confidence. To complicate matters further, the precise timing of the retreat of the glaciers from the Irish Sea Basin is unclear, and the exact nature of the post-glacial landscape of the region thousands of years ago is apparently also not known with certainty (Pedreschi and Mariani, 2015).

The better question, in this case, may not be about the native or non-native status of the northern pike in Ireland, but why should this matter so much for the management of this species? Whether the northern pike was brought to Ireland by people, or it arrived there by itself, this fish has been an integral part of the Irish freshwater ecosystems for thousands of years. After all this time, perhaps, as Pedreschi and Mariani (2015) put it, these fish "should now be considered naturalized to Ireland".

In a study published in 2005, van Leeuwen et al. analyzed ancient spores preserved in sediments taken from the bottom of some lakes in the Azores Islands. The spores belonged to the plant species *Selaginella kraussiana*, a relative of club mosses. The status of this species in the Azores was not clear, since some authors considered the species to be native to these islands, while others regarded this plant as either introduced, or even invasive, or of uncertain geographic origin. The analysis of the fossilized spore samples indicated that *S. kraussiana* had been present in the Azores Islands for at least the last 6,000 years, long before the arrival of people in this region. Therefore, the study by van Leeuwen et al. (2005) confirmed the native status of this species in the Azores, and settled a long debate about this issue.

Another case study of a difficult-to-assess conservation situation is that of the pool frog (*Rana lessonae*) in the UK. We know that this frog was introduced in the early 1800s from southern Europe, so it was assumed this species was non-native in the British Isles. However, in a major 2005 study, Beebee et al. reported that some pool frogs were in fact native to the UK. This team of researchers arrived at this conclusion based on various sources of evidence – among them, examining historical records, DNA data showing that these frogs were not directly descended from the introduced frogs, and the discovery of pool frog remains in the UK dating back 1,000 years or so, long before the southern European frogs were introduced. These native pool frogs were more closely related to conspecifics (members of the same species) from Sweden and Norway than to the southern European frogs. Since there is no evidence of any historical introductions from Scandinavia, for now at least, some *R. lessonae* frog populations in the UK are considered native while others are considered exotic. The difference between the two may seem purely academic, but this is not a trivial matter. Native status may afford a certain level of protection for a species, whereas an "introduced" designation may label it, in the words of the study's authors, "an undesirable alien". Should some ancient historical records come to light one day indicating that a Viking warrior carried a pool frog or two to Britain more than 1,000 years ago, the conservation status of this species may be revised yet again…

The pool frog may be relatively small and inconspicuous, but debates about the native or introduced status of species can go on for years, even when it comes to some of the largest animal species on Earth, such as the Bornean pygmy elephants. While these elephants are indeed smaller than most other living elephant species or subspecies, adding the word "pygmy" to their name is rather misleading, since males of this species can still reach 2.5 meters in height and a weight of 3,000 kg. It was believed that these elephants had been introduced into Borneo from mainland South-East Asia or Sumatra by people, possibly in 1750. As a result, the elephants of Borneo were considered a low priority from a conservation standpoint. However, more recent genetic studies suggested that the elephants of Borneo may represent a distinct lineage (*Elephas maximus borneensis*) which may have split from other Asian elephant groups somewhere between 18,000 and 300,000 years ago. The Bornean pygmy elephants are now considered an "evolutionary significant unit". However, the exact time when these elephants colonized Borneo remains hard to determine. It is possible that the Bornean pygmy elephants may have first reached their current island home by walking across land bridges exposed during periods when the sea levels of the region were much lower than today. After the subsequent rise of the sea levels, the Bornean elephants may well have become separated from conspecific groups in Sumatra and mainland South-East Asia (now known as *Elephas maximus sumatrensis* and *Elephas maximus indicus*, respectively). On the other hand, there is no historical evidence indicating the presence of these elephants anywhere except the north-east region of Borneo, where some of them may have been introduced by people in the 18th century, and, to complicate matters further, the bones of a very large elephant were found in a swamp in Borneo. It is not clear how this find may influence our understanding of the evolution of elephants in this region. Although the exact taxonomic status and origin of the Bornean elephants are still being debated, it does appear that these elephants have some unique morphological and behavioural features – such as tameness, for instance – and, as a result, their conservation status is of much greater concern than before, also because of the ongoing destruction of their forest habitats (Garbutt and Prudente, 2006).

Another controversial example is represented by the Falkland Islands dog (*Dusicyon australis*), also known as the Falkland Islands fox or the warrah. In addition to the dispute about whether this was a type of fox or a type of wolf, the initial geographic origin of this now extinct canid species is also controversial. Charles Darwin described it as "a large wolf-like fox", native to the Falkland Islands and found nowhere else (Darwin, 1845). However, it is not known with certainty whether *D. australis* was really native to the Falklands or not. Some experts suggested that *D. australis* evolved in the Falkland Islands, and was related to some fox species found only on the South American mainland. It is possible that the ancestors of this species reached the Falkland Islands on their own, thousands of years ago, during an Ice Age, when the sea levels were much lower than today. Other scientists hypothesized that the ancestors of *D. australis* may have been brought to the Falklands as pets by South American native people a long time ago. Like many isolated island species not accustomed to the presence of people, *D. australis* was apparently quite tame, and initially approached the first permanent European settlers without fear, and with a great deal of natural curiosity. Unfortunately, this behaviour did not help its survival chances, and the Falkland Islands dog, or fox, was destroyed by fur hunters and sheep farmers. The last recorded individual of this species was killed in 1876 (Flannery and Schouten, 2001; Piper, 2009). However, if the warrah would have survived until the present day, there would undoubtedly have been debates about its native or introduced status, and, therefore, about its overall perceived conservation value.

Common Names Don't Help

Common names may also be misleading with respect to the native or non-native status of the species they describe. For example, the Canada thistle (*Cirsium arvense*), despite its common name, is a plant species which was introduced into North America from Europe, probably in the 1700s. This species is now very common and widespread in North America, where it can be found from Newfoundland in the east, to Alaska in the north-west, and to the US state of New Mexico in the south (Peterson and McKenny, 1968; Kershaw, 2002; Dickinson et al., 2004). The Canada bluegrass (*Poa compressa*) is also a perennial plant of European origin, while the Norwegian cinquefoil (*Potentilla norvegica*) is native to North America, the Boston ivy (*Parthenocissus tricuspidata*) is native to parts of China and Japan, and the Mexican fireweed (*Kochia scoparia*) has been introduced into Ontario from Europe and Asia (Petrides, 1972; Royer and Dickinson, 2004; Del Tredici, 2010; Dickinson and Royer, 2014; Booy et al., 2015).

Similarly, Bermuda grass (*Cynodon dactylon*) actually originated in Africa, while the Kentucky bluegrass (*Poa pratensis*) was native to Europe or the Middle East. Both of these grass species have been used in modern times to form lawns in residential areas. Since these species were very common in North America for a long time, and they sometimes spread rapidly even ahead of European settlements, these grasses were considered by many to be native to America. The Kentucky bluegrass, for example, spread widely on its own over large areas of North America, after its initial introduction from Europe. There is speculation about the possible dispersal routes, but the exact patterns of dispersal are not known in detail (Jenkins, 1994). I have encountered local homeowners in the Toronto region, who insisted that they were killing "invasive" dandelions (*Taraxacum officinale*) in order to protect their "native" lawns, made up largely of Kentucky bluegrass. When I casually pointed out that Kentucky bluegrass and dandelions originally came from the same place (Europe), I was either not believed or was regarded with deep suspicion and even hostility, which, to me, shows that many people are unwilling to accept that the species they like could possibly be introduced – since they have been repeatedly told that non-native species have to be invariably nasty and disreputable.

Despite its name, the Norway rat probably originated in north-eastern China, and, in this case, the scientific name of this species (*Rattus norvegicus*) is just as misleading as the common name (Feinstein, 2011). Another species closely associated with human settlements in many parts of the world, the German cockroach (*Blatella germanica*), is not native to Germany, and probably originated in Africa, as did the also misnamed American cockroach

(*Periplaneta americana*) (Eaton and Kaufman, 2007). Apparently, the so-called German cockroach was not even present in Europe until after the end of the medieval period, and then spread on the European continent from the east to the west. Thus, as *B. germanica* kept spreading from one part of the world to another, its common name kept changing as well. For example, when this insect arrived in Germany for the first time, it was known as the Russian cockroach, and, later on, after its arrival in North America, in the late 1700s, it was called the English cockroach. The great Swedish naturalist and taxonomist Carl Linnaeus (1707–1778) started the system of binomial nomenclature, in which each biological species, including our own (*Homo sapiens*), receives a scientific name made up of a unique combination of a genus name and a species name. Linnaeus named many species, and he was also responsible for naming the German cockroach, simply because he apparently received these cockroach specimens from Germany, and was unaware of the true geographic origins of this species (Wiggins, 2009).

The Colorado potato beetle (*Leptinotarsa decemlineata*) is actually native to Mexico, and this species only reached Colorado around 1822. The story of the range expansion of this beetle is quite complex and interesting, and raises intriguing questions about what exactly can be considered a "natural" range expansion. Apparently, the misnamed Colorado potato beetle originated in south-central regions of Mexico, where it used to feed not on potatoes but on a related local plant called the buffalo bur or burweed (*Solanum rostratum*). After the Spaniards introduced domesticated cattle and horses to the Americas, the barbed seeds of the buffalo bur plant were inadvertently carried by these domestic animals northward (the seeds got easily tangled up in the hair of horses, for example), and the potato beetle simply came along for the ride, following its main food source. After the buffalo bur plants reached Texas, it appears likely that they might have been carried northward by migrating American bison herds, as well, which also led to a rather natural additional range expansion for the potato beetles in North America (Lu and Lazell, 1996; Marshall, 2006; Eaton and Kaufman, 2007; Mann, 2011).

The ancestral Mexican populations of *L. decemlineata* appear to have little interest in potatoes (*Solanum tuberosus*) and continue to feed on the buffalo bur. However, once the beetles encountered potato plants in the American Midwest, during their northern range expansion in the 19th century, it seems that a mutation may have occurred which allowed some of the beetles to expand their dietary preferences and start feeding on cultivated potato plants as well. The potatoes had been initially cultivated by the Incas, in western South America, and the Spanish conquistador Francisco Pizarro introduced this crop to his fellow Spaniards. The crop became popular in parts of Europe, and was eventually cultivated in England as well, and from there it was sent to the United States (Marshall, 2006; Eaton and Kaufman, 2007; Mann, 2011). So, both the potato and the Colorado potato beetle are exotic species in Europe and the USA, and therefore, their presence in these regions can be regarded as equally unnatural.

The crayfish *Cambarus bartonii bartonii* is also known by the common name of Appalachian brook crayfish, but the wide region of eastern North America where this species is considered native extends far beyond the Appalachian Mountains, into areas such as central and northern Ontario, for example (Taylor et al., 2005; Guiașu et al., 1996). In fact, among the approximately 100 species of its genus, *C. bartonii bartonii* has the northernmost distribution and can be found in the remote James Bay drainage area of northern Ontario, which is obviously very far away from the Appalachians (Guiașu et al., 1996; Guiașu, 2002a and 2009). So, clearly, the common name of this crayfish does not give any indication of its currently perceived native status in northern and central Ontario lakes and rivers.

"Weeds"

According to the Oxford dictionary and thesaurus (Tulloch, 1995), a weed is "a wild plant growing where it is not wanted". No mention of native or introduced status for such a plant is made or implied. Baker (1974) admitted that there were numerous definitions for this term, before offering his own version of a weed as a plant species whose populations, in any given area, may grow entirely or predominantly in places "markedly disturbed by man". Baker's definition, which did not apply to deliberately cultivated plants, is also fairly vague, as we are left having to decide at what point we can start considering an ecosystem "markedly" affected by human activities, for example. All plants adjusting well to, and growing freely in, human-modified habitats would be at risk of being classified as weeds, according to this definition. Since most ecosystems around the world have been clearly affected by the presence of people, to some extent, this category can incorporate a very large number of species indeed. Theoretically, such a definition could also be applied to very rare plants, which may grow mostly or exclusively in habitats changed by human actions. It should be noted that this definition does not make any reference to the native or introduced status of potential weeds either.

Interestingly, some important crops such as rye, oats, turnips, radishes, beets, leeks, and lettuce apparently started out as weeds a long time ago in regions such as the Fertile Crescent of Southwest Asia (more commonly known as the Middle East). These plants initially started growing opportunistically, and unwanted, in fields intended for different crops (Diamond, 1999).

In the introduction to their book *Weeds of the northern US and Canada*, Royer and Dickinson (2004) mention a few possible descriptions for the term weed, including "a plant of little value" and "a species that competes with crops" (despite the fact that, as we saw, some weeds became valuable crops over time), before offering a definition which is almost identical to the one found in the Oxford dictionary: "a weed might be best defined as any plant growing in an area where it is not wanted." Again, there is no connection made here between the term weed and the native or introduced status of the plants the term may be applied to. Indeed, the long list of weeds listed in Royer and Dickinson's book includes both plants which are native to North America and plant species introduced to the North American continent from elsewhere. The vague and arbitrary nature of this definition is quite obvious. All that is required is for somebody, for whatever reason, to consider a certain plant species unwanted at a particular location.

However, despite its very subjective nature, the term "weed" was not only adopted by some invasion biologists, but the use of this term was actually expanded. As a result, within the field of invasion biology, the terms "weed" or "weedy species" are applied not only to some exotic plants, but also to many introduced animal species (Baskin, 1998; Zimmer, 2001; Meyer, 2006; Quammen, 2008). In his essay, "Planet of weeds" (included in a collection of essays published in 2008), the well-known science writer David Quammen included among such "weeds" not only species that have initially travelled to North America from other continents, such as zebra mussels (*Dreissena polymorpha*) or rock doves (*Columba livia*), but also native North American species that have recently expanded their ranges on this continent. Quammen (2008) lists "the coyote in New York, the raccoon in Montana, the whitetail deer in northern Wisconsin or western Connecticut" and the brown-headed cowbird as some of the indigenous North American species which have become "weeds", because they (according to Quammen) "have increased grossly in abundance or expanded their geographical scope without having to cross an ocean by plane or by boat." These species, and others like them – for instance, the rusty crayfish (*Orconectes rusticus*), which will be discussed in subsequent chapters – are stigmatized essentially for being resourceful and adaptable, from an ecological standpoint, and for enlarging their ranges, often naturally, on the North American continent.

The brown-headed cowbird (*Molothrus ater*), for example, is blamed for having expanded its range (naturally, as far as we can gather) from one region of the US to other parts of the North American continent, by taking advantage of some human activities. Initially, this mainly seed and insect-eating bird species was apparently found primarily in the open grassland ecosystems typical of the Great Plains region, which is located (more or less) in the middle of North America. However, as a result of human-caused deforestation, and the accompanying growth of human settlements, the brown-headed cowbird has gradually expanded its range into eastern and south-western regions of North America. This is not a particularly recent phenomenon, since the range expansion into the south-western US areas took place in the late 1800s, for instance, and the gradual colonization of the eastern US must have occurred well before that (Stutchbury, 2007; Vuilleumier, 2009). The first reference to this species in southern Ontario is from the 1870s (Falk, 2007). The presence of *M. ater* outside the strict boundaries of its presumed "native" North American range was labeled as "unwelcome" by Stutchbury (2007), who also described some drastic control programs against this cowbird.

The common name of the species comes from this bird's habit of following bison herds to feed on insects scared off or startled by these large ungulates. Given the origin of this name, it may have been more appropriate for this species to be known as bisonbird, but, as also discussed in a different section of this chapter, the common names of many species can be quite misleading.

FIGURE 1 *Juvenile brown-headed cowbird* (Molothrus ater)
PHOTO BY DORIS WONG

It may be worth noting that, if North America had not lost its elephant species (and other large herbivores) during the human-caused Pleistocene megafaunal extinction, perhaps those North American elephants would have cleared some forested areas, in the same ways as modern day elephants do this in Africa, and kept more open grassland habitats available for species such as cowbirds. So, while people deforested parts of the continent hundreds of years ago, the prehistoric human destruction of the elephants may have led to a progressive reduction of the grasslands and an increase in woodland areas thousands of years ago.

The main "crimes" the cowbird is held responsible for are its ability to thrive in human-modified habitats, such as agricultural pastures, orchards, and suburban gardens, and the fact that cowbirds are brood parasites which lay their eggs in the nests of many songbird species. Stutchbury (2007) referred to cowbirds as "serial killers who attack dozens of host species". By that logic, I suppose we can describe all parasitic species in this manner, since they have negative effects on their hosts, and we can apply the "serial killers" label to all the predators (lions, wolves, lady beetles, etc.), as well, since their survival obviously requires the death of their prey. And then, following this train of thought, we can discriminate not only against cowbirds (why stop there?), but also against all the hundreds of thousands of species, or more, whose survival may lead to the death of individuals from other species. Imagine the kind of (disastrous and misguided) direct human interference in the natural world that would result if such "principles" were consistently applied, and if we always decided to intervene on behalf of species some of us may like, for whatever subjective reasons.

The cowbird's range expansion is simply a predictable result of the anthropomorphic changes in the landscapes of North America. One of the rare times I have seen brown-headed cowbirds in the Toronto area, in southern Ontario, where I live, a few of these birds were feeding on insects found in the grass of an empty soccer (or football) field, in a neighbourhood park. Since the eastern forests were cut down following the initial European settlement of the continent, this grassland bird species simply moved into the newly available suitable habitat. It is perfectly logical that, if we replace forests with grasslands, we will alter the species composition of these transformed regions. Can we really blame cowbirds, or raccoons, for being some of the few species that can do well around our settlements, in habitats we have altered to suit our own needs? The rapid growth of the North American suburbs may well be a very valid environmental concern, but persecuting cowbirds will not adequately address this problem. As the eastern forests of North America are now regenerating, the amount of suitable habitat for cowbirds is declining naturally in these regions.

In fact, Falk (2007) reported a significant recent decline in cowbird populations, at a rate of 3.9% per year in Ontario and 3.5 % per year in Canada (overall), between 1981 and 2005. In Ontario, this species is restricted to agricultural and developed areas, and the current decline of the cowbird in this province is attributed primarily to the abandonment and loss of farmland because of increased urbanization as well as agricultural intensification which has led to the destruction of woodlots, hedgerows, and wetlands, and, therefore, to a reduction in habitat diversity in this part of the world. While *M. ater* does require open fields and crop and pasture lands for feeding on seeds and insects, as mentioned, this species also needs some woodlots as high-quality breeding habitats. Thus, the loss of small woodlots and pasture lands due to agricultural intensification is thought to have had a negative effect on the abundance of the cowbird in Ontario. At the same time, the dense, widespread forests typical of central and more northern areas of Ontario have always been relatively unsuitable for this mainly grassland species (Falk, 2007).

In addition, as the climate changes, the relative percentages of grassland-dominated regions and forest-covered areas will inevitably change as well, and this will have an obvious effect on the types of species we can expect to find in these respective habitats. In other words, again, we have to keep in mind that the natural world is constantly changing, with or without human interference, and we are unlikely to find the exact same species and species associations and interactions in the same area for very long.

Coming back to the brown-headed cowbird example, brood parasitism may not be endearing to songbird lovers, but it is a perfectly legitimate way for a species to make a living in the natural world, and getting rid of brood parasites seems just as misguided, from an ecological standpoint, as previous attempts to destroy top predators. Trying to eliminate species some people don't like, presumably for the purpose of maintaining biodiversity, can be a very questionable and short-sighted way to proceed, particularly from a (supposedly) scientific point of view. I will discuss this topic further in subsequent chapters.

Changing Perceptions

Furthermore, our perception of "non-native" species can change over time. Species perceived by many people as native to, and representative of, a particular location, may in fact have been introduced to that area hundreds or thousands of years before. Tulips, for instance, have been a beautiful symbol of The Netherlands for quite some time, but they are actually plants native to central Asia, which were brought to Holland around 1570 (Dash, 2000). The ubiquitous tumbleweeds (*Salsola kali*) seen in cowboy movies about the Old

West are native to Russia and were introduced to North America in the early 1870s (Martin, 1987; Thacker, 2004).

Another such example is represented by the presence of the European rabbit (*Oryctolagus cuniculus*) in Britain. This rabbit species has been accepted as an integral part of the "original" fauna of the British Isles for a long time, and has become a major character in beloved English children's books such as *Winnie-the-Pooh*, *Alice in Wonderland* and, of course, *Peter Rabbit*. If asked, many British citizens would probably regard the rabbit as a "native" species, and an essential and very familiar part of the local natural environments. It would be hard to imagine the British countryside without rabbits. In fact, the rabbits may have been introduced to Australia to remind nostalgic early British settlers of the fauna of the country they had left behind.

However, the European rabbit is actually native to south-western Europe, in particular southern France and some regions of Spain, as well as, perhaps, north-western Africa (Clutton-Brock, 1999; Burnie and Wilson, 2001). The oldest fossils – about 500,000 years old – and the most reliable and abundant ancient records for this species were found in the Iberian Peninsula (Corbet, 1986; Dickenson, 2014; McLeod and Saunders, 2014). As is the case with many other species, there are controversies regarding the exact extent of the native range of *O. cuniculus*. Corbet (1986) could not find reliable evidence indicating that European rabbits were also native to North Africa, and suggested that this species may have been introduced there from the Iberian Peninsula by people during the Neolithic period. The same author also mentioned the discovery of very old fossil European rabbit specimens in Britain. These fossils, which were approximately 200,000 to 300,000 years old, placed *O. cuniculus* in Britain during a previous interglacial period. Although this would seem to bolster the case for awarding some type of native status for this rabbit species in the UK, we should remember the previously discussed examples of *Rhododendron ponticum* and the *Tsuga* and *Pterocarya* trees. These plant species, just like the European rabbit, were once found in Britain long ago, during a previous interglacial period. However, they are still considered alien there now, because they did not travel to Britain on their own yet again, after the end of the latest Ice Age. For that apparently unforgivable reason, the rabbit is widely regarded by experts as a non-native species in the British Isles (Flux, 1994; Dickenson, 2014).

This rabbit species was introduced in many other parts of continental Europe by the Romans, and there is speculation that, if the Romans had not helped *O. cuniculus* greatly expand its initial European range, this rabbit may have become extinct, just like the related Mediterranean pika (*Prolagus sardus*), for example (Clutton-Brock, 1999). Thus, by introducing the rabbit elsewhere, the Romans may actually have saved this species from extinction. Sykes and Curl (2010) report some evidence which shows that the European rabbit may have been brought by people to Britain for the first time during Roman times. In fact, the debate about the exact status of the rabbit, and the general historical period when it first arrived in the UK, continues. However, the most comprehensive available evidence suggests that the European rabbit was introduced into Britain in medieval times, possibly as early as the 12th century (Attenborough, 1984; Ponting, 2007; Sykes and Curl, 2010). Some may regard all the wild European rabbits living outside their initial "native" range of Spain and southern France as "feral", since they are the descendants of individuals which escaped from captivity, generally a very long time ago. However, Clutton-Brock (1999) considered the use of the term "feral" to be inappropriate in this case, since the rabbits escaped from captivity in Europe as long ago as the time of the Roman Empire, and they are now well-adapted to their new environments and living there as a fully wild species.

This information is not meant to imply that the presence of the rabbit should not be considered perfectly natural today in Britain. Far from it. I am simply mentioning this classic example to highlight the fact that, in the long run, some non-native species become accepted by many people as natural and important components of the ecosystems they have been (or may have been)

FIGURE 2 *A domesticated European rabbit* (Oryctolagus cuniculus) *in England*
PHOTO BY ADRIAN LEU

introduced to long ago. Perhaps in a few hundred years, for example, no one will make a fuss about the presence of the rusty crayfish – a species suspected of having been introduced from the US decades ago – in parts of Canada, and the current debates about this issue will be regarded simply as some sort of historical curiosity.

Name Calling

Another problem worth mentioning in this chapter is the frequent use of a variety of subjective and derogatory terms to describe non-native species. The list of such obviously inappropriate terms, sometimes used not only by certain people in non-scientific fields but also by some scientists (who should know better), includes: biological pollution, nuisance or pest species, ecological abnormalities, environmental "malignancies" or "diseases", aliens, natural enemies, monsters, the previously mentioned "weeds" or "weedy" species, and so on (as discussed or mentioned by Lodge, 1993; Bright, 1998; Chew and Laubichler, 2003; Coates, 2006; Wonham, 2006; and Larson, 2008, among others). Biological "pollution"? Surely, bringing a dandelion seed or a rusty crayfish into a new area is not in any way equivalent to dumping mercury into a lake, or burying plutonium into the ground, and any rational person should be able to grasp that obvious fact.

These viciously negative terms form the basis of a previously mentioned propaganda against introduced species (or even native ones that may be expanding their ranges), which often goes far beyond what we actually know about these species and their impacts on their new environments. Such terms have no place in any rational scientific debate or discussion, and this overblown and hateful rhetoric against introduced animals and plants richly deserves, in my view, to be singled out for particularly strong condemnation. Thoroughly demonizing "the enemy" – in this case, some exotic animal and plant species – can sometimes also hide the fact that one may have a weak case for going to war in the first place.

Meyer (2006) insisted that the term "weedy species" was "not pejorative, but descriptive". However, only two pages later in the same book, the author explained that certain "weedy species have become so comfortable living among us that we now consider many of them to be pests, requiring active control measures: the common (Norway) rat (*Rattus norvegicus*) and the white-tailed deer (*Odocoileus virginianus*) come immediately to mind." So what are we to make of such a statement? Is the perhaps unintended, but nevertheless fairly obvious, implication that we should only appreciate and value species that are not comfortable living among us, or near us? This way, we would not have to see any other species at all, except perhaps on television or the computer screen... Can we really say that white-tailed deer are generally "comfortable" among humans? The deer live in forests, for the most part, after all, and they usually tend to flee as soon as people get near. Also, perhaps this is due to my tolerant and easy going nature, but when I saw white-tailed deer in the wild, control programs never came to my mind at all. In fact, I watched the deer, briefly, before they quickly disappeared into the woods, with a sense of wonder and curiosity, and I felt privileged to have seen them and happy to know they were there. These wonderful encounters with wildlife make my day. I enjoy hiking and walking in the woods, and during many years of engaging in both recreational walks and field research, I only saw white-tailed deer in Ontario a few times, and I never thought for a second that I came across a species that can possibly be classified as a "pest".

As I was searching the Internet for some references related to the term "alien species", I came across a fascinating, and a bit surreal, mix of both invasion biology and science fiction books and articles. One of the more interesting titles was *Star Wars – the new essential guide to alien species*, a nicely illustrated book where a variety of fictional species such as Ortolans, Hutts, and Kowakian monkey-lizards were described in detail (Lewis and Keier, 2006). In fact, while I am on the subject of works of science fiction, the general attitude of some invasion biologists to introduced species can perhaps best be summarized by the title of a 1954 movie about giant radiation-mutated ants – *Them!* One of the images on the movie poster shows a police officer, or possibly a military man, speaking into a microphone and letting the public know that "This city is under martial law until we annihilate 'THEM!'." (As can be seen, the research I had to do for this book led me to a fairly wide range of sources...)

Sometimes, articles about real species read very much like horror or science fiction novels. For example, an article titled *Frankenfish raises fears*, which deals with the possible arrival of the northern snakehead fish (*Channa argus*) in Canada, begins this way: "They are frighteningly resilient with ferocious appetites. This bold air-breathing, land-crawling fish has the head of a snake, the jaws of a piranha and can live out of water for three days" (Enberg, 2002).

In an attempt to summarize the terminology used in invasion biology, and separate relevant terms from less appropriate ones, Wonham (2006) promoted the use of the terms introduced, non-native or non-indigenous when referring to species found outside their "native" ranges. At the same time, terms such as alien, nuisance, pest, weedy, and, interestingly, the word "invasive" itself, were regarded as ambiguous, poorly defined, and essentially non-scientific, and the use of such terms in the study of species invasions was therefore not recommended. Even though I agree with these suggestions, I am still using the word "invasive" in this book, simply because this term is very widespread and commonly used in many of the studies in this field. Therefore, any discussion of such studies, even a critical analysis, pretty much requires the use of this rather pejorative term. In fact, the field of invasion biology is, fundamentally, based on the negative connotations associated with the words "invasive" or "invader". Such words can, and are probably meant to, instill a sense of fear, anxiety, and foreboding in members of the general public, and this makes it easier for certain invasion biologists to get their message across, and justify the need for control programs against some species suspected of having come from elsewhere, possibly with initial human assistance.

Some of the ideas and examples analyzed in detail in this book were first discussed in a cover story I wrote for the *Royal Ontario Museum Magazine* (Guiașu, 2008a). One of the predictably negative, and in my view also quite defensive, reactions by invasion biologists to my article came in the form of a letter to the editor published in the next issue of the same magazine (Mandrak, 2008). In his letter, Nicholas Mandrak suggested that the terms "non-native", "exotic", "introduced", "alien", and "invasive" should not be used interchangeably, since they all mean something different to biologists. The implication seemed to be that biologists know exactly what each of these terms means, and agree on all these distinct definitions. Contrary to Dr. Mandrak's claim, however, a thorough survey of the relevant scientific literature shows that this is simply not the case, and that no such consensus exists. In fact, there are many on-going debates about the precise meaning and use of these terms in various relevant fields and even (or perhaps especially) among invasion biologists themselves (as shown, for example, in studies or comments by Richardson et al., 2000; Shrader-Frechette, 2001; Slobodkin, 2001; Davis and Thompson, 2001, 2002; Colautti and Mac Isaac, 2004; Falk-Petersen et al., 2006; Wonham, 2006; Warren, 2007; Guiașu, 2008b; Davis, 2009; Young and Larson, 2011; and Humair et al., 2014, among others). Mandrak (2008) considers the term "invasive species" to be the most relevant, and offers a particular definition for this term. According to this definition, invasive species have a significantly negative impact on the ecosystem in which they are introduced. However, Colautti and MacIsaac (2004) consider the term "invasive" to be most confusing, and list five different definitions used in the invasion biology literature by various authors for this one term. Colautti and MacIsaac (2004) also give a non-exhaustive list of 32 adjectives commonly used in the English language literature on invasive species (including, of course, the aforementioned "invasive", "alien", "exotic", "weedy", etc.), and mention that many of these terms have been used interchangeably and inconsistently in various studies in this field. In a worthy effort to provide some much needed order and logical consistency to the terminology used in invasion biology, these authors propose their own reasonably complex framework, or flowchart, for defining important terms in the field (Colautti and MacIsaac, 2004). However, despite the authors' hope that a consensus on terminology can be reached among invasion biologists based on this particular framework, I think this is unlikely, at least in the foreseeable future, given the nature and intensity of the continuing debates on these issues.

In their survey of 422 invasion biologists (chosen because they acted as peer reviewers for the journal *Biological Invasions*), Young and Larson (2011) found almost equal support for two alternative definitions of the term "invasive" – one of these definitions suggested that invasive species have a negative impact on the environment and the other one did not.

Interestingly, Charles Elton, the man credited with starting the field of invasion biology, writes on page 118 of his book *The ecology of invasions by animals and plants* that: "Introduced animals often do replace or reduce the numbers of native ones." So, Elton himself refers to "introduced" species having a negative impact, rather than "invasive" ones. Furthermore, Elton's statement stands in contrast with the general current recognition by most ecologists that the vast majority of introduced species do not cause problems in their new environments. The impact of introduced species will be discussed in much more detail in the fourth and fifth chapters of this book.

In a 1993 article, freshwater ecologist David Lodge explained that biological invasions are normal and common events in nature, and seemed to suggest that terms such as ecological malignancies should not be used, in general, while describing exotic species (Lodge, 1993). However, a few years earlier, Lodge himself had titled one of his co-authored articles on the introduced rusty crayfish, *Lake-bottom tyrant* (Lodge et al., 1985). Such anthropomorphic terms, which have really nothing to do with the ecology and characteristics of the targeted animal and plant species, are not difficult to find in the invasion biology literature and related publications.

For example, Barlow (2000), in an otherwise wonderful book on cichlid fishes, referred to an introduced aquatic plant as a "botanical gangster", even though this plant had not been involved, to my knowledge, in any bank robberies, muggings, or other types of crimes.

On the front page of the *Toronto Star*, on Sunday, June 17th, 2012, there was a brief note about invasive plants in the Toronto region. This note, which was linked to a two page article titled *Invasive plants taking over the landscape* (Scrivener, 2012a), provided the following worrisome information for interested readers: "Periwinkle, English ivy, winged euonymus – their names are dainty, but their intentions are dark. Invasive species moving in on our turf." So, you see, according to this note, invasive plants are not just possibly causing some ecological problems – they also somehow have dark intentions, implying that such plants are deliberately up to no good, and, furthermore, in a description perhaps more suitable for a story on gang warfare, they are "moving in on our turf".

One recent book about invasive plant species, authored by a professor of ecology, was titled *Demons in Eden* (Silvertown, 2005). A brief quote from the book helps to identify one of the familiar targeted non-native species:

> Plenty more Darwinian demons from Europe and Asia have now been unleashed upon the Americas. One of the most widespread of these is purple loosestrife (*Lythrum salicaria*), which arrived on the eastern seabord of North America in the early 1800s.

Far from being a "demon", purple loosestrife was actually introduced due to its beautiful flowers, which made it an attractive garden plant, and medicinal qualities (Botkin and Keller, 1998; Kershaw, 2002). Slobodkin (2001) hinted that purple loosestrife may not even be a non-native species in North America after all, but provided no additional details. There will be more information about *Lythrum salicaria* later in the book, since the impact of this species in North America is considerably less clear than paragraphs such as the one cited above may suggest.

Obviously, one is entitled to gather and present evidence, and express general opinions, against introduced species, and the evidence and opinions can then be scrutinized and analyzed by other ecologists. However, the use of terms such as "demons" and "unleashed" seems somewhat excessive in this case, at least to me, not to mention decidedly unscientific. How does one "unleash" a wetland plant, anyway? I suppose one could make the case that such books are written for a more general audience, and, therefore, it is all right to use more colourful and evocative terms than would normally be found in regular peer-reviewed scientific articles. On the other hand, I would argue that the main role of such books, written by ecologists for the general public, should be to educate and inform interested readers, and not to demonize species the authors may not like, for any number of more or less subjective reasons.

Overall, in my view, such derogatory and inflammatory terms serve mainly to brand introduced species as inherently bad, or even somehow "evil", "outsiders" that should be controlled or eradicated in order to restore some vaguely defined balance and order to our fine native ecosystems. Never mind that we frequently have a tough time distinguishing native from non-native species, and that the distinction may often be basically irrelevant, from an ecological or logical standpoint. Never mind that the impact of introduced species in their new surroundings may be quite complex and barely understood. Never mind that ecosystems and the species associations present within them are constantly changing, and past ecosystems can never really be recreated. This anti-exotic species propaganda is a way to desensitize members of the general public, as well as, perhaps, some of the ecologists themselves, and create the impression that species suspected of having been introduced are somehow less important and less valuable than other species, and that, ultimately, "non-native" species are essentially disposable, and can therefore be controlled or destroyed without any need for second thoughts or remorse, or even, in some cases, a coherent fact-based justification.

CHAPTER 2

The Dynamic Distributions of Species and the Static Concept of Native Range

> Because the history of evolution is that life escapes all barriers. Life breaks free. Life expands to new territories. Painfully, perhaps even dangerously. But life finds a way.
>
> DR. IAN MALCOLM (MATHEMATICIAN AND ONE OF THE MAIN CHARACTERS IN MICHAEL CRICHTON'S 1990 BESTSELLER *JURASSIC PARK*)

> Ranges of obscure species tend to center on institutions employing systematists who specialize on those groups.
>
> WALTER K. DODDS

The statement above, made by a biologist and taken from an otherwise serious book titled *Laws, theories, and patterns in ecology* (Dodds, 2009), was meant to be on the lighter side, of course, but, as all good satire, it contains an uncomfortable element of truth. (Dodds also stated, jokingly, that "The only absolute law in ecology is that there are no laws.") Having done some work on the distributions of species, I learned from my own experiences just how incomplete and potentially unreliable the data on the changing ranges of species can be, and just how difficult it is to infer the displacement of one species by another based on such data.

Some years ago, when I was gathering the distribution records for various crayfish species in Ontario, based on the analysis of museum specimens and my own fieldwork records, I realized that numerous data points indicating the presence of *Cambarus bartonii* were clustered within and around the general area of Algonquin Provincial Park. Of course, this could have been in part explained by the fact that this protected region provided some great habitats, such as waterfalls, rushing cool waters, and so on, for this crayfish species. However, such suitable habitats, and relatively pristine rivers and lakes, also existed in some other parts of Ontario. Therefore, it seemed quite clear that the unusually high number of *C. bartonii* records from the Algonquin Park region was at least in part related to the intensity of collection efforts by various biologists, over time, in this popular area for research.

 | DOI 10.1163/9789047426134_004

Cattle Egrets, Blue Sharks and Long-Range Dispersal

Species can find various interesting and surprising ways of reaching distant regions of the planet, and entire books can be written about this fascinating topic alone. In his *On the origin of species* book, Charles Darwin dedicated two chapters to a discussion of the geographical distribution of species, and mentioned intriguing examples of dispersal, such as the transport of plant seeds on the muddy feet of birds. In one particularly interesting example, Darwin removed hardened soil from the leg of a bird. This hard "ball" of earth was kept in storage for three years. Even after this very long period of time, when the soil was finally watered, 82 plants belonging to several species sprouted from seeds which had been present in the soil sample all along. Darwin also found that some freshwater molluscs could survive out of the water, on the feet of ducks, for as long as twenty hours. Thus, in addition to plant seeds encased in mud, birds can also transport some small animals, such as freshwater molluscs, on their feet over long distances (Darwin, 1859). Despite their lack of tolerance for saltwater, it seems that land snails can travel to distant islands, across large sections of oceans, by attaching themselves to the feet and the plumage of certain migratory birds. This is one of the dispersal methods which accounts for the impressive abundance and biodiversity of such snails on many tropical and subtropical islands around the world (Holland, 2009).

Migratory birds, fish, and even insects can spread other species to various parts of the world, sometimes over very large distances. In addition, even non-migratory species can travel quite a bit, on a regular basis, while carrying a variety of small organisms as "passengers". Humans are obviously not the only wide-ranging species capable of affecting the distributions of other species. To pick just one out of many possible examples, the blue shark (*Prionace glauca*) can cross the Atlantic, and tagged individuals of this fast-swimming open ocean species have been found to travel between such far apart destinations as the coasts of England and Brazil, or between the waters near New York and the coast of Spain (Paxton, 1995). Who knows how many small invertebrates, or other species, get a ride from these sharks, and other large open ocean species, across the Atlantic? So, theoretically, if one day we find a small marine invertebrate species native to, say, European waters, off the east coast of North America, we may assume that the species was brought across the ocean in the ballast water of ships, when, in fact, sharks or whales may have helped the invertebrates complete their long journey.

Highly mobile organisms, such as birds, are capable of very impressive dispersal feats. The cattle egret (*Bubulcus ibis*) is native to Africa, but has also managed to cross the Atlantic and reach northern South America entirely on its own in the late 1800s. Since that time, *B. ibis* has not only become a successful resident of the Americas, but has also managed to gradually expand its range northward into Central America, the Caribbean region, and the United States (Lomolino et al., 2010). Cattle egrets were first seen in Florida in 1941, and about three decades later they were breeding in Minnesota and becoming established in some Canadian provinces. In 1981, this species reached southern Alaska (Fitzpatrick, 2004). Given the scope of this range expansion, if we did not know that the egret arrived in the New World by itself, we might have been tempted to believe that this was yet another invasive species. However, sailors on ships regularly observe cattle egrets crossing the Atlantic (Fitzpatrick, 2004), which shows that these birds are indeed able and willing to fly over the ocean and make the journey from western Africa to South America. This type of evidence apparently saved this species from being labelled as an "alien". Instead, in a field guide to the birds of Ontario, the cattle egret was referred to as "a recent immigrant" (Hughes, 2001).

However, it is entirely possible that cattle egrets could rest on ships from time to time, during their travels across the Atlantic, and they undoubtedly benefitted from human-made habitat transformations in the Americas, such as the clearing of forests and the related increased availability of pasture land for grazing cows. Just like the brown-headed cowbirds, cattle egrets tend to follow large herbivorous mammals around, and feed on small animals (usually insects and other invertebrates) exposed by the movements of the ungulates. In Africa, cattle egrets tend to be associated mainly with rhinos and African buffalo herds, but after their arrival in North America, these birds have adjusted to local conditions and are now feeding beside cattle herds, just like the cowbirds (Temple, 2004). It is not clear to what extent these two species may be competing directly with one another for food, but, since North America has plenty of cattle, and both birds are currently expanding their ranges, this does not seem to be a significant factor, at least for now. What is clear is that the spectacular range expansion by *B. ibis* in the New World, mainly during the second half of the 20th century, was also facilitated by human activities. So, while this species is not strictly speaking "invasive", since humans did not take it across the Atlantic, it nevertheless took advantage of anthropogenic environmental changes to its new environment. This is a common feature of numerous currently successful species, which is not at all surprising, given that humans have modified so many of our planet's ecosystems in fundamental ways. Therefore, it can be argued that such range expansions are, in fact, entirely natural, even though they may occasionally be assisted by people, since humans are a major and very influential and ubiquitous component of life on Earth at the moment.

Green Iguanas, Storms and Natural Rafts

Even animals which, at first glance, may not appear capable of such long-range travel can sometimes arrive at surprisingly far away destinations, due to powerful natural dispersal agents such as winds and storms. Censky et al. (1998) reported the arrival of at least 15 individuals of the green iguana (*Iguana iguana*) on the beaches of the Caribbean island of Anguilla. Although this particular iguana species is widely distributed in an area that includes parts of Mexico, Central America, South America, and some other Caribbean islands, *I. iguana* was not present on the island of Anguilla before (as far as we know). The green iguanas arrived at this new destination on an impressive natural raft made up of huge logs and uprooted trees which had been pushed for long distances over the water, from one island to another, by hurricane-force winds. Censky et al. (1998) mentioned that both male and female green iguanas "invaded the island" in this way and repeatedly referred to this event as an "invasion", even though this arrival had nothing to do with human activities. Since the green iguanas were healthy and survived on the island, it appears that this new species may have a good chance of perhaps establishing itself in Anguilla. Clearly, this new arrival can be expected to change things for some of the other species on the island, but these types of changes are obviously natural, and have always been an integral part of life on islands around the world. The arrival of the green iguanas on natural rafts due to hurricanes could easily have been missed (such events are rarely witnessed, after all), and the discovery of the new species in Anguilla may well have been attributed to transport by people.

Even though, as we have seen, the green iguanas arrived on Anguilla without human assistance, Lever (2003) reported that it was decided to eradicate *I. iguana* on this island. This seems to be another example of the inconsistency and lack of clear standards which are sometimes typical of our attitudes and reactions towards newly arrived species. Are we supposed to prevent new species from colonizing islands at all costs now, even when these species arrive on natural rafts pushed by winds and ocean currents? This is one of the main ways in which numerous species have reached islands all over the world for millions of years. After somehow beating the odds, and surviving a dangerous long journey on floating logs during major storms, it seems to me that the green iguanas, and other similarly successful and resourceful species which colonize islands on their own, have earned their right to become established on Anguilla the hard way. By trying to deny them the opportunity to do so, we are, in fact, artificially interfering with a natural island colonization event, and, therefore, changing the future course of evolution of life on this island.

The clear implication is, once again, that some invasion biologists and wildlife managers are trying to keep species compositions the same in perpetuity – an elusive and impossible goal. No new arrivals are allowed, even when we know they came on their own, without our help, or we are uncertain about their exact means of transportation.

The green iguana is considered to be an introduced species in Florida, and the introduction of this species there is thought to be linked to the pet trade. However, since, as discussed, green iguanas can travel for fairly long distances on natural rafts, I suppose it is not inconceivable to assume that perhaps some of these lizards can make the journey, gradually, to an area such as the Florida Keys on their own, therefore complicating matters regarding the exact status of this species. For example, although we know that some of the green iguanas in Florida are former pets, which have either escaped or were released, or the descendants of former pets, what if some other green iguanas arrived on their own, undetected, on a natural raft? Such populations of multiple origins are bound to exist in nature for several species suspected of having been introduced.

Lever (2003) mentioned the debates and uncertainties regarding the status of *I. iguana* on many Caribbean islands. Some authors think that this species may have been introduced by native people to these islands as far back as about 2,500 years ago, while others believe that the green iguana is essentially a native species in the Caribbean region, whose distribution from island to island may have been occasionally assisted by indigenous people in prehistoric times.

Case Study: The Rusty Crayfish and the Fuzzy Issue of "Native Range"

Since 1991, a good part of my research has focused on the ecology, conservation, biogeography, and behaviour of the crayfish species of Ontario, Canada, as well as on the diversity of the many crayfishes of North America, in general. I have spent many years learning about crayfish, in the field, in the laboratory, and in museum collections. As a result, not surprisingly, there will be no shortage of crayfish-related examples and information in this book. North America is a good place to be if one wants to study crayfish diversity, since approximately 77% of the world's more than 500 or so crayfish species are found on this continent, primarily in the south-eastern areas of the United States. These crayfish species occupy a wide range of aquatic and semi-terrestrial habitats, and can be found in surface streams and lakes, as well as in underground rivers

FIGURE 3 *The much maligned rusty crayfish* (Orconectes rusticus)
PHOTO BY PAUL MOORE

flowing through caves, and even in mud burrows sometimes located away from any permanent water sources (Taylor et. al., 1996, 2007; Guiașu, 2007a, 2009).

Orconectes rusticus, commonly known as the rusty crayfish, is the most reviled crayfish in North America today. Considered invasive in many parts of the continent, including Ontario, it is currently the target of a number of control programs that aim to eradicate it from certain lakes and streams (Hein et al., 2006, 2007). However, the exact native range of the rusty crayfish is unclear. According to some leading experts, the initial or native range of *O. rusticus* may well have included southern Ontario (Hobbs, 1989; Hobbs et al., 1989). Hobbs et al. (1989) mentioned that the native range of this species "is somewhat difficult to determine", but probably consisted of southern Ontario, Michigan, Ohio, Indiana, Kentucky, and Tennessee. Hobbs (1989) also includes southern Ontario, Michigan, Kentucky, and Tennessee within the initial range of *O. rusticus*. In this case, the author gives a rather vague original range for the rusty crayfish, stating that this species is found from "Southern Ontario and Michigan to Kentucky and Tennessee" and thus leaving it up to the readers' imagination to fill in the blanks and decide which regions located between Ontario and Michigan, on one hand, and Kentucky and Tennessee, on the other hand, should be included as parts of this "natural" distribution.

The native range of *O. rusticus* is reduced only to southern Ontario, Michigan, Kentucky, and Tennessee – apparently with nothing in between – in a newly published book by Booy et al. (2015). Interestingly, this is, of course, a disjunct distribution, since Kentucky and Tennessee are not adjacent to, and therefore do not share a border with, either Michigan or Ontario.

Presumably, at some point, the native range of *O. rusticus* would have had to include regions which bridge the gap between Kentucky and Tennessee in the south, and Michigan and Ontario in the north.

Other authors give somewhat different initial ranges for the rusty crayfish. For example, Mather and Stein (1993) state that this crayfish species "naturally occurs in Indiana, Kentucky, and western Ohio". A "natural" occurrence presumably refers to the regions where this species is considered to be native, while the presence of the rusty crayfish in aquatic habitats outside this initial range would probably be regarded by some as "unnatural". Butler and Stein (1985) agree that Indiana and western Ohio are indeed part of the native range of *O. rusticus*. However, these authors add the state of Illinois to, and remove the state of Kentucky from, this range. Butler and Stein (1985) are also quite specific about the native range of this crayfish within the US state of Ohio, for example, by listing *O. rusticus* as native in western Ohio, as mentioned, but invasive in east-central and southern Ohio. Some other authors are more vague, and simply mention that *O. rusticus* is native to the Midwestern United States (Maude, 1988), or, a bit more specifically, to the southern Midwest region of the US (DiDonato and Lodge, 1993; Hill and Lodge, 1994), while Kuhlmann (2008) states unequivocally that the rusty crayfish "is native to the Ohio River" (and presumably cannot be considered native in any other stream, lake, or pond).

Crocker and Barr (1968) considered *O. rusticus* to be an introduced species in Ontario. Despite some more recent updates for the distributions of certain Ontario crayfish species (for example, Guiașu et al., 1996; Hamr, 1998), the book on the crayfishes of Ontario written by Crocker and Barr nearly fifty years ago remains the most comprehensive and influential source of information available on the crayfish species of this Canadian province. This highlights the need for more research on the current distributions of the Ontario crayfishes. It is reasonable to expect that these current distributions may be quite different from the ones recorded by Crocker and Barr (1968) several decades ago.

Hamr (1998) stated that *O. rusticus* "is thought to be native to the Ohio River Basin and the states of Ohio, Kentucky, Michigan, and Indiana", and gave a long list of North American regions where this crayfish may have been introduced. This list includes the Canadian province of Ontario and sixteen US states (Connecticut, Iowa, Illinois, Maine, Massachusetts, Minnesota, Missouri, New Hampshire, New Jersey, New Mexico, New York, Pennsylvania, Tennessee, Vermont, West Virginia, and Wisconsin). Hobbs (1989) writes, cautiously, that *O. rusticus* was "presumably introduced in one or more of the following: New England, Illinois, Iowa, Minnesota, New Mexico, West Virginia, and Wisconsin". Jezerinac et al. (1995) mentioned that *O. rusticus* "is most likely an introduced species" in West Virginia, where the presence of this crayfish was recorded for the first time in 1978.

This is only a sample of some of the expert opinions found in the scientific literature about the native range of the rusty crayfish, as well as about the possible regions where this species may have been introduced. An analysis of the ranges given above reveals several clear discrepancies. For example, Butler and Stein (1985) consider *O. rusticus* to be native to Illinois, while Hamr (1998) thinks that the rusty crayfish was introduced in that state. Tennessee is listed as a state *O. rusticus* is native to by Hobbs (1989) and Hobbs et al. (1989). However, Hamr (1998) disagrees, and considers that this crayfish species was introduced in Tennessee. Hamr (1998) and Hobbs (1989) agree that Michigan is within the native range of the rusty crayfish, but some other authors (for instance, Butler and Stein, 1985 and Mather and Stein, 1993) do not include Michigan among the states where *O. rusticus* may be considered as a native species. And so on...

The controversy regarding the exact extent of the "native range" of the rusty crayfish continues. Dresser and Swanson (2013) described two rather different possible native ranges for this species. According to the United States Geological Survey's version, *O. rusticus* is native to Indiana, Illinois, Kentucky, Ohio, and Tennessee. However, the alternative native range cited for this same crayfish species encompasses Ohio, Kentucky, Indiana, and Michigan. According to Dresser and Swanson (2013), this latter range is "accepted by researchers of invasive crayfish". Three particular references are cited in support of that contention (Taylor, 2000; DiStefano et al., 2009; and Peters and Lodge, 2009), but, as we have seen, quite a few other studies can be cited which show that no such consensus exists among researchers with an interest in North American crayfish. In fact, it turns out that none of the three studies cited by Dresser and Swanson (2013) in support of the revised native range of the rusty crayfish deal mainly with this issue. Furthermore, in one of these studies, Taylor (2000) actually mentions "a less than clear understanding" of the ranges of several closely related and hard to tell apart *Orconectes* species, including *O. rusticus*, and admits that for his study "only a small number of specimens of *O. rusticus* were examined from outside Kentucky, and its range in adjacent states is far from being completely known."

This discussion is not meant as a possible remedy for insomnia, for readers without a keen interest in the minute details of crayfish distributions, but simply as an illustration of the fact that not even the experts can agree on the extent of the so-called native distributions of species and on all the regions where some species are supposed to have been introduced. This is an important point to emphasize, in my view, because, again, invasion biologists and wildlife managers may target introduced species and use control programs against them.

The certainty with which some invasion biologists seem to separate "native" from "non-native" species often amazes me, given the frequently ambiguous, incomplete, and insufficient information available about the geographic origin of these species, as well as their subsequent means of dispersal.

To illustrate this point, we can also consider the case of the crayfish *Orconectes limosus* in Canada. In his report on the conservation status of Canadian crayfishes, Hamr (1998) mentions that "little is known about the history and distribution of this species in Canada" and acknowledges that "It is also not certain whether the records for this species in Canada are natural populations or the results of introductions."

Again, in this case, the word "natural" is used as a synonym for "native". As we have seen, terms such as "natural populations" or "natural range" are restricted to species perceived as native. This strongly implies that non-native species are somehow not "natural" and therefore do not belong anywhere outside their hypothetical native range. The well-known environmental historian William Cronon (1996) stated that:

> As soon as we label something as "natural", we attach to it the powerful implication that any change from its current state would degrade and damage the way it is "supposed" to be. But in fact we are rather selective about the parts of nature we choose to view in this way.

Furthermore, when we talk about a crayfish species, or any other species, being native to a particular state or province, we have to keep in mind that these borders are simply artificial lines on a map, drawn for political reasons. These lines have no biological meaning, and are irrelevant to any other species but our own. A rusty crayfish obviously has no idea that the moment it moves across a state boundary, by walking along the bottom of a stream or a lake, it may be at risk of becoming an invasive species, or breaking certain laws. When somebody says that this crayfish is native to Kentucky, or Michigan, or Indiana, to name just some of the states in this so-called native range, what does this really mean, anyway? Surely, *O. rusticus* is unlikely to be native to *all* of Kentucky, for example, or to every little bit of aquatic habitat found in Michigan. Obviously, the crayfish must have originated in a small section of this range, long ago, and then spread from there. The truth is that nobody really knows the exact limits of this native range, and we cannot expect this species not to move beyond these boundaries by any means necessary, if environmental conditions are suitable for further dispersal. And if the species is native to certain states bordering the Great Lakes, such as Michigan, or Ohio, then what would prevent it from expanding its range naturally into Ontario, for instance?

In addition, there is also the serious problem of mistaken species identity, which will be analyzed in more detail later in the book. As a quick preview of that discussion, it may be worth noting that Wetzel et al. (2004) found that in large areas, which included several river basins, of Iowa and southern Minnesota "many records attributed to *O. rusticus* are in fact *O. luteus*." Therefore, since at many of the locations in these two states, and possibly elsewhere, the presumably native *O. luteus* was mistakenly believed to be the similar looking and supposedly invasive *O. rusticus*, the ranges of these species have to be adjusted accordingly. Wetzel et al. (2004) also stated that "some records of *O. rusticus* from portions of Wisconsin and northern Illinois also could be *O. luteus* or another species." Such identification mistakes are probably quite common both in the field and among labelled museum specimens. For example, during my own extensive work with crayfish museum specimens, conducted at the Royal Ontario Museum (ROM) in Toronto, I found that a significant number of *Cambarus* crayfish specimens from Ontario had been consistently misidentified over the years. In this case, the confusion involved mostly the *Cambarus robustus* and *Cambarus bartonii* species. For the purposes of my research, I did not look at the numerous *Orconectes* specimens in the ROM's vast collections, but, if I had, there is no reason to believe that identification mistakes would not have been found there as well, particularly since the number of *Orconectes* species found in Ontario is higher than the number of local *Cambarus* species (according to Hamr, 1998, Ontario has three *Cambarus* and six *Orconectes* species, assuming *O. limosus* also exists in this province).

Case Study: The Rusty Crayfish in Ontario

Often, some biologists assume the rusty crayfish was introduced into Ontario by fishermen from the US who used it as bait. However, for the most part, we are just not sure how this crayfish species arrived in Canada. It is very likely that some of the range expansions of *O. rusticus* in Ontario were natural—in other words, after the initial possible (but undocumented) introduction, the rusty crayfish increased its range in this province without human help. This resourceful crayfish species can survive in a variety of aquatic habitats, from small ponds to fast streams, and is able to spread quite well naturally (Guiașu and Guiașu, 2003; Guiașu, 2008a). Although this is primarily a river and lake species, *O. rusticus* (unlike many other typical stream-inhabiting crayfish, such as, for example, *Cambarus robustus*) is able to survive for extended periods of time (at least a couple of weeks) even in areas with mostly wet mud but very little water available above the ground (Guiașu, 2007a). The rusty crayfish is also able to dig fairly deep and extensive burrows in streams with clay substrates, although this behaviour is not frequently observed in this species (Hamr, 1998). However, semi-terrestrial habitats with relatively little water are clearly not optimal for this crayfish species, and *O. rusticus* is not expected to thrive in such conditions.

I recently analyzed the rusty crayfish records found in the Royal Ontario Museum (ROM) database and I was struck by just how little information we really have about the distribution of this species in Ontario. The ROM houses by far the largest and most comprehensive collection of Ontario crayfishes, and so an examination of the museum's crayfish database should give access to most of the information available about the history of scientific crayfish collecting efforts in this province. As can be seen in Figures 4 and 5, *O. rusticus* specimens were collected during five decades in Ontario – the first record is from 1963 and the last one from 2003. However, in these five decades, rusty crayfishes were collected in only 13 different years, for a total number of just 25 collection days. So, let us think about this for a moment. Almost everything we know about the range and historical presence of this notorious crayfish in Ontario is the result of about three and half weeks of rather uneven collection efforts, distributed over the span of five decades.

A closer look at the data is even more unsettling. Figure 4 reveals that 16 (or 64%) of the total 25 collection days were back in the 1960s, with the greatest collection efforts (11 days, or 44% of the total) made in only a single year – 1964. In fact, 1964 was the only year when we had more than 2 collection days for this crayfish species in Ontario. The reason why a majority of the collection efforts were made in the early to mid-1960s is obvious – this was the period of time when most of the data were collected in the field for the previously mentioned book on Ontario crayfishes by Crocker and Barr (1968). Other than the more sustained fieldwork done mostly in 1964 for this specific purpose, the *O. rusticus* collection efforts in this province can be characterized as sporadic, rare, and obviously very incomplete. There was only one collection day in the 1970s (in 1976), followed by 5 collection days in the 1980s, 2 collection days in the 1990s, 1 collection day in 2003, and nothing since then. There have been only 9 collection days since the end of the 1960s, and in 10 of the total 13 years when this crayfish was collected in Ontario there was only 1 collection day per year, which was still better than the 28 years between 1963 and 2003 when no *O. rusticus* were collected at all. Clearly, this is not the basis for any informed judgement on the changing distribution of the rusty crayfish in this part of the world. There have been some crayfish collection efforts by other individuals and institutions, as well,

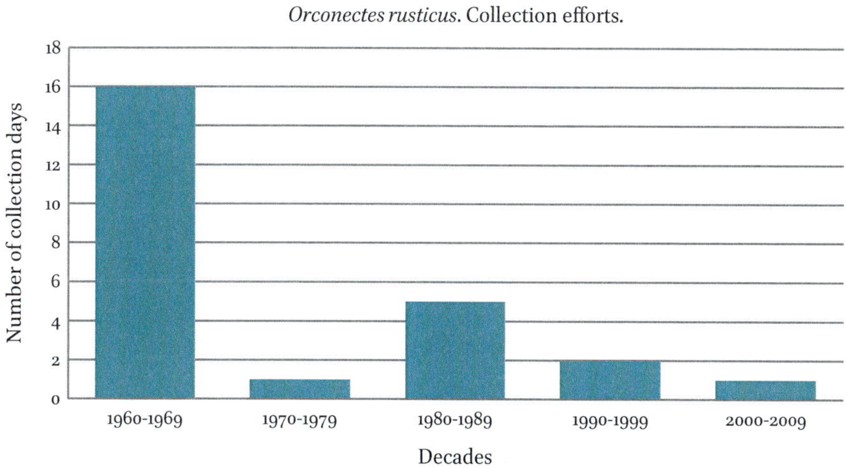

FIGURE 4 *Collection efforts related to the rusty crayfish in Ontario per decade (based on the Royal Ontario Museum crayfish database)*

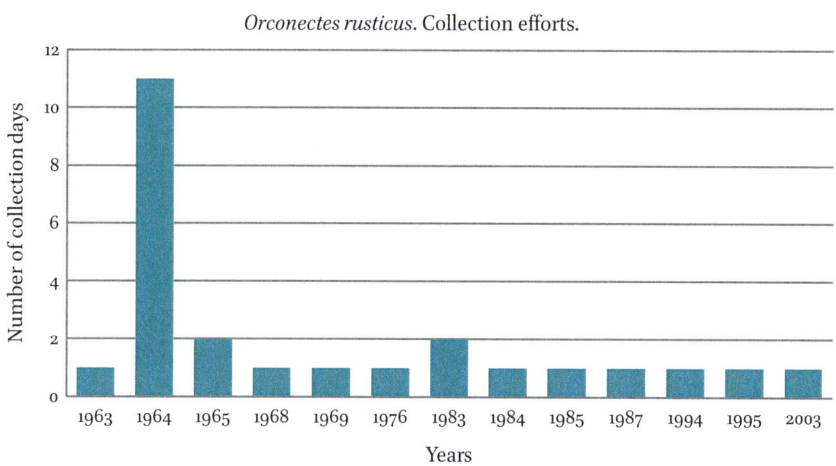

FIGURE 5 *Collection efforts related to the rusty crayfish in Ontario per year (based on the Royal Ontario Museum crayfish database)*

but the ROM has the most comprehensive historical records of this type in the province.

The oldest Ontario crayfish records found in the ROM database are, in chronological order: 1903 for *Orconectes virilis*; 1908 for *Cambarus robustus*; 1912 for *Orconectes propinquus*; 1913 for *Cambarus diogenes* and *Orconectes immunis*; 1915 for *Cambarus bartonii*; 1919 for *Fallicambarus fodiens*; and 1963 for both *Orconectes obscurus* and, as we have seen, *Orconectes rusticus*.

So, what do these oldest Ontario crayfish museum records really tell us? Perhaps not very much. The only obvious conclusion is that we have no information at all about the presence and ranges of these species in Ontario prior to the early years of the 20th century. For all we know, several of these crayfishes, including species presumed to be "native", may have been introduced into this province, and then moved around extensively, by fishermen or other people. Does the fact that the oldest *O. virilis* record predates the oldest *F. fodiens* record by 16 years really indicate that *O. virilis* arrived in Ontario before *F. fodiens*? No, of course not. The records may simply reflect collection priorities or crayfish abundance at the relatively few initial locations where the species were captured. Burrowing, semi-terrestrial species, such as *F. fodiens*, are difficult to collect and wouldn't normally be encountered if someone decides to catch crayfish in a local river, for instance.

At first glance, the list of oldest records would appear to show that *O. obscurus* and *O. rusticus* must have been introduced more recently, because they were first recorded in this province in 1963 – several decades after the other, presumably native crayfishes. However, given the scarcity of the data, we cannot even be sure of that, and alternative explanations exist. It is possible that *O. rusticus* and *O. obscurus* were present in Ontario well before 1963, but were not found, either because the species were very rare in the recent past, or due to the fact that nobody bothered to look for them at the specific locations where they might have been present. I know from personal experience that collecting crayfish can be a rather unpredictable pursuit. I sometimes collected crayfish on consecutive days, at the same locations, and on some days I saw plenty of specimens of several species, while on other days I found nothing, even though everything else seemed to be the same. I once searched for crayfish, together with several dedicated and capable field assistants, for hours in a particular stream and did not catch anything, and then we came back the next day to the exact same location and caught a record number of specimens using the same methods as before. There was no clear explanation for this, but it seems likely that crayfish migrate along the stream sometimes, from one area to another, for reasons unknown.

Coming back to the oldest ROM crayfish records, it is also possible that rusty crayfish may have been present in Ontario in previous centuries, but then disappeared before the early 1900s and were then either reintroduced or returned on their own. Perhaps the current distribution of *O. rusticus* in Ontario is the result of local extinctions and relatively isolated remnant populations that were only "discovered" in 1963, when more systematic collection efforts were made prior to the publication of the Crocker and Barr (1968) book.

It is obvious that Ontario crayfishes didn't just magically appear in local waters in the early 20th century. That is simply when scientists first started to look for them and store them in museum collections in this region. So, the assumption seemed to be that the species first recorded here in the early 1900s must have been the only native ones, when, in fact, we really don't know anything about the history of crayfishes in Ontario before the 20th century. How can we make decisions on the long-term historical range changes of these species when we only have a few decades worth of very incomplete and scattered records? We are assuming a species arrived in Ontario on its own thousands of years ago, following the end of the last glaciation, because it happened to be found by a scientist in the early 1900s. And if another species wasn't seen until decades later, we assume it must have been introduced. However, both assumptions may be wrong, of course, especially when collection efforts were so meager. It is easy to miss crayfish species in a vast province, if only a handful of researchers look for them once or twice every few years at a small number of locations. Apparent modern day range "expansions" may be nothing more than the result of more intensive collection efforts in certain areas, which can lead to the discovery of species that may have been there, undetected, all along.

Huston (1994) defined invaders as "species that have spread far beyond their original distribution, to distant continents or sometimes around the entire globe." Such a definition would not seem to apply to the spread of the rusty crayfish within North America, and certainly not to the relatively modest range expansion of this species from its original range in parts of the US states of Michigan and Ohio to the neighbouring Canadian province of Ontario. As mentioned, currently, this species is considered native in parts of Ohio, but invasive if one drives only a few hours (perhaps four or five hours) north and finds the rusty crayfish in Ontario lakes or streams. This seems particularly arbitrary and perhaps unfair, because, whether *O. rusticus* was brought into Ontario by US fishermen or not, it seems very likely that, sooner or later, this species would have expanded its range into Ontario naturally anyway from neighbouring regions where it is considered native. So, if some US fishermen did indeed bring this crayfish to Ontario, they inadvertently prevented the possibility of ever having this species considered native in this province. Even if some *O. rusticus* individuals may cross the border and colonize the aquatic habitats of Ontario completely on their own, in the near future, they will probably always be considered as members of an invasive species by the invasion biologists working in this jurisdiction, because of the assumption that these crayfish were once introduced by people.

Although Simon and Thoma (2006) state quite explicitly that the rusty crayfish "is considered a non-indigenous invasive species in the Great Lakes", and this view seems to be widely shared among many crayfish researchers in this region of North America, the situation may not be that straightforward. First of all, as we have seen, the rusty crayfish is regarded as a native species in some rivers and lakes found in states that border the Great Lakes (such as Ohio, Michigan, Indiana, etc.). At least in some of these cases, it is certainly possible to contemplate that *O. rusticus* may have expanded its range naturally, and gradually, along certain streams, all the way to the Great Lakes. Perhaps those mythical fishermen are not solely responsible for every range expansion of the rusty crayfish after all… I find it interesting that, on the one hand, some invasion biologists keep mentioning just how good *O. rusticus* is at expanding its range, but, on the other hand, the same biologists insist that this crayfish couldn't possibly have reached any areas outside its presumed "native" range without human assistance. In any event, those who always assume that fishermen must have been the only ones who helped increase the range of this crayfish species generally do not have any concrete evidence to back up those suppositions (these often repeated claims are obviously very convenient for those who want to control the spread of *O. rusticus*, by the way).

In addition, in a recent study, Peters et al. (2014) mention that the rusty crayfish was present in Lake Erie for more than a century (as far as we know – in fact, this crayfish may have been there, undetected, for a much longer period of time, of course). According to these authors, "there is debate regarding whether *O. rusticus* was originally native to southwestern Lake Erie". Well, if this crayfish is indeed native to Lake Erie, and has been present there for a long time, shouldn't we at least consider the possibility that it might spread on its own to other parts of this vast aquatic region? After all, the Great Lakes of North America are interconnected bodies of water, and there is no reason why *O. rusticus* would not be able to move about freely, and largely undetected by us, within this huge freshwater aquatic system.

So, branding species with native or non-native status is not straightforward. Such labels are often dependent on individual local jurisdictions in charge of species management. For example, the crayfish *Orconectes virilis* is considered native in places such as Wisconsin and Ontario (Hamr, 1998; Olden et al., 2006), but invasive in North Carolina (Cooper and Armstrong, 2007), while another crayfish species, *Orconectes propinquus*, is listed as native in southern Wisconsin, but invasive in the northern part of that state (Olden et al., 2006). Following this line of reasoning further, we can envisage a scenario in which a species can be native on one side of a road, but invasive if it ever crosses over to the other side. In this field, the answer to the question: "Why did the species cross the road?" would always be: "To become invasive, of course."

Naturally changing crayfish distributions, in response to changing environmental conditions, may in fact present a challenge to invasion biologists and ecologists in charge of monitoring crayfish populations. If, for example, a crayfish species expands its range naturally from one US state to another, or from the US to Canada, would such a range expansion be recognized as a natural process, or would it be regarded as an "invasion" that should perhaps be stopped or controlled?

On the Natural Heritage Information Centre web site of the Ontario Ministry of Natural Resources (2008), for instance, there is a note stating that the semi-terrestrial, burrowing crayfish *Cambarus diogenes* may have invaded Ontario's Niagara Peninsula from adjacent New York State. However, since *C. diogenes* is considered both a native species and a very rare crayfish with a small distribution in Ontario, using the term "invasion" to describe a local range expansion for this species is perhaps inappropriate (Guiașu, 2009).

In my view, it is important to recognize the obvious fact that the distributions of species are dynamic – in other words, these distributions are not frozen in time or space, and quite often we don't really know exactly if a particular range expansion may have been directly assisted by humans or not (Guiașu, 2008a). This problem is even more difficult, because we frequently have very incomplete and outdated information about the distributions and range changes of many species. This is certainly true for the distributions of many of the North American crayfish species (Taylor et al., 1996, 2007). For example, in a detailed report, Dubé and Desroches (2007) mentioned that it is not known if the current presence of the crayfish species *Cambarus robustus* in the Saint Lawrence river, within the Canadian province of Quebec, is due to a natural range expansion by this species, or to deliberate, or accidental, introductions of this crayfish by humans. The same authors also discussed the incomplete current knowledge of the limits of the range of the crayfish species *Cambarus bartonii* in Quebec (Dubé and Desroches, 2007), and I have already discussed the uncertainty regarding the native or introduced status of *Orconectes limosus* in Canada (Hamr, 1998). Overall, Taylor et al. (2007) estimated that basic ecological data and current distributional information are lacking for approximately 60% of the hundreds of crayfish species found in the US and Canada. Given the lack of such essential information, it is difficult to accurately assess the conservation status of these species.

The persuasive recent information showing that the ranges of several species are changing in response to human-induced climate change was summarized by Parmesan (2005). The potential impact of global warming on crayfish species is yet to be thoroughly explored. However, it is reasonable to assume that a warming climate may allow at least some of the many crayfish species currently restricted to the south-eastern regions of the US to gradually expand their ranges northward. Eventually, this may lead to an increase in the crayfish species richness found in the north-eastern US states and in southern and south-central areas of Canada (including southern Ontario) (Guiașu, 2009).

For example, at this time, most of the North American semi-terrestrial burrowing crayfish species seem to prefer the warmer climate and the milder and shorter winters currently found in south-eastern areas of the US About 25% of the numerous North American crayfish species found south of 35 degrees N latitude are burrowers that occupy semi-terrestrial or terrestrial habitats. Such burrowing crayfishes are usually prevented from expanding their ranges further north in part by the occasionally frozen soil and longer winters (France, 1992). However, global warming may, theoretically, allow more semi-terrestrial burrowing crayfish to expand their ranges into areas such as southern Ontario, and perhaps other regions of southern Canada as well. On the other hand, global warming may also reduce the extent of wetland areas in Canada (Schindler, 2001), and this additional loss of wetlands would, of course, have a negative impact on all wetland species, including burrowing crayfishes.

At the same time, as the climate warms up, some southern areas of the US may become less suitable for crayfish species that prefer cooler waters, such as *Cambarus bartonii* (Cox and Beauchamp, 1982; Guiașu, 2002a).

Some of the inevitable future changes in the distributions of North American crayfish species will lead to new species moving into certain regions, sometimes, perhaps, at the expense of previously established crayfishes. However, the new species may succeed in their new surroundings mainly because they are better adapted to the changing environmental conditions. Thus, the attempts of some invasion biologists to resist or reverse such natural changes may be both misguided and futile (Guiașu, 2009).

All the species that live in the province of Ontario, and most of the rest of Canada, now arrived relatively recently, within the last 10,000 years or so. This is, of course, almost the evolutionary equivalent of a blink of an eye, taking into account that life on Earth goes back about 3.5 billion years, and considering the many climatic, geological, and biological changes that happened in this almost unimaginably vast period of time. At the height of the last Ice Age, approximately 20,000 years ago, all of Ontario and almost the entire area of modern-day Canada, as well as the northern halves of Europe and Asia, were covered by huge ice sheets hundreds of meters thick (Storck, 2004). As the ice retreated northward, starting about 12,000 years ago for the southern regions of Ontario, species from ice-free regions to the south began moving gradually into the province to occupy the newly available habitats. Thus, even the earliest post-glaciation arrivals moved into Ontario from glacial refugia in the United States only within the last few thousand years. Since the ice sheets retreated gradually, over a fairly long period of time, the colonization of central and northern areas of Ontario, after the last Ice Age, took place even more recently than the arrival of the first post-glacial species into southern Ontario. It is estimated that the glaciers disappeared completely from the Great Lakes region of North America only approximately 9,000 years ago (Grady, 2007).

Crayfish species such as *Cambarus robustus*, which presumably reached Ontario on its own after the last glaciation, are considered native, while species such as *Orconectes rusticus*, which may have arrived as recently as several decades ago, perhaps initially with some human help, are assumed (by some ecologists) to have been introduced, as previously discussed (Guiașu et al., 2006; Guiașu, 2008a). I was at a Canadian scientific meeting a few years ago, where one biologist referred to the rusty crayfish as "the American crayfish". The implication was that this "foreign" crayfish invaded Ontario from the United States, and would have a negative effect on our local, homegrown crayfishes. However, in a sense, all our current "Canadian" crayfishes are, in fact, fairly recent immigrants from the US, and there are no crayfish species endemic to Canada, so appealing to Canadian nationalism in this context seems somewhat far-fetched. Besides, as discussed, *O. rusticus* is already quite unpopular in many US jurisdictions as well.

Thus, essentially, the earliest "native" species now found in Ontario, and a good part of the Northern Hemisphere, have only arrived in these regions a few thousand years ago. The re-colonization of the regions formerly occupied by ice is a work in progress, and it is likely that this complex and lengthy process continues slowly and naturally today.

For example, there is evidence that the black locust tree (*Robinia pseudoacacia*) lived in Ontario before the last Ice Age but did not have enough time to get back to this province again naturally just yet. To date, this tree species has expanded its range northward on its own only as far as central Ohio and Pennsylvania (Eckenwalder, 1999). However, human reintroduction programs have brought the species back to Ontario to reforest polluted habitats where few other trees can survive. This very useful ability allows the black locust to prosper in urban and industrial areas and it is an asset in many of Toronto's wooded ravines because its roots are also particularly good at preventing erosion. Yet, despite its usefulness and history of being native to the province (before the last Ice Age), the black locust tree is still referred to in certain field guides as an aggressive invader in Ontario – simply because it was reintroduced by humans

FIGURE 6 *The black locust tree* (Robinia pseudoacacia) *in Ontario*
PHOTO BY RADU GUIAȘU

(Kershaw, 2001). A similar field guide advises readers to pick introduced flowers freely but not to touch native ones (Kershaw, 2002). This kind of advice can encourage the mistreatment of certain species and perhaps even interfere with the public's enjoyment of the beauty and diversity of nature. It may be more difficult to enjoy a walk in the woods if you are constantly on the lookout for species that are not allowed to be there.

CHAPTER 3

Speciation, Biodiversity, and Introduced Species

> From so simple a beginning endless forms most beautiful and most wonderful have been, and are being, evolved.
>
> CHARLES DARWIN

The relentless fight against invasive species is undertaken presumably in the name of the preservation of the planet's current biodiversity. However, the term biological diversity was not defined until 1980. At the time, the term included genetic diversity (the amount of genetic variation among the individuals within a species) and ecological diversity (the number of species in a given ecosystem or the species richness of that ecosystem). Biological diversity was first abbreviated to the now famous and fashionable term biodiversity in 1986. This term became more and more widely used, both in the specialized scientific literature and in related newspaper articles, and its meaning continued to expand, to include more and more things. Thus, in the 1980s, ecosystem diversity was added as a third level of biodiversity, to complement the already mentioned genetic diversity and species diversity. By the early 1990s, the ecological and evolutionary processes resulting from all the interactions among species in ecosystems were also included in a much more comprehensive and all-encompassing definition of biodiversity (Farnham, 2007).

As a result, biodiversity now includes all life on Earth, the genetic diversity shown by the sum of all the individuals within each species, all the biological communities made up of the species of our planet, all the ecosystems these species and communities are a part of, and all the ecological and evolutionary processes that take place within these ecosystems. Certainly, preserving all of this, under the ever-expanding notion of biodiversity conservation, would be a very tall order indeed. As Farnham (2007) put it: "there are those who believe that the expanding quality of the concept has rendered it useless as a conservation objective." Or, as Lautenschlager (1997) wrote in a paper provocatively titled *Biodiversity is dead*: "Biodiversity is so all-inclusive that it has become meaningless." Referring to biodiversity, Lautenschlager (1997) went on to write that:

> The problem with the word is not a lack of an acceptable definition, but that its definition is too all-inclusive. Hence, the word commonly fails to

 | DOI 10.1163/9789047426134_005

provide understanding when used in communications or in directions for research or management.

In other words, preserving everything may sound noble in theory, but would be impossible, from a practical standpoint. Thus, it is a lot easier to keep adding elements to the definition of biodiversity, than to actually start saving all the many components of this now very broad concept in practice. The more the definition of biodiversity expands, the more vague and confusing it may become, and, therefore, in reality, such a greatly expanded definition may take us farther away from (rather than bringing us closer to) reaching certain concrete conservation goals.

By including ecological and evolutionary processes in this definition of biodiversity, ecologists made sure that the concept of biodiversity became a dynamic one, where everything constantly changes and evolves. If biodiversity includes all life on Earth, then it surely also includes non-native species, as well as native ones, which means that all species should be worthy of conservation, irrespective of their geographic origins. And if all ecological and evolutionary processes within ecosystems should be respected and maintained, then the adaptations of introduced species to their new ecosystems, as well as the interactions between native and non-native species should also be given due consideration, from a conservation standpoint. Finally, since we, as human beings, are also one of the living species on Earth, the human-assisted movements of species around the world, and the novel ecosystems that resulted at least in part from human activities, over thousands of years, are also part of the biodiversity concept, and, therefore, should also be protected.

Species as the Main Units of Conservation

Fortunately, our understanding of the species which largely define the world's biodiversity has greatly expanded over the centuries, although much remains to be learned. For instance, Aristotle (384 B.C.–322 B.C.), a pioneer in the field of taxonomy, mentioned only about 550 types of animals in his writings, while Carl Linnaeus listed no more than a few thousand species of animals and plants in his most famous and enduring works (Mayr, 1982). Thus, in his well-known book *Species plantarum*, which was published in 1753 and contained 1,200 pages, Linnaeus listed approximately 5,900 species of plants. This book is generally regarded as the starting point of modern botanical nomenclature and classification, and it is probably safe to assume that nobody could list and name more plant species than Linnaeus in the 1750s. The widely recognized beginning of modern zoological nomenclature can be traced to the publication of the tenth edition of Linnaeus' *Systema naturae* in 1758, and, in this book, about 4,000 species of animals were listed (very few of which were marine animals, for example) (Mayr, 1982; Blunt, 2001). So, during the second half of the 18th century, the foremost expert of his time in the field of taxonomy listed only about 10,000 species of animals and plants, overall, in his works. Stearn (1971) gives the slightly higher figures of approximately 4,400 animal species and 7,700 plant species which had been scientifically named by Linnaeus, but these numbers are in the same general ballpark as the previously shown estimates.

In the 17th century, Antoni van Leeuwenhoek, the inventor of one of the very first microscopes, opened the door to an entirely new part of the living world by describing microscopic organisms – which included various types of bacteria, protozoa, and rotifers – for the first time (Magner, 2002; Goddard, 2010).

The discoveries of numerous species, big or small, continued in the last few centuries. Despite its enormous size, the largest fish in the world, the whale shark (*Rhincodon typus*), came to the attention of scientists for the first time only in 1828, when a specimen was captured off the coast of South Africa (Ellis, 2009).

Given the current knowledge of the world's biodiversity, the most plausible places where new species could be found would be species-rich tropical forests and coral reefs, or largely unexplored areas, such as the ocean floor. Although many newly discovered species are likely to be relatively small and inconspicuous – for example, insects – new species of rather big fish, reptiles, and even mammals are still being identified from time to time.

For example, some large animals were not known to science until the 20th century. A list of such species, with the corresponding year of formal scientific discovery between brackets, would include: the okapi (*Okapia johnstoni*) – a relative of the giraffe (1901), the Komodo dragon (*Varanus komodoensis*) – the world's largest living lizard species (1912), and the megamouth shark (*Megachasma pelagios*) – a huge filter-feeder (1976) (Ellis, 2009; Goddard, 2010).

By now biologists have identified, and assigned scientific names to, roughly 1.8 million species, and currently about 20,000 new species (twice the total number of species known to Linnaeus two and half centuries ago) are formally described every year (Mayr, 2001; Primack, 2008). The majority of all living species currently known are insects (there are about 1 million insect species), and, among the insects, the beetles are the most diverse group with more than 350,000 species or so (Evans and Bellamy, 2000). Tropical rainforests are particularly rich in species, because some of the most diverse groups – the insects and the flowering plants (there are about 240,000 species of flowering plants) – are particularly abundant there (Wilson, 1989; Primack, 2008).

Species are generally the main units of conservation (although vulnerable or endangered populations can also be important), since, as can be seen in the previous paragraphs, and throughout the book, the world's biodiversity tends to be divided into species of various main types (approximately 4,500 species of mammals, 9,500 species of birds, 40,000 species of crustaceans, etc.). The well-known biologist Edward O. Wilson, one of the pioneers in the study of biodiversity, wrote that: "life has divided into millions of species, the fundamental units, each playing a unique role in relation to the whole" (Wilson, 1992). This idea is reinforced in numerous publications, including, for instance, a book titled *Species: the units of biodiversity* (Claridge et al., 1997). The importance of species in conservation biology tends to be also reflected in the names given to various key pieces of relevant legislation – *The Endangered Species Act* passed in the United States in 1973, the Canadian *Species at Risk Act* (SARA) (2002), influential lists, such as *The IUCN Red List of Threatened Species*, and so on.

While we discover more and more species all the time, our knowledge of most of these species remains very limited. For example, only 2.5% of the species described in the scientific literature had been assessed with respect to their conservation status for the 2004 *IUCN Red List*. Furthermore, most of the species which had been evaluated for this much cited list were terrestrial vertebrates and plants from biologically well-known parts of the world (such as Europe or North America, presumably) (Baillie et al., 2004). This, of course, leaves us with significant gaps in our knowledge of huge groups, such as the vast majority of insects, and large regions containing most of the world's biodiversity (such as tropical rainforests and coral reefs).

According to the 2004 *IUCN Red List*, all of the 9,917 described species of birds and the 5,743 species of amphibians had been evaluated with respect to their conservation status. However, some other vertebrate groups were much less well-known from this standpoint, since only a small minority of fish (1,721 out of 28,500 species) and reptiles (499 out of 8,163 species) had been assessed. Perhaps not surprisingly, the situation was even more dire for invertebrate groups, as the conservation status of all but a relatively very small number of invertebrate species is largely unknown. As a clear example of that, it may be sufficient to mention that of the 950,000 described insect species mentioned in the 2004 *IUCN Red List*, only 771 (or 0.08%) were assessed (Baillie et al., 2004). Although more species have been evaluated since then, the November 2014 version of the *IUCN Red List of Threatened Species* only includes about 4%

of the world's formally described species, and groups such as the insects, fish, and reptiles remain poorly known, overall, from a conservation standpoint (Monastersky, 2014).

The lack of knowledge about the conservation status of most species can also be highlighted if we focus on a very widespread and ecologically important group, such as the ants. In 2009, Bert Hölldobler and Edward O. Wilson, perhaps the world's foremost experts on ants, stated that: "About 14,000 species of ants are known, for example, but twice that number or more are likely to exist. Of those known, fewer than 100 can be said to have been well studied." This lack of knowledge extends to the precise geographic origins of many ant species. In a detailed book on the ants of New England, Ellison et al. (2012) acknowledge that some of the ant species considered native to this region may in fact have been introduced by people long before scientific studies of local ants began. This seems particularly likely, since, according to the authors, systematic surveys of the ants only started early in the 20th century. In other words, we don't know exactly how many ant species are really "native" and how many may have been introduced as a result of human activities. The same is true for other groups of organisms as well. Arthur Shapiro – one of the greatest experts on North American butterflies (Thacker, 2004) – stated in an article published in 2002 that: "The butterfly fauna of the Sacramento Valley was unstudied before the middle of the 20th Century; there are neither old records nor old specimens." As a result, Shapiro concluded that in this region, when it comes to butterflies, "the composition of the pre-European fauna is thus unknowable." And, as we have seen, the oldest crayfish records found in the vast collections of the Royal Ontario Museum are from the early 1900s, and, therefore, we know nothing about the distribution and presence of crayfishes in Ontario before the beginning of the 20th century.

While evolutionary biology focuses primarily on, and natural selection operates mainly at the level of, the individual or even the gene (as discussed, for example, in the classic book by Dawkins, 1976), conservation biology and, by extension, invasion biology deal mainly with species. In addition to the already discussed thorny fact that we know next to nothing about the initial geographic origins and ecology of many (probably most) species, another obvious problem is that the species concept itself remains quite controversial and incompletely defined in modern biology. Entire books – for instance, Claridge et al. (1997), Wilson (1999), Stamos (2007), Wilkins (2009), Richards (2010) and Kunz (2012) – have been recently dedicated to the analysis of this tricky, and yet fundamental, concept, which sits at the very core of both conservation biology and invasion biology. If the study of biodiversity is based on the numbers and types of species, then, if we cannot clearly define what a species is, we are having what has come to be known as "the species problem" (Mayr, 2001; Richards, 2010). Obviously, in order to even begin the analysis of which species may be native and which may be introduced, and which species may be endangered and which may be relatively safe, we should know, first and foremost, where to draw the lines between the different species found in any given ecosystem. This can be a very difficult job even for the genuine experts, with plenty of time and well-equipped laboratories available, and yet, often, when it comes to making quick judgements about the identity of very similar native or non-native species in the field, amateurs are in charge.

The Diversity of Species Concepts

Wiley (1981) discussed the long-standing debates among biologists about the reality of species. In other words, are species real, distinct, and natural units of biodiversity, or are they simply artificial and somewhat subjective human-generated "units of convenience" used by biologists to make sense of the natural world? Furthermore, even if we come to the conclusion that species are, by and large, real units of biodiversity, finding suitable and universally acceptable definitions for these units has proven to be a formidable challenge. As Isaac et al. (2004) put it: "There are almost as many concepts of species as there are biologists prepared to discuss them."

In fact, more than 20 different species concepts, none of them perfect or all encompassing, are currently being used, and the numbers of species of various groups, which define our very idea of biodiversity, vary widely depending on which particular species concept is applied. As Richards (2010) mentioned, "often disagreements about species counts are due to the fact that different researchers use different ways of defining and conceiving species."

The biological species concept, popularized and promoted by Ernst Mayr (1982, 2001), among others, states that species are groups of interbreeding or potentially interbreeding organisms, which are reproductively isolated from other groups. Unfortunately, breeding information regarding many species is often not available. Sometimes, all we have are a handful of preserved specimens of a particular presumed distinct species, and no idea of how such individuals may be interacting in the wild. Furthermore, this definition does not really apply to the significant number of organisms which do not reproduce sexually.

The morphological definition of a species may sometimes appear to be more practical, since it groups species according to perceived morphological similarities which can be studied both in the field, using live specimens, and in museum collections, looking at preserved specimens. However, there can be great morphological differences sometimes among members of the same species, which may cause researchers to place such individuals in different species, particularly if the species and the full range of forms and structures among its members are poorly known. As Coyne (2009) noted, if the many different breeds of domestic dogs, obtained through artificial selection by humans, would only be known as fossils, paleontologists would undoubtedly classify many of these dog breeds as different species, due to the great morphological differences among them. But, as we know, even though breeds such as the Pekingese, or the Saint Bernard, or the greyhound, look very different from each other, they all actually belong to the same species (*Canis familiaris*, or subspecies, *Canis lupus familiaris* – controversies persist even when it comes to the exact status of this well-known taxon), and are all likely descended from the gray wolf (*Canis lupus*).

On the other hand, sometimes, members of different species may look very similar from a morphological standpoint. This can obviously make it even more difficult to distinguish native species from non-native ones. For example, invasive cane toads in Australia carried with them parasitic nematode lungworms from their South American native range. However, initially, it was believed that the cane toads were infected with a different species of nematode lungworm, which was native to Australia. The identification error was apparently made in part because of the morphological similarities between the Australian and the South American species of parasitic nematode lungworms – both of which belong to the genus *Rhabdias*. Eventually, the error was corrected by a genetic analysis of the lungworm species (Dubey and Shine, 2008). In this case, misidentifying the parasite would also lead to a different interpretation of the interactions between the introduced host species and its parasites. For instance, an infection with a native Australian lungworm would suggest that these parasites were capable of a host shift from native Australian frogs to the non-native cane toads. However, as discussed, this turned out not to be the case. The introduction of the cane toad to Australia will be discussed in more detail later on.

Sometimes, even widespread and ubiquitous species, such as the dandelion, can be surprisingly difficult to identify with certainty, using either the biological or the morphological species concepts. In most field guides, the common dandelion, which is regarded as an invasive species in North America, is listed

FIGURE 7 *The common dandelion* (Taraxacum officinale)
PHOTO BY DORIS WONG

as a single species – *Taraxacum officinale*. However, in the 19th and the 20th centuries, botanists found small differences in the leaves, flowers, and seeds of these plants, which suggested that dandelions could be divided into quite a few distinct groups. As a result, the number of presumed separate dandelion species kept on climbing. According to some estimates, there may be as many as 2,000 different dandelion species, although dandelion taxonomy and the relationships among the many different dandelion lineages remain rather complicated, controversial, and confusing. The problem appears to be that, while some older dandelion species, or lineages, still reproduce sexually, many other dandelion lineages do not. Quite a few dandelions reproduce asexually, by parthenogenesis, which means that members of such lineages are clones of the parental plant. There are many such lineages of clones, sometimes coexisting in the same areas, and they have probably evolved independently numerous times from sexually reproducing ancestral dandelions (Schilthuizen, 2001). If we become fixated on preserving genetically distinct lineages found only in relatively restricted areas, then classifying the various types of dandelions out there and trying to figure out which lineage may be native to which area should keep us very busy for a very long time…

The phylogenetic species concept, which links species according to their hypothesized genealogical descent from common ancestors, is based largely on the methods initially developed by Hennig (1950, 1966). According to this concept, species are the smallest possible groups whose members are descended from a common ancestor, and all members of a species have derived characteristics that distinguish them from individuals of other species. Therefore, these derived characteristics define each species and allow us to uncover the unique evolutionary history of each species (Zimmer and Emlen, 2013). Actually, Wilkins (2009) describes several distinct phylogenetic species concepts, but there is only so much hair-splitting that can be done on this subject in this section, and, furthermore, this book is not mainly about the various species concepts in existence. In any case, the phylogenetic species concept has been used with increasing frequency, but the application of this concept, for instance, tends to greatly increase the numbers of species in most of the groups analyzed using this method (Agapow et al., 2004; Richards, 2010). So, biodiversity appears much greater, in most cases, if we apply the phylogenetic species concept, instead of other species concepts.

Isaac et al. (2004) referred to this process as "taxonomic inflation". According to these authors, the numbers of species in several groups are increasing rapidly mostly because known subspecies are raised to the level of full species, as a result of the increasing use of the phylogenetic species concept, instead of the biological species one. Isaac et al. (2004) estimated that the use of the phylogenetic species concept has led, on average, to the recognition of 48% more species than the application of the biological species concept. Clearly, this is a substantial difference, and it is due entirely to the use of different methodologies and species concepts, and not to the discovery of new species.

In addition, species numbers for various groups are notoriously influenced by the propensity of taxonomists and systematists to be either "lumpers" (with a tendency to generally place several potentially distinct groups into a single species) or "splitters" (with a strong inclination to readily consider all such groups as distinct species). Apparently, there are also different taxon-specific classification traditions among researchers working with different groups. For example, it seems that ant taxonomists may often act as "splitters", by describing morphologically different ant populations with distinct distributions as separate species, while biologists specializing in the classification of butterflies tend to be "lumpers", since they generally consider locally distinct butterfly populations as subspecies of widely distributed species. As Isaac et al. (2004) stated: "The result is that regional or global species counts of butterflies and ants are not comparable."

As an additional example of the uncertainties regarding the numbers and exact identity of currently recognized species, we can look at land snails on islands. Despite their limited active dispersal abilities and poor saltwater tolerance, terrestrial snails have become very widely dispersed on oceanic islands. The snails have reached these islands through a number of passive dispersal mechanisms, such as possibly being carried by hurricanes and birds (dispersal of land snails by birds has been documented and is also discussed briefly elsewhere in the book). Once the snails reached islands which offered suitable habitats, the diversity of snail species increased, sometimes dramatically, through the process of speciation. It appears that several thousand snail species evolved on islands, throughout the world. As a result, many tropical and subtropical islands and archipelagoes have an impressive number of land snail species. The Hawaiian Islands, for instance, shelter between 750 and 800 such species. While the vast majority of these snail species have apparently either arrived on these islands without human assistance (although it is often hard or impossible to be sure about such things, as discussed) or are descended from species that reached the islands on their own, through passive dispersal that did not involve humans, there are also some land snail species which are considered aliens. It should be noted that, well before the arrival of European settlers, the Polynesians routinely and inadvertently helped a number of snail species, as well as certain lizards and other types of organisms, reach Hawaii and some other Pacific islands (Ziegler, 2002; Holland, 2009). Thus, humans have played a crucial role in the dispersal of many inconspicuous wild species – not just the ubiquitous, domesticated ones – for a very long time. Once again, it is probably impossible to ascertain, with any confidence, exactly how many species of animals and plants were carried to distant islands by people, storms, natural rafts, birds, or all, or several, of the above.

However, while the great diversity of terrestrial snails on many islands is obvious, the exact numbers of snail species on these islands are rather debatable and controversial. Rough estimates of the overall global numbers of such species range widely, and figures such as 24,000 and even 80,000 species have been given. As Holland (2009) put it, the lists of island snail species "are notorious for taxonomic inaccuracy, including the fact that many species remain undescribed, whereas others have been described multiple times as different species." Obviously, such confusion regarding the numbers and exact identity of land snail species can be expected to have an impact on the usefulness and effectiveness of any conservation initiatives intended to preserve the diversity of these groups on islands. If you don't even know how many species you have in a particular ecosystem, or how these species may differ from one another, it is obviously difficult to understand the myriad of interactions among them and make well informed decisions about their conservation status.

Although there are continuing controversies regarding the best ways to measure biodiversity, as well, all the major indices of biodiversity currently in use take into account the number of species, or the species richness (Guiașu and Guiașu, 2010a, 2010b, 2011, 2012, and 2014).

The various ways of defining species can have an impact not only on the way we quantify species richness, but also on our perception of some of these species. For example, if we use the biological species concept, hybrids between different species, including a native and a non-native species, should not pose a problem. If individuals of such species can readily interbreed, and produce viable and fertile offspring in the wild, then, according to this perhaps most widely used of all species concepts, the distinction between these species should be basically irrelevant.

In an informative and entertaining article titled *What is a species*?, Zimmer (2008) mentioned that there were at least 26 published species concepts in the scientific literature, and that the variety of currently available such concepts can have an influence on which species are listed (or not listed) as endangered. So, clear and logical definitions for essential terms such as species are not just bureaucratic exercises, but fundamental cornerstones for conservation biology. In order to make sense of the biodiversity all around us, we have to know how to group organisms into species, using the same criteria in each case, as much as possible, and, once we have done that, we should also have clear, consistent, and comprehensive standards for separating species into native or non-native ones, if we think that such a distinction is important, and provides a sound basis for conservation programs.

As the number of species known to us becomes greater and greater, it would be logical to assume that the number of question marks regarding the native or non-native status of many newly discovered species would tend to greatly increase, as well. For example, if we discover a new species of beetle at a particular location, how can we be sure that this beetle species is indeed native to that area, and was not brought there inadvertently by people, at some point during the last several thousand years? What if some of the beetle's ancestors hitched a ride in the belongings of some indigenous people, and were therefore transported by humans for hundreds or perhaps even thousands of kilometers to the location where they were recently found for the first time by scientists? And if the beetle had disappeared in the initial region people brought it from, and its remains did not get preserved in the fossil record there (perhaps due to unsuitable conditions for fossil preservation), we would never know it was ever present elsewhere. This is a hypothetical scenario, of course, but it is entirely plausible. We know very little about most of the new species we discover, and we often lack clear information about the geographic origins of species, which is the type of often elusive knowledge that fuels the field of invasion biology.

Furthermore, our assumptions about the geographic origins of species also tend to be subjective, and depend to a large extent on our preconceived notions about the places whose biodiversity we are exploring. For example, as we have seen earlier in the book, a plant species discovered in the Galápagos Islands as recently as 1972 was automatically considered (erroneously, as it turned out) to have been introduced there by people, even in the absence of any relevant evidence, because we did not expect to find new native plant species in the Galápagos Archipelago, and the biologists working there were conditioned to think that those islands need to be protected from exotic species. On the other hand, any new species discovered in the Amazon rainforest, for instance, is generally assumed to be native, even though we now know that the biodiversity and the ecosystems of the Amazon region were modified and influenced by the activities of indigenous people to a much greater extent than previously thought. For example, Heckenberger et al. (2008) reported that archaeological studies conducted in the Upper Xingu region of the southern Amazon (in Mato Grosso, Brazil) show evidence of "a highly self-organized anthropogenic landscape of late prehistoric towns, villages, and hamlets, with well-planned road networks across the region." It is clear that pre-Columbian, prehistoric human civilizations in the Amazon region have altered the forested environments of the area, and human-caused changes in the distribution and abundance of local species must have taken place for thousands of years. In most cases, identifying the exact nature and full extent of such ancient changes would probably be impossible. Therefore, we may well have to ask ourselves if there are still any truly pristine ecosystems, entirely untouched by the impacts of human activities, left anywhere on Earth. Increasingly, we are finding out that many of the ecosystems assumed to have been "pristine" and "uncontaminated" by people had been in fact repeatedly modified and disturbed by humans many times throughout history and prehistory. So, often, the diverse "untamed" ecosystems first discovered by Europeans in various parts of the world were just at a particular stage which resulted at least in part from previous human-caused or human-influenced changes. Thus, not only is the quest for pristine ecosystems probably futile and utopian, but the so-called restoration of such ecosystems would only mean a partial restoration, at best, to a previous version of a human-modified environment.

Exotic Species, Great Voyages of Discovery, and the Growth of Natural History

Mayr (1982) stated that "the early naturalists knew only the limited fauna or flora of their home country." From the context of the book section this quote was taken from, it is clear that Mayr is referring to naturalists who lived several hundred, and perhaps even a thousand or more, years ago. In a way, it could be argued that modern day invasion biology is bringing us back to that ancient state, by encouraging a positive focus only on the native species found in each country, province, county, or neighbourhood. Species coming from elsewhere are not only unwelcome, but the main interest of invasion biology, as far as exotic species are concerned, is to find ways to get rid of them and limit their spread and ecological success. At least the early naturalists had the excuse of not being aware of the much greater biodiversity found beyond the borders of the relatively small parts of the world they were familiar with.

Ironically, in light of the currently negative attitudes towards non-native species, the fascination with exotic species was one of the key factors which drove the great European voyages of exploration of the 1700s and 1800s. In those days, the interest in non-native species was much more positive in nature, and remarkable naturalists and biologists were doing their best to study these species and bring them to Europe. Both the scientists and the members of the general public were very much interested in seeing these exciting foreign species and learning about them. As a result, some of the world's great botanical and zoological collections, made up of both live and museum specimens, were born in those days, and the general awareness and knowledge of the extent and value of the world's biodiversity grew enormously.

The famous London Zoo started receiving visitors in the late 1820s, and played an important role in introducing the public to many fascinating and previously little known exotic species. The number of animal species on display expanded quickly, and the Zoo became an important and popular attraction. The animals were housed in diverse and innovative pavilions, such as the world's first Reptile House (1849), the very first public Aquarium (1853), the first Insect House (1881), and so on (Barrington-Johnson, 2005; Ito, 2014). The British Museum (Natural History), better known as the Natural History Museum, opened in London in 1881 and became one of the most comprehensive and well-respected institutions of its kind (Whitehead and Keates, 1981). The impressive Smithsonian's National Museum of Natural History, in Washington, D.C., USA, opened in 1910 and now welcomes about 7 million visitors per year (Barry et al., 2008).

There are now many collections of both living and preserved animals and plants in gardens, zoos, aquaria, and museums throughout the world.

I remember fondly my own childhood visits to the wonderful Grigore Antipa National Museum of Natural History (named after a distinguished Romanian biologist) in Bucharest, and the many hours I spent there, watching the dioramas and preserved specimens from all over the world, were decisive in leading me eventually towards the pursuit of an academic career in biology. Several years later, I also had the opportunity to visit the Smithsonian's National Museum of Natural History, for research purposes, only a few months after starting my graduate studies at the University of Toronto. I was given a "Visiting Scientist" badge, which allowed me free access to the public exhibits, and, as a result, I could take my time and explore the museum at a leisurely pace, over several days. The only slight drawback was that, upon seeing my official looking badge, some visitors asked me to direct them to the cafeteria or the nearest washrooms, undoubtedly assuming that this is why I was there.

Mayr (1982) referred to the 18th century as "the great age of natural history". During this time there was a remarkable expansion of the scientific knowledge of the world's biodiversity. For example, the German naturalist and physician Georg Wilhelm Steller accompanied Captain Vitus Bering during his second voyage, which was completed between 1740 and 1742 and was intended to explore the northern Pacific region (Macleod et al., 2010). In 1741, Steller was the first European to set foot in Alaska, and during his brief stay there he made notes on several previously undescribed species including the salmonberry. Bering died on this voyage, as his ship was returning to Russia, but Steller survived and wrote an account of this trip which was eventually published. Although many of his original scientific papers were lost, some of Steller's descriptions of plants and sea animals were preserved, and he is given credit for the scientific discovery of several new species both in Siberia and in north-western North America (Frost, 1988; Williams, 2013). Some of the species described by Steller, which have been subsequently named by others in his honour, include birds (Steller's jay – *Cyanocitta stelleri*, Steller's sea eagle, Steller's eider), mammals (Steller's sea lion, and the now extinct Steller's sea cow) and plants (the Alaskan common alpine heather – *Cassiope stelleriana*) (Littlepage, 2006). In particular, Steller's notes are the only scientific record of the giant sea cow (*Hydrodamalis gigas*), which became extinct due to hunting in 1768. Virtually everything we know about the behaviour, anatomy, and ecology of this large relative of the manatee comes from the the observations made by Steller in 1741 (Frost, 1988; Flannery and Schouten, 2001; Williams, 2013). Steller's work was also known to, and appreciated by, Linnaeus, who incorporated some of Steller's descriptions of plants into his own writings (Littlepage, 2006).

Almost three decades later, the botanist and naturalist Sir Joseph Banks accompanied Captain James Cook during his first voyage around the world, which took place between 1768 and 1771. During this time, Banks collected thousands of plant specimens, including some belonging to more than 1,000 plant species previously unknown to science (Lomolino et al., 2010). On this voyage, Banks was accompanied by Daniel Solander – a botanist who had received training from Linnaeus – as well as several other assistants and artists, whose job was to preserve and illustrate the specimens collected. In fact, Botany Bay – the first place where the crew managed to land during the pioneering exploration of Australia – was named by Captain Cook in honour of the spectacularly successful work performed by the ship's naturalists. Together, Banks and Solander collected not only many plants, but also quite a few species (including new ones) of birds, fish, molluscs, crustaceans, insects, and jellyfish at various locations throughout mostly the Southern Hemisphere. One of the more unusual and spectacular species described by Banks in 1770 in Australia was a kangaroo, and the skin of one specimen was brought back to England, together with some sketches of a living animal (Williams, 2013).

Such voyages of scientific discovery, which greatly enhanced our knowledge of the natural world, continued in the 19th century, culminating with Charles Darwin's famous journey around the world on the HMS *Beagle* between 1831 and 1836. During this voyage, Darwin collected thousands of specimens, including some of the well-known finches he found on the Galápagos Islands, which influenced his views on the changing or evolving nature of species. Darwin's finches, as they are now known, were sorted out by ornithologist John Gould after Darwin's return to England (Macleod et al., 2010; Williams, 2013). Fourteen species of these finches are now recognized – thirteen of these species are found in the Galápagos Islands, and one is found on the Cocos Island (or Isla del Coco), which is located in the eastern Pacific Ocean, about 800 kilometers from the Galápagos Archipelago. Genetic studies revealed that all of these finches form a monophyletic group, and they are all descended from a common ancestor which colonized the Galápagos Islands from the South American mainland two to three million years ago. There is speculation that the South American ancestors of Darwin's finches may have been pushed out towards the remote Galápagos Islands by powerful winds (Zimmer and Emlen, 2013). In other words, this was a rather unusual and random colonization event, which led to several speciation events, as the birds adapted to the specific conditions found on various islands. It may be worth noting that such a successful colonization event may be regarded with suspicion, and possibly even fought against, these days by invasion biologists. If a few South American birds would somehow manage to make the long journey to the Galápagos Islands today, by unknown means, there is a real chance that these birds would be regarded as invasive, and eradicated, in order to "protect" the "pristine" archipelago from outsiders and resist changes to the current species compositions and interactions. If invasion biologists have their way, speciation events such as the ones that led to the formation of the fourteen species of Darwin's finches may not be allowed to happen from now on. Instead of keeping an open mind towards exotic species, the way the early naturalists had done, many ecologists now simply see them as nothing more than a threat.

Hybridization between Native and Non-native Species: Something to Fear or a Possible Major Source of Future Biodiversity through Speciation?

The so-called "species problem" is clearly not the fault of invasion biologists, but it is an additional complication many of them do not seem particularly aware of. This should be of some concern, however, especially when dealing with closely related and very similar species, some native and some non-native, which are now occurring in the same areas. Not only is it often very difficult to tell such species apart, but, in addition, the lines between them may often be blurred even more by extensive interbreeding.

For example, a few years ago, someone from the Ontario Ministry of Natural Resources (OMNR) sent me a few crayfish specimens for analysis, since I am, for what this may be worth, considered to be one of the relatively few crayfish experts in this province. My assignment was to determine the species identity of these specimens, and, in particular, to try to find out if the crayfish were hybrids between the so-called invasive species *Orconectes rusticus* (the rusty crayfish) and another, presumably native, species, *Orconectes virilis*. After carefully examining the specimens in the lab, I concluded that no hybrids were present. Some of the specimens I had received were *O. rusticus* and the others belonged to the *O. virilis* species. The news probably came as a relief to the OMNR employee who had sent me the specimens, but, as I reflected further on this matter, I found myself wondering why potential hybrids between two *Orconectes* species should cause such concern. Presumably, some biologists feel that interbreeding between individuals from a native and a non-native species would somehow dilute the genetic "purity" of the native species. Understandably, this can be a potentially delicate and uncomfortable subject, with rather unseemly implications. If two individuals of two different species freely choose to interbreed in the wild, and, as a result, we get healthy and fertile offspring, well adapted to their environment, why should we mind? Why the obsession with making sure that native species never get any new genetic contributions from individuals of other very closely related species, which may happen to be relatively recent arrivals to the area?

Sometimes, hybrids may be better adapted to the current environmental conditions than individuals of either parental species. As a result, such hybrids will be stronger and survive better than the parental species – a phenomenon known as hybrid vigor (Primack, 2008). Eckenwalder (1999) analyzed the tree species growing in the Toronto region, and found that some exotic and native species can hybridize successfully, and the resulting hybrids are fairly common (such as the ones formed as a result of hybridization between the Siberian elm and the native slippery elm). Hybridization among introduced and native tree species in the urban forests of Toronto can also be expected in genera such as *Salix* (willows) and *Tilia* (linden trees), among others. Apparently, many such hybrids can be difficult to recognize and may be much more common than previously believed. Eckenwalder (1999) concluded that "hybrid vigor may make such crosses particularly successful in coping with the stresses of an increasingly urbanized environment."

This is evolution in action, and, in some cases, it may lead to hybrids better adjusted to our changing ecosystems, and eventually even to speciation (the formation of new species), and, therefore, to a gradual increase in biodiversity in the long run. In a recent one page comment published in the journal *Nature*, Chris Thomas (2013) convincingly argues for the usefulness of speciation driven by hybridization between native and introduced species in generating increased future biodiversity. Mallet (2007) suggested, based on genetic evidence, that hybrid speciation "is more common in plants and also animals than we thought" and gave several examples of hybrid species formation in plants, insects, and fish. Thomas (2015) expanded his argument in a new study, and offered evidence and analysis which showed that the rate of speciation due to human activities – including agriculture, horticulture, and the transport of species throughout the world – and subsequent hybridization between native and non-native species may be higher than the rate of extinction at least for some groups at certain locations. For instance, although the relevant databases are admittedly incomplete, it appears that in the last 300 years or so (between 1700 and 2010), the number of new plant species formed through hybrid speciation in Great Britain and in the rest of Europe is higher than the number of plant species which became extinct. As a result, plant biodiversity clearly increased overall in this region, due to the hybridization process taking place between introduced and native species (Thomas, 2015), and this is likely happening in other parts of the world as well. Thus, perhaps we should not only accept, but also celebrate, such natural processes, rather than fighting against them.

Sometimes, the species obtained as a result of such a hybridization process can also be of great importance to people. For example, the most widely cultivated wheat species today is the bread wheat (*Triticum aestivum*), which is the result of the hybridization process between Emmer wheat (*Triticum turgidum*) and the wild grass species, *Aegilops squarrosa* (Mindell, 2006).

Hybridization also seems to have been responsible for a small part of the genetic heritage of certain modern human populations. After members of our own species (*Homo sapiens*) left Africa for the first time, somewhere between 60,000 and 40,000 years ago, some of them eventually encountered the Neanderthals (*Homo neanderthalensis*) in parts of Europe and possibly also Western Asia. Recent studies suggest that interbreeding between these two closely related species of the genus *Homo* did occur and resulted in some hybrid offspring. This may explain the fact that an average of 2.5% of the genes of modern humans living in parts of Europe and Asia today are of Neanderthal origin (Zimmer and Emlen, 2013). Neanderthals and modern humans are now widely regarded as distinct but very closely related species. In fact, Neanderthals are seen as the closest hominid relatives, or the sister group, of *H. sapiens* (Stringer, 2002; Noonan et al., 2006). However, according to the biological definition of species, Neanderthals and modern humans should not be considered as separate species at all, since it appears that they managed to interbreed naturally and produced viable and fertile offspring.

Neanderthals were native to Europe and parts of Western Asia. Despite their remarkable ability to survive in the harsh climate of Europe during the last Ice Age, Neanderthals became extinct about 30,000 years ago, roughly 10,000 years after the arrival of modern humans on the European continent. The exact nature of the interactions between these two closely related species is uncertain and hotly debated, but it appears likely that *H. sapiens* may have contributed to the extinction of *H. neanderthalensis* possibly through competition for resources and perhaps even direct conflict (Mellars, 2004; Stringer and Andrews, 2005).

Early in my research career, I studied the phylogeny and evolution of the surgeonfishes (family Acanthuridae) – a group of tropical, coral reef fishes which are particularly diverse, or species rich, in some regions of the Indian and Pacific Oceans (Guiașu, 1991; Guiașu and Winterbottom, 1993 and 1998). The common name of these fishes comes from the scalpel-like fixed or movable (depending on the genus) spines found on each side of the caudal peduncle, which can be used as effective defensive weapons. Randall (2001) reported several examples of naturally occurring hybridization between various acanthurid species. Hybridization was observed in waters near Hawaii between the common species *Acanthurus achilles* and the locally rare *Acanthurus nigricans*. Apparently, some *A. nigricans* individuals can sometimes join spawning groups of *A. achilles*. Within the genus *Acanthurus* – the largest of the surgeonfish genera with respect to number of species – other hybrids were reported between *Acanthurus pyroferus* and *A. tristis*, *A. olivaceus* and *A. nigricauda*, as well as *A. olivaceus* and *A. tennentii*.

During my initial work on the evolution of the six acanthurid genera (*Acanthurus*, *Ctenochaetus*, *Naso*, *Paracanthurus*, *Prionurus*, and *Zebrasoma*) (Guiașu, 1991; Guiașu and Winterbottom, 1993), I also uncovered some potentially interesting interspecific differences among the species of the genus *Zebrasoma*, which were analyzed further in a subsequent study (Guiașu and Winterbottom, 1998). According to our preferred hypothesis of the phylogenetic relationships among the six *Zebrasoma* species (*Z. flavescens*, *Z. gemmatum*, *Z. rostratum*, *Z. scopas*, *Z. veliferum*, and *Z. xanthurum*), *Z. scopas* and *Z. flavescens* are sister groups (that is, they are each other's closest relatives) and *Z. rostratum* is the sister group of a monophyletic group formed by *Z. scopas* and *Z. flavescens* (Guiașu and Winterbottom, 1998). Interestingly, Randall (2001) reports the presence of hybrids between the sister groups *Z. flavescens* and *Z. scopas*, as well as between the closely related *Z. rostratum* and *Z. scopas*. Thus, it appears that these species are still closely related enough to each other to be able to interbreed and generate viable hybrid offspring. Hybrids among acanthurids, and no doubt some other coral reef fishes, can occur if two or more closely related species engage in spawning activities side by side at the same time (Randall, 2001).

Among some invasion biologists, hybridization seems to be a dirty word, as long as one of the parental species is considered to be a non-native one. Again, this often struck me as a bizarre attitude, which seems to reflect a static, stagnant, and thus non-evolutionary view of the living world. According to such an attitude, we not only have to preserve the exact same species we now have, everywhere around the world, with no new additions or other changes, but we also have to make sure that even very rare and barely surviving species, with only a handful of individuals left, never receive any fresh genetic input from closely related newcomers which have arrived in the neighbourhood. Far from restoring a "natural" order to our already much and often modified ecosystems, this seems like a rather unnatural way to proceed – essentially, we are getting in the way of normal interbreeding among individuals of some closely related species, and trying at all costs to preserve some vulnerable species, basically by keeping them in a sort of bubble, where they are not allowed to interact naturally with other species in their environment.

This typically negative invasion biology approach to hybribization between native and non-native species can also be illustrated by the attitude towards the introduced white mulberry (*Morus alba*) tree in Ontario. Because the white mulberry can form hybrids with the closely related native, and locally rare, red mulberry (*Morus rubra*), the presence of the white mulberry is regarded by some as a problem. The verdict regarding this matter given in a comprehensive recent field guide on the plants of southern Ontario was unequivocal: "This hybridization is a threat to native stands of red mulberry." (Dickinson and Royer, 2014).

I was at an ecology and evolutionary biology conference held at the University of Guelph, years ago, and I remember attending a presentation on the vulnerable red mulberry populations of Ontario. At the end of the talk, a member of the audience suggested, apparently in all seriousness, that we should perhaps remove all white mulberries from the province to prevent any further hybridization between the two local mulberry tree types. This suggestion seemed to frighten even the speaker, whose main topic was the conservation of the red mulberry tree. As I recall, he quickly said something to the effect that the data presented were preliminary, and we should not jump to conclusions and do anything drastic. But it was clear that there were at least some ecologists in the room for whom this idea did not seem entirely crazy.

Case Study: The Ruddy Duck in Britain

Hybridization, or even the possibility of hybridization, between an exotic species and a rare native one can lead to a death sentence for many individuals of the introduced species, and even full-scale eradication attempts. A rather unsettling example of this is provided by the anti-ruddy duck campaigns conducted in parts of Europe – particularly in Britain and Spain.

The ruddy duck (*Oxyura jamaicensis*) is native to North America. This species was deliberately introduced to the UK in 1948 by Sir Peter Scott, a conservationist and bird painter, and a few individuals escaped from wildfowl sanctuaries in the early 1950s. As a result, relatively small wild ruddy duck populations became established in Britain, over the next few decades. The total ruddy duck population in the UK probably contained approximately 3,500 individuals by the 1990s. This was not considered to be a problem, initially, but then conservationists realized that the ruddy duck can migrate from Britain to Spain at the beginning of winter, and, once in Spain, these introduced ducks can interbreed with the rare native white-headed duck (*Oxyura leucocephala*). The first hybrid obtained as a consequence of interbreeding between these closely related duck species was first observed in 1991. Apparently, the hybrids do not have mostly white heads, and there were fears that interbreeding with the introduced ruddy duck may destroy the perceived uniqueness of the remaining white-headed ducks of Spain (Milton, 2000; Rawles, 2004; Vidal, 2012; Carrington, 2014; Carter, 2014). It should be mentioned that only the adult males of both *O. leucocephala* and *O. jamaicensis* have conspicuous partly white heads. The main difference is the percentage of the head covered by white plumage – as expected, based on the common name of the species, adult male white-headed ducks have mostly white heads, with some black feathers on top, while only the lower half of the heads of male ruddy ducks is white. In other words, the differences between the two species are subtle, and male hybrids may presumably be identified in part by the amount of white feathers on their heads – a rather trivial distinction.

To make matters even more complicated, there is a fair bit of intraspecific variation in the amount of white and black plumage on the heads of males of both the *O. jamaicensis* and the *O. leucocephala* species (Johnsgard and Carbonell, 2010). The females of the two species are less colourful than the males, and can be even tougher to tell apart on the basis of differences in feather colours (Jonsson, 1993).

Furthermore, hybridization among various duck species is apparently a common phenomenon, and some experts even doubt that the ruddy duck and the white-headed duck are truly distinct species (Milton, 2000; Rawles, 2004). Mallet (2005) estimated that 16 of 21 (76.2%) duck species native to Britain were capable of hybridizing with at least one other species in the wild. Since Mallet (2005) excluded hybridization between native and non-native species from his analysis, this very high percentage of hybridizing species of British ducks obviously does not include hybrids between ruddy ducks and other local duck species. This study clearly shows that hybridization among ducks of various species is widespread and should be considered a perfectly natural phenomenon. The boundaries between these species do not appear to be as clear-cut as some conservationists may think.

The range of the white-headed duck – which once included wider areas in the Mediterranean region, as well as parts of Eastern Europe and Central Asia – became much reduced during the 20th century, and the numbers of these ducks plummeted particularly due to hunting and the destruction of wetlands. There was uncertainty about the exact size of the global population of white-headed ducks – estimates ranging from 19,000 to 5,000 birds were found in the relevant conservation literature in the 1990s. The attempts to increase the white-headed duck population of Spain were relatively successful, since the numbers of these ducks increased from a low of only 22 birds in 1977 to more than 1,000 twenty years later (Jonsson, 1993; Milton, 2000). Spanish conservationists clearly saw the newly arrived ruddy ducks as a threat to the white-headed ducks from the very beginning, and, therefore, ruddy ducks and the hybrids between the two species were shot in Spain shortly after their discovery in 1991. Since the ruddy ducks were coming from Britain, conservationists in Spain and some other parts of Europe asked for the implementation of control programs against this species in the UK as well. As a result, under international pressure, the British government and the relevant conservation organizations (such as the Wildfowl and Wetlands Trust and the Royal Society for the Protection of Birds) approved an anti-ruddy duck campaign which eventually led to the killing of thousands of birds. Apparently, the shooting of female ruddy ducks while they are sitting on their nests is a key part of the control program. This program was generally supported by conservationists in Britain, but opposed by some local landowners and several other members of the public (Milton, 2000; Rawles, 2004; Vidal, 2012). In fact, landowners only gave permission for control programs to be undertaken on their properties about 50% of the time (Smith et al., 2005).

Despite the concerns and the controversy, the killing of ruddy ducks went ahead at 132 sites and may soon achieve the goal of virtually eradicating this species from Britain. There may have been about 6,000 ruddy ducks in Britain in the year 2000 (Carter, 2014). The eradication campaign against them started in 2005 (Henderson, 2009). An estimated 2,689 ruddy ducks were destroyed between September 2005 and January 2007, according to the country's Department for Environment, Food and Rural Affairs (Defra) (Vidal, 2012). In a progress report on the UK ruddy duck eradication program, Henderson (2009) stated that more than 6,200 of these ducks were killed in England, Scotland and Wales, which led to a population reduction of almost 90% by the winter of 2008/2009. By 2012, as many as 6,500 ruddy ducks were killed and the costs of the control program added up to more than £5 million. At that time, no more than 100 ducks of this species were left in the UK, and the government announced the allocation of an additional £200,000 to eliminate the remaining local *O. jamaicensis* individuals (Vidal, 2012). By 2014, only approximately 50 ruddy ducks were left in the UK, and Defra allocated another £120,000 to finish them off as well, which meant that the government was willing to spend £2,400 per bird for the killing of each remaining British ruddy duck. This rather extravagant expense was described as "wasteful" by some experts, while others have described these eradication efforts as "obsessional" (Vidal, 2012; Carter, 2014). However, as reported in a recent newspaper article, a spokesperson for Defra reaffirmed her organization's commitment to the complete elimination of the ruddy ducks by stating: "We will continue to fund ongoing work to eradicate this non-native species from the UK to protect the white headed duck." (Carter, 2014).

Possible Speciation in Progress Due to the Introduction of a New Species: The Apple Maggot Fly Case Study

The introduction of apple trees (*Malus pumila*) to North America by Europeans about 400 years ago provided a useful new plant resource not only for the human settlers, but also for a native insect, *Rhagoletis pomonella*, currently known as the apple maggot fly. This species is a fruit parasite whose life cycle revolved mainly around the small red fruits of the hawthorn tree (*Crataegus* spp.). The hawthorn tree and this particular parasitic fruit fly are both considered native to North America (Zimmer and Emlen, 2013).

However, the arrival of the apple trees on this continent changed the life cycle and the course of evolution for some of these flies. In the early 1860s there were published reports about some *R. pomonella* flies which were using apples instead of hawthorn fruits in orchards located in the Hudson River valley in New York State. Apple maggot flies have now been observed on apples throughout the north-eastern and central regions of North America, and have even made their way to apple orchards on the West Coast of the continent (Prokopy and Roitberg, 1984; Bush, 1994; Schilthuizen, 2001).

The adult *R. pomonella* flies mate on the host fruit, and the females of this species lay their eggs, using the sting-like ovipositor, inside this fruit. The males establish territories on the fruits, and will vigorously fight other conspecific males to keep or gain control of these vital resources. When they are not defending their territories against rivals, the males will also try to attract the females with particular wing-waving displays. The courtship, mating, and egg-laying activities take place while the fruits are still attached to the branches of the tree. Afterwards, the overly ripe fruits eventually fall to the ground. The fly eggs hatch inside these rotting fruits. The larvae develop there as well, and feed on the fruit. During the winter, the larvae move into the soil, where they spend the winter in a lengthy dormant stage known as diapause. The next generation of adult flies will emerge during the next summer and start their search for the same species of host fruit they were born in. Adult flies can live about one month, and therefore have a relatively short time to locate suitable host fruits and reproduce (Prokopy and Roitberg, 1984; Schilthuizen, 2001; Zimmer and Emlen, 2013).

These fruit flies recognize, and are attracted by, the odour of their favourite host fruit. They also apparently identify the fruits by colour and size (Linn et al., 2003; Zimmer and Emlen, 2013). The preference for different species of fruits separates the hawthorn flies from the apple ones, even though the two varieties of flies – also known as host races – live in the same general areas.

Over time, in the approximately 150 years or so since some *R. pomonella* fruit flies switched from native hawthorn fruits to introduced apples, genetic differences have emerged between the two host races. Since their ranges overlap to a great extent, these host races are sympatric, of course, and, therefore, their genetic divergence is considered a textbook case of emerging sympatric speciation (the opposite of allopatric speciation, which requires geographic isolation). The reproductive isolation between the hawthorn and the apple races of *R. pomonella* is also maintained because the two types of flies come out of diapause at different times. Apple fruits usually become ripe several weeks before the hawthorn fruits, and, as a result, the host race dependent on apples will emerge two to three weeks earlier than the host race which favours hawthorn fruits. Since the adult flies only live for a few weeks, this difference in the timing of the end of the diapause ensures that adults from the two host races are not usually seeking mates and host fruits at the same time. This clearly reduces the chances of potential encounters between members of the two fly races even further, which keeps the sympatric speciation process on track. It should be noted that adults of the two host races are capable of mating with each other if they are artificially brought together in the laboratory. However, in the field, each host race normally spends most of its active life on a different species of host fruit, as discussed, and therefore the separation between these races is maintained naturally outside the lab (Bush, 1994; Ridley, 1996; Schilthuizen, 2001; Pennisi, 2006; Zimmer and Emlen, 2013).

It is often said that introduced species are successful because they don't have any parasites and pathogens in their new surroundings. In this case, however, it clearly didn't take very long for a parasitic fruit fly to "discover" apples, and take advantage of this valuable new resource in the environment. The case study illustrates that it is quite possible for a native species to be the "pest" which exploits a beneficial exotic species. In other words, the overly familiar "alien species bad, native species good" scenario generally promoted by invasion biology does not apply here. This classic example of sympatric speciation in progress also shows that the introduction of non-native species into a particular region can bring about the evolution of new varieties within the native species. In the long run, this should lead to the formation of new species, and, therefore, to increases in local biodiversity.

Some people may dismiss the evolution of new types of fruit parasites as undesirable, and that is undoubtedly true from the point of view of orchard owners. However, parasitic insects are also an integral and abundant part of the planet's biodiversity and complex ecosystems, obviously. Therefore, if we are really interested in understanding the mechanisms that lead to increased biodiversity, then surely the formation of new varieties of *Rhagoletis* fruit flies, and possibly numerous other herbivorous insects (which represent an

FIGURE 8 *The apple maggot fly* (Rhagoletis pomonella)
PHOTO BY ANDREW FORBES

impressive percentage of the global species count), as a result of adaptations to novel plant hosts, would have to be considered both important and fascinating. Otherwise, if only species we deem "desirable" or "useful" are allowed to exist, we would end up with nothing but domesticated plants and animals, and a few rare species kept alive on small pockets of rather artificial habitats, maintained at great expense, mainly as exhibits of a living world which can never again exist on its own, without our constant supervision and care.

In a very interesting study, Forbes et al. (2009) showed that the sympatric speciation, in this case, was not restricted to the *Rhagoletis* fruit flies. Not only were these parasitic flies undergoing sympatric speciation in response to the introduction of novel fruit trees in the environment, but, in addition, the evolution of different fruit fly varieties, or host races, led to possible sympatric speciation in the parasitic wasps which attack the fruit flies. The wasp *Diachasma alloeum* lays its eggs in the bodies of *Rhagoletis* larvae or maggots. As the wasp's own larvae develop, they devour the fly maggots from the inside and then burst out of their hosts' bodies – a gruesome concept somewhat reminiscent of the *Alien* movie series. So, the fruit flies make their living as parasites of the fruits, and the parasitic wasps, in turn, exploit the fruit flies.

Since the parasitic wasps obviously present a lethal danger to the flies, the host shift from hawthorn fruits to apples was also seen as a way for the apple maggot flies to leave their parasites – the dangerous wasps – behind. With the parasitic wasps still affecting the fruit flies dependent exclusively on the hawthorn fruits, the *Rhagoletis* flies living on apples could presumably enjoy a relatively safe and carefree existence. However, this state of affairs did not last, as some of the parasitic wasps adjusted and followed the apple maggot flies on the journey to the apple trees. Over time, both the fruit flies and the wasps adapted to a particular fruit tree and avoided or ignored other types of fruits.

Diachasma alloeum wasps now attack the hawthorn and the apple dependent varieties of *Rhagoletis pomonella*, as well as the closely related fruit fly species *Rhagoletis mendax*, which lives on blueberries, and *Rhagoletis zephyria*, which prefers snowberries. However, as the wasps adjusted to their specific *Rhagoletis* host, genetic differences started to emerge among the parasitic wasp populations as well. Thus, over time, the wasps which parasitized the apple maggot flies became genetically distinct from the wasps which attacked fruit flies dependent on hawthorn fruits. Tests showed that, just like the fruit flies, the wasps preferred the smell of the fruit they were raised on or near. The wasps did not feed on the fruit directly, of course, but they did consume the larvae of the flies which ate the fruit, so, in order for the wasps to find their particular variety of fly hosts, they had to orient themselves towards the right type of fruit as well. It appears that the same mechanisms which contributed to the reproductive isolation of the *Rhagoletis* varieties which feed and mate on different host fruits also led to the separation of the *D. alloeum* wasps which attack the various types of flies. As we have seen, the different fruit fly varieties are genetically distinct and each of them reach the adult stage at different times of the summer, in order to be able to lay eggs on their specific host fruit. Therefore, the emergence times of the various types of fruit fly adults have to be synchronized with the fruit ripening times of their host fruit trees. The parasitic wasps, in turn, have to synchronize their life cycles with those of their particular fruit fly hosts. As a result, since the adult flies and wasps only live for a few weeks, the opportunities for mating with conspecifics of other host races are minimized (Forbes et al., 2009). The different types of *Rhagoletis* fruit flies and their corresponding wasp parasites are becoming genetically distinct, in each case, by adjusting to different ecological niches, and, therefore, they may be undergoing a process known as ecological speciation (Zimmer and Emlen, 2013).

Sympatric speciation by host shift may be quite common in insects. Ridley (1996) mentioned the possibility that this mechanism may explain, at least in part, the existence of 750 species of fig wasps, each of which is associated with a different species of fig. Also, since the number of insect species feeding and mating on plants is enormous, and parasitic wasps may represent 20% of all insects, these types of linked sympatric speciation events may be very frequent and important for the growth of global biodiversity (Forbes et al., 2009).

CHAPTER 4

The Controversies Regarding the Perceived Negative Impacts of Non-native Species

> A man should look for what is, and not for what he thinks should be...
>
> ALBERT EINSTEIN

> Even if deleterious effects have not occurred, invasion involves change, and ecologists tend to oppose change.
>
> LAWRENCE B. SLOBODKIN

Upon reading some of the articles and books describing the terrible or potentially catastrophic impacts of introduced species in their new surroundings, we are often confronted with an apparent paradox. On the one hand, we are told that native species are more valuable because they are better adapted to the ecosystems in which they originated, but, on the other hand, we are also informed that the newly arrived introduced species are taking over by outcompeting native species. Obviously, there is a contradiction here. If native species are indeed better adapted to their environments, then clearly there is nothing to fear from competition with the "out of place" and poorly adapted exotics.

When confronted with this conundrum, some invasion biologists will respond that native species may lose "the battle" with non-native ones because their environment has been "disturbed" and changed by humans. This is especially true in and around large urban or agricultural areas, for example. So, what would a practical solution to this perceived problem be? Reversing urbanization? Reducing the extent of the areas set aside for farming? Removing any trace of human impact in these regions? This would be highly unlikely to work out very well, given the rapidly rising global human population, the need to produce more and more food to feed everybody, and the fact that more and more people are moving into the cities. Under these circumstances, does it make sense to insist on favouring native species over introduced ones everywhere, under all conditions? This seems like a rigidly ideological, largely utopian, and potentially destructive and expensive endeavour, and the scientific basis for such aggressive promotion of "native" species over "alien" ones is often simply not there. After all, from an evolutionary standpoint, if a species is thriving in a particular environment, how can that species be "in the wrong place"?

 | DOI 10.1163/9789047426134_006

In a sense, isn't the use of the term "invasive", in this context, another way of stating that a species is particularly well adjusted to its current surroundings?

Proponents of the notion that native species should be encouraged and promoted exclusively, even in urban settings, often state that native species will do well with minimal care or maintenance in their presumably native environments, because, after all, such species are supposedly perfectly adapted to local conditions. This type of reasoning is frequently used in support of programs favouring native species. Plant only native species in your garden, the slogan goes, because they'll require little to no maintenance (unlike those pesky exotics), and besides, you'll never have to feel guilty about assisting and abetting plant invasions that could ruin our fragile native ecosystems... Well, if native species are indeed so well adapted, and require so little care and maintenance, then why do we even need to become involved and promote them, and why do we have to first remove and suppress introduced species in order to allow the native species to flourish? In essence, some invasion biologists and restoration ecologists are doing all they can to put obstacles in the path of introduced species naturally thriving in their current environments (where they may have been introduced many generations ago), while promoting supposedly native species, which either are no longer that well adapted to current conditions, and are therefore not doing particularly well on their own (for a variety of reasons, sometimes unrelated to the presence of introduced species), or continue to be well adapted and thus require no particular assistance from us. Also, by removing or inhibiting certain introduced species, we may just be weakening those native species which have developed, or may develop, new and mutually beneficial relationships with the introduced species we are trying to eliminate in the name of conservation.

Sometimes, ecologists, and the bureaucrats and members of the public influenced by ideas generated by invasion biologists, may be guilty of tunnel vision. Only the particular species they are trying to preserve and promote matters, to the exclusion of virtually everything else. If you have to remove, say, ten or one hundred introduced species and manipulate the environment in various drastic ways to make conditions more suitable for the one native species you are trying to safeguard at all costs, then so be it. So, you end up promoting "conservation" and "biodiversity", and other such rather vague and lofty goals, by engaging in questionable species eradication campaigns which, for all practical purposes, end up making the local ecosystems obviously less diverse and probably more fragile. But that is O.K., because, according to some invasion biologists at least, introduced species do not provide "real biodiversity" (whatever that may be) anyway. That is just an illusion for those of us who have not yet assimilated the invasion biology dogma.

This is a very slippery ideological slope, of course, as we are asked to aggressively interfere in natural ecosystems to promote some species and eliminate others simply because of their suspected places of origin (which, as we have seen earlier in the book, are often far from clear). By doing this we are also not allowing our current ecosystems to reach their own new temporary balance or equilibrium, involving the species present here and now. We do not give introduced species a reasonable chance to fully contribute to their new surroundings, and we do not give ourselves, as ecologists, a fair shot at studying the many complex and often positive contributions introduced species may bring to new ecosystems. The impulse to study and analyze new and little understood phenomena objectively, which is so vital to healthy and genuine scientific investigation, is replaced by the much lower human instinct to eradicate and suppress what is new and poorly known. In other words, shoot (and control) first, and ask relevant questions much later, if at all.

Guilty until Proven Innocent

Not only is it often very difficult, if not impossible, to determine if a species is really native to a particular region or not, but we may also have a very tough time proving conclusively that a native species may be driven to extinction because of the recent appearance of an introduced species. As I sometimes mention to students in my Conservation Biology course at York University's Glendon College, if a student leaves the room shortly after I entered the classroom, does this automatically mean that the student went out because I came in? This is just one among many possibilities. The student may have needed to get a drink of water, or call someone about a family emergency, or had to use the washroom, etc. Theoretically, it is possible that the amount of oxygen in the lecture hall is only sufficient for, say, 30 people, so when a 31st person walks in, one individual has to walk out, to avoid running out of breathable air. Practically, however, this is highly unlikely, and one would have to prove that air is indeed a limiting factor, which does not allow the addition of any more people to the room. We would also have to show that no other variables, with the exception of air availability, have any significant impact on the number of people the room can accommodate. Often, such clear evidence does not seem to accompany the claims made against newly introduced species. The usual assumption seems to be that if a new species arrived somewhere, it will likely have a negative impact, and native species will suffer as a result. According to an article by Ruesink et al. (1995), for instance, all non-indigenous species should be considered "guilty until proven innocent". These authors advise that

FIGURE 9 *Mute swan* (Cygnus olor) *in Lake Ontario*
PHOTO BY ILEANA GUIAȘU

"the most effective way to reduce harm from new nonindigenous species is to assume that all are problematic".

Studies conducted under the influence of such an assumption may carry a strong bias from the beginning, since, if one sets out to "prove" that a species has a negative impact simply because the species exists (in other words, it takes space, feeds, etc.), one can prove that about any species anywhere, of course. As an example, we can consider a passage about mute swans (*Cygnus olor*) – a species of Eurasian origin, introduced into North America in the mid-1800s – from an article by Kerasote (2005). The author interviewed a graduate student who was studying the effects of these swans on the subaquatic vegetation of Chesapeake Bay, in the US state of Maryland. The study had been funded by Maryland's Department of Natural Resources. Maryland is one of the US states where control programs targeting the mute swan have taken place (Ciaranca et. al, 2007). Upon observing the mute swans swimming with their heads often underwater, a common activity in many waterfowl species, Kerasote asked if the swans "were eating the seagrass native waterfowl depend on", an obviously leading question. The graduate student responded: "I don't know what else they'd be doing with their heads down." Well, there you have it. A crime has been discovered. The mere fact that swans were engaging in a perfectly normal, common, and peaceful activity is used against them, in this context, simply because they are considered to be an introduced species. The question was not: "Do you think the mute swans may be eating seagrass?", but "Are the swans eating the seagrass *native waterfowl depend on*?" Or, to put it differently, is that seagrass reserved only for native waterfowl? When native waterfowl eat, they are only doing what comes naturally to sustain themselves. But when mute swans are doing the same thing, they must be stealing food from native species (after all, what else could they be doing?), even if no native species happen to be around at the time. Do we know if the so-called seagrasses are in short supply in this area? Maybe there is enough seagrass for everybody. Are native waterfowl going hungry because a few mute swans are feeding here? Is there clear evidence for this? These questions are not addressed in the article. So, it is obvious that we often have one set of standards for species we consider native and quite another set of standards for introduced species. Native species can go about their lives, but non-native ones will be kept under careful observation, and each one of their activities will be regarded with deep suspicion or even open hostility. We can imagine this type of biased reasoning being extended to all types of regular activities. For instance, if a native duck defecates, the droppings will add valuable nutrients to the soil, but if a mute swan does the same thing, it is polluting the environment, making it unsuitable for native species, and bringing possible diseases to our formerly pristine shores... Furthermore, if one starts to conduct a study about the impact of a species with this type of approach, we can predict beforehand what the conclusions of the study will be.

As seen during the discussion of the anti-ruddy ducks campaigns in the UK, control programs against waterfowl species can be controversial and may widen the gap between conservationists and members of the public. In the United States, concerned citizens have initiated legal challenges to stop wildlife management organizations from killing the graceful mute swans. Control programs against these swans in states such as Rhode Island and Maryland involved destroying eggs and removing adult birds (Kerasote, 2005; Ciaranca et al., 2007; Guiașu, 2008a). In New York State, wildlife officials recently labelled the mute swans as "an aggressive, invasive species that poses a threat to people and native animals", and made plans to kill these birds. However, faced with massive public protests over this proposal, New York's Department of Environmental Conservation had to back down and announced changes to the initial plans. The changes may appease the public by allowing these popular mute swans to survive (*The Globe and Mail*, 2014). Mute swans can sometimes behave aggressively towards people, but that is simply because the birds are protecting their territories and their nests from intruders (McKeating, 1990; Sandilands, 2005; Guiașu, 2008a). In a way, the accusation that mute swans can be aggressive while defending their nests reminds me of the French maxim: “Cet animal est très méchant. Quand on l’attaque il se defend.” (cited in Henderson, 2012), which can be translated as: “This animal is very mean. When it is attacked, it defends itself.” Besides, large waterfowl species native to North America, such as trumpeter swans (*Cygnus buccinator*) and Canada geese (*Branta canadensis*), can also defend their territories vigorously and even attack people who get too close to them, from time to time (Kerasote, 2005; Sandilands, 2005; personal experience).

Furthermore, mute swans can also become quite tame around people, which is one of the reasons why these beautiful birds were commonly seen on ponds in various country estates and city parks (McKeating, 1990). I have encountered mute swans once in a while during my fieldwork, including a pair of adults and their cygnets on the slow-moving waters of a large Ontario river, and I did not detect the slightest sign of aggression on the part of these majestic birds.

In another example of an apparent rush to judgement, Wilcove et al. (1998) claimed that 57% of imperiled plant species in the USA are threatened by alien species – presumably mainly due to competition with alien plant species. However, Davis (2009) pointed out that, as of 2007 (nine years after the publication of the Wilcove et al. study), there was still no evidence that any native North American plant species had become extinct, or even disappeared from a single US state, due to competition with non-native plant species. Therefore, Davis (2009) labelled the implication that non-native plants are an important threat to a majority of imperiled US native plant species as “quite unjustified” and “even reckless”. The study by Wilcove et al. (1998) will be discussed in more detail in another section of this chapter.

In a recently published study, based on the analysis of the extensive British Countryside Survey Data, Thomas and Palmer (2015a) found no support for the hypothesis that introduced plants may cause the extinction of native plants through competitive exclusion, and concluded that “non-native plant species are not a threat to floral diversity in Britain.”

In a more specific example, garlic mustard (*Alliaria petiolata*) was introduced to parts of central and eastern North America from Europe in the 1800s. This species has been labelled as “highly invasive” and is considered a major threat to native North American plants (Davis et al., 2012; it should be noted that in this article Davis et al. did not endorse this view but simply reported it). Dickinson et al. (2004) mentioned that garlic mustard spreads in forested areas and outcompetes many native spring plants in Ontario. In a brochure titled *The landowner’s guide to controlling invasive woodland plants*, garlic mustard

FIGURE 10 *Garlic mustard* (Alliaria petiolata) *plants growing on the edge of a Toronto woodlot*
PHOTO BY DORIS WONG

is labelled as "a major threat to native woodland flora and the wildlife that depend on it" (Pridham, 2012). In this same brochure, we are informed that "garlic mustard populations double in size every four years, and can displace native woodland wildflowers such as trilliums and trout lilies within ten years." The brochure was published with the support of the Ontario Federation of Anglers and Hunters, the Ontario Invasive Plant Council, Ontario's Invading Species Awareness Program, the Victoria Stewardship Council, and the Ontario provincial government. As a result of this perceived threat to native plants, control and eradication programs have been undertaken against garlic mustard in North America. Such programs included the use of herbicides, fire, and the physical removal of the plants (Davis et al., 2012; Pridham, 2012), which sometimes involves members of the public as well (more on these types of plant removal campaigns in the last chapter of this book).

In my own experience, during frequent walks through southern Ontario woodlots, garlic mustard does not replace trout lilies, at least in the forested areas I have observed. First of all, trout lilies (*Erythronium americanum*) and the garlic mustard do not emerge at the same time in this part of the world. The trout lily comes out first, usually in late April or early May, and then garlic mustard appears a few weeks later – in mid- to late May – after the trout lily vanished for the year on its own. Secondly, I have been keeping notes on the plant life of a few local woodlots for several years, and during this time I did not notice any obvious changes in the populations of trout lilies, despite the fact that garlic mustard has been also present along the edges of these forests throughout my observation period. What I did notice, however, is that garlic mustard seems to have become less abundant and widespread in some woodlots in the last few years.

Recent formal studies back up my observations. In a series of detailed long-term studies, Mark Davis and teams of researchers investigated the population dynamics and the role of garlic mustard in a Minnesota oak woodland. The researchers found that *A. petiolata* does not have a significant negative effect on native plants. The native plants included both wildflowers and tree seedlings. Furthermore, the changes in abundance and distribution of garlic mustard populations appear to be quite independent of similar types of changes taking place in the populations of native plants found at this location (Davis et al., 2012, 2014, and 2015). Davis et al. (2014) concluded that:

> In four years of study, we have not been able to document any substantial effects of *A. petiolata* on other plant species. Overall, our findings are not consistent with the common claim that *A. petiolata* is a noxious invasive species responsible for the decline of many North American native forest herbs.

Leaving Enemies Behind?

Another related common assumption is that introduced species do well because they don't have any (or too many) predators and/or parasites in their new environment. Presumably, the idea behind the assumption is that invasive species left their traditional enemies behind when they moved away from their native range – a concept known as the Enemy Release Hypothesis or ERH – (Davis, 2009), and are therefore free of most of the former constraints they had to face back home. This general explanation for the perceived success of some exotic species is found in quite a few articles (for example, Clay, 2003 and Torchin et al., 2003), and has been much discussed. It is the type of explanation which adds to the perception of introduced species as somehow different than the native ones, not only with respect to geographic origins, but also in terms of some other key features related to survival and expansion in a new environment. In one recent book, Hamilton (2010) referred to particularly successful non-native species as "super species". In other words, these are somehow "otherworldly" and invincible species, and the normal rules of nature don't apply to them. This is usually a rush to judgement, which may simply reflect incomplete data sets and observations and the fact that the new species has just arrived in the area. This is somewhat like saying that an individual who has just moved into a new neighbourhood two days ago has no friends (or enemies) on that street. Given enough time and opportunities, the fellow may make some enemies, or some friends, and, just as importantly, perhaps his new neighbours will get to know him and establish various relationships with him. Similarly, if a new species becomes established in a particular area, it is quite likely that, given sufficient time, individuals of some native species will learn to adjust to this introduction, and interact with the members of the introduced species. Thus, over time, the new species may well have new predators, and parasites, as well as engage in commensal or mutually beneficial relationships with local species.

House sparrows (*Passer domesticus*) were introduced to North America in the New York area, from Britain, in the mid-1800s. The first introduction attempt, in 1850, was not successful, but in 1853, after sustained efforts to help the birds adjust to their new home, a few sparrows survived and eventually became established. Only two decades later, native North American predatory birds, such as owls and northern shrikes, were seen preying on house sparrows in New York, Boston, and other places in the eastern United States (Todd, 2012).

In a comparative study of native and exotic plant species found along the coasts of Rhode Island and Massachusetts, in the United States, Heard and Sax (2013) found that herbivorous insects were much more likely to eat and damage exotic plants than native ones. This directly contradicts the prediction of the Enemy Release Hypothesis, of course. However, from an evolutionary standpoint, it is reasonable to expect that native plants would have had much more time to develop effective defenses against local insect predators than newly arrived exotic plants.

The number of new parasites acquired in the area where a species is introduced would be expected to grow over time. For example, the rainbow trout (*Oncorhynchus mykiss*) – a species native to western North America – was introduced to the Patagonia region of Argentina more than a century ago. During this relatively long period of time, the rainbow trout has accumulated thirteen species of fish parasites native to Patagonia. On the other hand, the common carp (*Cyprinus carpio*) was brought to Patagonia only approximately ten years ago, and during this considerably shorter period of time, the non-native carp acquired only three native parasitic species (Waicheim et al., 2014).

Often, the time scale of our ecological studies will affect the conclusions we draw about the nature of the interactions among various species. If we study such interactions for one day, or one week, or even one year, we may form one set of opinions, but if we study these relationships over the long term, for,

FIGURE 11 *Information about invasive species at the Korthright Centre for Conservation in Ontario*
PHOTO BY RADU GUIAȘU

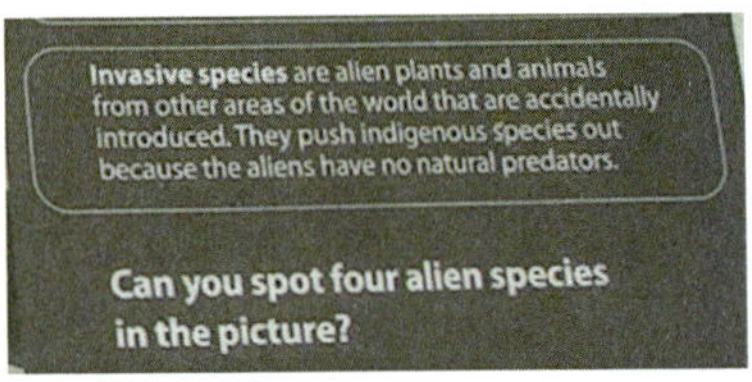

FIGURE 12
Close-up of a section of the sign at the Kortright Centre for Conservation. As can be seen, this section promotes the Enemy Release Hypothesis as an explanation for the success of invasive species
PHOTO BY RADU GUIAȘU

perhaps, a few decades, or centuries, the conclusions we reach may well be very different indeed. Too often, it seems to me, invasion biologists are overly eager to jump to conclusions soon after a new species is introduced, and the initial impressions they form may be premature, and certainly not a reliable basis for definitive statements or control programs.

Unfortunately, research grants and even academic careers are usually considerably shorter in duration than many long-term ecological and environmental phenomena associated with changing species diversity and distributions. So, if you are applying for a five-year major research grant, your chances of success will likely be better if you propose a series of short-term projects, yielding quick results and publications, rather than long-term work stretching over several decades. And if you are fortunate to receive such a grant to look at the effects of, say, the rusty crayfish on the "native" crayfish of Ontario, you will probably, from a practical standpoint, study the impact of this "invasive" crayfish in the field for at most two or three field seasons (likely to take place from late May to about mid-September, in this part of the world), at a limited number of locations. You will also likely have to limit the analysis to only one or two native crayfish species, and, overall, you will predictably end up with only a few frames from a much longer and more complicated movie, with many other characters and much unseen and unrecorded action.

In addition, there is the tendency of researchers to perhaps overstate the importance of their own research, consciously or sub-consciously. This is somewhat understandable – after all, most researchers tend to believe strongly in the significance of their work, and this is what sustains them during their careers – but it can sometimes make such scientists lose some of their objectivity and sense of proper context. When it comes to invasion biology, the use of hyperbole and blatant exaggeration is not restricted only to journalists, and can drive the very negative public perception of exotic species. Just as I was working on this section of the book, on January 19th, 2012, I came across an advertisement for a presentation in the Biology Department Colloquium Series, at my university. The title of the talk, by a senior biology professor, was: *The spiny water flea: the worst threat to the biodiversity of our lakes since acid rain*. Although I don't know the professor personally, I am aware of the fact that he is a respected and experienced researcher in his field, and I am sure he sincerely believes that spiny water fleas are very bad indeed (particularly when they live in *our* lakes). However, in general, and without disputing the professor's research and motives, I am a bit wary (and weary) when I read titles such as this one (particularly for scientific talks or articles). To me, there appears to be something a bit self-serving and self-aggrandizing in such titles, even if one really believes in their validity. Whenever I read that some animal or plant species is the greatest threat since acid rain, or since the last huge meteorite struck the Earth, or since Attila the Hun, etc., I can't help thinking that what the researcher is actually saying (perhaps sub-consciously) can be translated as: "this problem is the greatest one of its kind, at the moment, and, since I am the person studying this, my studies are more important than any others, and deserve prominent funding and attention." Some may say that is a cynical view, but even cynics can be right from time to time.

I could probably fill at least a whole chapter with similar examples, from my voluminous files on the subject, but I think the point was made, and readers can easily find their own such examples, during a survey of the relevant literature.

One problem with a field that was created to constantly address "crises" is that every species introduction may be treated as a "crisis", at least by some, and drastic actions may be taken, before anybody bothers to collect and carefully analyze the necessary relevant evidence. The danger of overreacting is always there. Another problem is that during a crisis there is little time for considering alternative points of view or engaging in thorough and reasoned debates about the nature of the evidence and the pros and cons of taking action against perceived threats.

Furthermore, a crisis atmosphere is not easy to maintain in the long run. People tend to get tired of repeated dire warnings, when they see that the sky has not fallen after all. So, the temptation may be there to ramp up the propaganda against invasive species by exaggerating the negative features and impacts of the well-known ones and finding new "invaders" to get outraged about on a regular basis.

As for the previously mentioned Enemy Release Hypothesis, the empirical support for this influential idea appears to be far from universal. In other words, some studies show a measure of support for the hypothesis and some do not (Davis, 2009). Not only do invasive species find new enemies in the areas where they were introduced, but sometimes, the old parasites are not actually left behind after all. As we have seen earlier, Dubey and Shine (2008) analyzed nuclear and mitochondrial genetic sequences to identify the lungworm parasites which affect cane toads (*Bufo marinus*) in Australia. Cane toads – a species of South American origin – are considered invasive in Australia, as we know. Dubey and Shine (2008) found that in Australia the cane toads were infected with parasitic nematode lungworms of the South American species *Rhabdias pseudosphaerocephala*. Despite the fact that the relatively small number of toads initially released in Australia did not come directly from South America, these amphibians still carried with them some of the major parasites from their original South American native range. The growth and survival of infected toads were negatively affected by the parasites, and individual toads in Australia sometimes carried more than 200 lungworms – a very high number of parasites for this host species. Thus, this highly successful invasive toad species of South American origin managed to spread rapidly over a large area of Australia, despite being negatively affected by parasites from its native range.

The fire ant (*Solenopsis invicta*), another invasive species of South American origin, came to North America with some of the pathogens from its native range as well (Cooling and Hoffmann, 2015).

As previously discussed, the common carp acquired three native parasitic species since its introduction to Patagonia, Argentina about a decade ago. However, this exotic fish species also apparently arrived in Argentina with at least three non-native species of parasites (Waicheim et al., 2014). Therefore, the common carp is an example of an introduced species which acquired new parasites in its non-native range, while at the same time keeping some of its old parasites as well.

A similar example is provided by the starling (*Sturnus vulgaris*), a species introduced into North America from Britain. Starlings arrived in North America with some of their original British helminth parasites, and then acquired even more species of such parasites in the decades since their arrival on the North American continent. In fact, it appears that more than half of the species of helminth parasites present in these introduced starlings (or the descendants of the initially introduced starlings) have been acquired in North America, and, as a result, starlings seem to be infected by a higher number of species of helminth parasites in North America than in their British native range (Colautti et al., 2005).

Furthermore, sometimes species are considered invasive even when they are not found all that far away from their presumed native ranges. If we take the rusty crayfish as an example, it is simply hard to believe that this species left its predators and parasites behind, when it moved from its native Ohio and Michigan to very similar freshwater aquatic habitats in nearby Ontario.

Invasive Species – The Second Most Important Threat to Native Species and Biodiversity?

The statement which makes up the title of this section, minus the question mark, has been frequently repeated in numerous invasion biology-related articles and books. The validity of the assertion that invasive species are the second greatest threat to native species after habitat destruction has been thoroughly questioned and debunked by Davis (2009). Nevertheless, given the main focus of this book, and the pervasiveness of this belief, I will still deal with this issue here, since this is an influential idea which seems to be used often in support of various anti-invasive species programs or in order to silence the critics of such programs.

A bit of historical background to this controversy may be useful. In a chapter titled *Biodiversity threatened*, within his 1992 book *The diversity of life*, Edward O. Wilson – one of the pioneers of conservation biology and a towering figure in the field – mentioned that in prehistoric times the main threats to biodiversity were "overkill and exotic animals", and "In recent centuries, and to an accelerating degree during our generation, habitat destruction is foremeost among the lethal forces, followed by the invasion of exotic animals." While the acknowledgement of the role played by people in changing the global distribution of species in prehistoric times may be useful, it is worth remembering that, as discussed in a previous chapter, species which moved to new regions before the beginning of recorded human history are considered to be native, regardless of their means of dispersal, by some biologists. The "introduced" label is reserved for species known or assumed to have expanded their ranges due to human assistance in historical times (Myers and Bazely, 2003). Since it is often difficult to determine the exact ways in which species distribute themselves around the world even these days, it would be logical to assume that the farther back in time we go, the harder it would be to reconstruct such ancient dispersal events and the exact ways in which they might have been completed.

Coming back to more recent times, Wilson (1992) clearly suggests that "the invasion of exotic animals" (introduced plants were not mentioned) is second only to habitat destruction among the major threats to biodiversity. This assertion is followed by a fairly general and rather brief example of the threats to the endangered, vulnerable, and rare freshwater fishes of North America. As far as I could tell, the data included 27 fish species which became extinct during the century which preceded the publication of Wilson's book, as well as 265 fish species (out of the total 1,033) described by Wilson as "liable to extinction". These fishes were selected because they are listed as either: extinct, endangered, vulnerable, or rare, according to the IUCN (the International Union for Conservation of Nature and Natural Resources, also known as the World Conservation Union). The "destruction of physical habitat" is considered to be the main problem for these particular species, followed closely by the almost equally serious (with respect to the percentage of species affected) "displacement by introduced species". In any event, the limited information presented about the freshwater fishes of North America appears to have been extrapolated to the threats to the biodiversity of the entire world.

On the front page of an article titled *Quantifying threats to imperiled species in the United States*, Wilcove et al. (1998) announced that: "Habitat loss is the single greatest threat to biodiversity, followed by the spread of alien species." This article, which had been cited 2064 times by November 10, 2013 according to a survey made using Google Scholar, proved to be very influential and instrumental in leading many to the conclusion that introduced species were the second greatest threat to endangered species, or biodiversity in general, pretty much everywhere around the world. However, as Davis (2009) pointed out in his careful analysis, the Wilcove et al. (1998) study had some problems which would suggest that the conclusion of this article regarding alien species should be interpreted with much more caution. First of all, as Wilcove et al. themselves admitted, the conclusions of their 1998 study were often not based on any data, but on the potentially subjective opinions of experts such as state Fish and Game employees monitoring the fate of endangered species in certain regions. These opinions may have value but they are clearly not a substitute for real quantitative data or experimental evidence. Also, as the title unambiguously indicates, this is a study of the threats to imperiled species in the USA, and, therefore, the conclusions of such a study should not be extrapolated to the rest of the world. Furthermore, the inclusion of data from Hawaii significantly affects the result. If one looks at the data for the continental United States, introduced species are a much lower overall threat to the biodiversity found there (Venter et al., 2006; Davis, 2009).

In a subsequent study on the threats to endangered species in Canada, Venter et al. (2006) found that introduced species were, in fact, the least serious threat among the six categories examined. These six categories were, in decreasing order of importance: habitat loss, overexploitation, native species interactions, natural causes (storms, droughts, limited dispersal ability, etc.), pollution, and introduced species. As can be seen, and as Venter et al. (2006) explicitly mentioned, interactions among native species and natural causes were more important threats to Canadian endangered species than either pollution or introduced species. It should be noted that interactions among native species and natural causes were not even listed as categories in the Wilcove et al. (1998) study. Venter et al. (2006) compared their own data to the findings of the Wilcove et al. (1998) study and concluded that without the Hawaiian data the levels of the threats to biodiversity posed by introduced species were quite similar in Canada and the continental USA. Although small in area in comparison to the rest of the United States, the Hawaiian Archipelago has a big influence on the results, because of the great number of threatened and endangered species found there. As mentioned elsewhere in the book as well, it is difficult, and probably often quite misleading, to apply conclusions based on the effects of introduced species on small islands to the analysis of the presence of such species on continents.

In fact, Davis et al. (2011) concluded that "recent analyses suggest that invaders do not represent a major extinction threat to most species in most environments."

Young and Larson (2011) found that only a minority (27.3%) of the 422 invasion biologists surveyed thought that invasive species were the greatest or the second greatest threat to biodiversity. So, despite all the on-going propaganda, almost three quarters of a substantial number of invasion biologists did not rank invasive species among the top two greatest threats. This shows that relentless propaganda prevails even when the vast majority of the scientists in this generally anti-invasive species field do not agree with the alarmist statements. Clearly, in this case, the minority view was more effectively communicated to the public and conservation organizations, and the media may have been more receptive to this claim, because of its simple and straightforward message. There is little room for scientific ambiguity in headlines and brochures. And the assumption is that people are only likely to pay attention to the items at the top of most lists. Few may care as much about the 3rd, 4th, or 279th-ranked threats to biodiversity.

More recently, in an article published in the December 11th, 2014 issue of the scientific journal *Nature*, an attempt was made to gather and present the best available data on the status of life on Earth. The information on the main threats to biodiversity was taken from the *WWF Living Planet Report 2014*. (WWF is the acronym for the World Wildlife Fund). According to this presumably most up to date and comprehensive data set, the main threats to biodiversity are, in decreasing order of importance: exploitation (through hunting and fishing and other such activities; 37%), habitat degradation and change (31.4%), habitat loss (13.4%), climate change (7.1%), invasive species and genes (5.1%), pollution (4%), and disease (2%). So, once again, invasive species are clearly listed among the lower ranked threats to biodiversity – in fifth place among the seven categories chosen, and well behind exploitation and habitat degradation and change (Monastersky, 2014; *WWF Living Planet Report 2014*).

Although, as we have seen, the argument that invasive species are the second greatest threat to native species seems to have no basis in fact, and should certainly not be applied universally to each case study involving invasive species throughout the world, this simplistic statement continues to be made all the time, rather casually, even by reputable scientists and in respectable publications. For example, an article published in the *Journal of Applied Ecology*, which presents a model for the eradication of the non-native ruddy duck in the UK, starts with the following sentence: "Non-native species are the second most important threat to global biodiversity after habitat loss." (Smith et al., 2005). Presumably, this definitive statement is used to prepare the readers for the argument that all the ruddy ducks found in Britain have to be killed.

And the opening sentence of the Introduction section of a recent scientific article about a new species added to the IUCN's "100 of the world's worst invasive species" list contains the statement: "invasive alien species (IAS) are regarded as the second largest threat to biodiversity worldwide" (Luque et al., 2014).

In a 2010 book about the Endangered Species Act published by the American Bar Association, the Wilcove et al. (1998) study is cited as a source for the statement that for most species: "The primary causes of imperilment are habitat loss and predation by or competition from nonnative species" (Goble, 2010). Please note the use of the term "nonnative species" in this case, rather than invasive species.

Also, in an even more recent book, published just a few months before I started writing this paragraph and titled *Letters to a young scientist*, Edward O. Wilson refers to alien species as the "enemies of native faunas and floras", before concluding that: "Invasive species are the second most important cause of extinctions of native species, exceeded only by the destruction of habitats through human activity" (Wilson, 2013). In other words, the same familiar refrain – another mantra of invasion biology. As if the thorough analyses by Venter et al. (2006) and Davis (2009) never took place.

This often repeated statement has practical implications at a local level, as well. For instance, in a letter to the editor published in the *Toronto Star*, the Coordinator of the Social Marketing Programs for the Toronto and Region Conservation Authority – an organization which manages local conservation areas and woodlots – identified herself as "a person who works to raise public awareness of invasive plant species" and mentioned in the very first paragraph that invasive species are considered to be "a significant threat to biodiversity, second only to habitat loss." The author of the letter went on to discuss the perils associated with having some non-native plants in private gardens and encouraged the public to grow native plants instead and "to garden in a more ecologically thoughtful manner" (Cirillo, 2010). In my reply, which was published in the same newspaper four days after the initial letter, I briefly explained some of the problems related to the article by Wilcove et al. (1998) and the findings of the Venter et al. (2006) study, which is much more relevant to southern Ontario, before adding that plant biodiversity in the Greater Toronto Area actually increased in the last couple of centuries due to the addition of exotic species and mentioning that introduced plants can also bring useful contributions to the new ecosystems they become an integral part of (Guiașu, 2010). As far as I know, there were no additional letters published in this particular exchange of views, but, overall, I doubt that anything will change, at least in the short run. The notion that invasive species, in general, are the second greatest threat to biodiversity, everywhere and at all times, is just the kind of convenient all-encompassing simplistic concept that is tailor-made for constant use in the propaganda campaigns linked to invasion biology.

Case Study: Purple Loosestrife in North America

We first encountered purple loosestrife (*Lythrum salicaria*), briefly, in the first chapter, when I discussed the fact that this wetland plant had been introduced from Europe to North America early in the 19th century, and was referred to as a "demon" in a recent book (Silvertown, 2005). This much maligned introduced plant species has received considerable bad press, and is almost universally regarded as a major threat to the well-being and ecological integrity of North American wetlands, and the native species found in these wetlands. In fact, since the negative information about *L. salicaria* is repeated so often, many people take it for granted that this species must have well-established and horrible negative impacts on wetlands, and, therefore, the need for control programs against purple loosestrife is obvious and no longer has to be analyzed or discussed.

A sample of the anti-purple loosestrife literature may be educational. For example, Botkin and Keller (1998), in their popular *Environmental science* textbook, write that purple loosestrife "has caused many problems for the biological diversity of native plants" and can replace native North American vegetation in wetlands. Clay (2003) refers to the "widespread ecological damage to native communities" caused by this invasive plant in the United States. Silvertown (2005) observes that this species does not cause problems in its native European range, but in North America, *L. salicaria* "has conquered millions of hectares of wetlands and displaced native plant species, reducing some to the brink of extinction." Bright (1998) mentions that purple loosestrife overruns North American wetlands, "where it has eliminated native vegetation and ruined the waterfowl forage base." In their book on the weeds of northern North America, Royer and Dickinson (2004) go even further, and state, categorically, that *L. salicaria* "spreads rapidly and replaces all native vegetation, destroying wetland areas." So, presumably, according to this statement, purple loosestrife acts like some sort of a bulldozer, eliminating all (not just some) native plants and leaving behind wetlands which are utterly devastated and completely destroyed. After reading this statement, I must admit that the first thing that came to mind was an image of a drained wetland, where all plant life had been removed, and nothing but dry mud remained. And the caption underneath this imaginary picture of total wetland devastation read: "Purple loosestrife was here."

Kershaw (2002) writes that purple loosestrife is an "aggressive weed", known as a "silent killer", which "thrives unmolested, rapidly choking out native plants and converting complex wetland ecosystems into silent monocultures that are largely unpopulated by insects, birds and frogs." This information is repeated in a more recent field guide by Kagume (2008), where, once again, the terms "silent killer" and "aggressive weed" are applied to purple loosestrife, and we are again informed that this introduced Eurasian species "turns wetland communities into silent monocultures by smothering native plants." Echoing this same theme, Lockwood et al. (2007) state that: "In North America purple loosestrife suffers little to no herbivory from native species and takes over wetland areas, creating monocultures that persist for generations." Bright (1998) refers to purple loosestrife as "an aggressive Eurasian plant that can overrun a North American wetland and reduce its wildlife value to roughly that of a parking lot". In an article mostly about other invasive species, Mathieu (2009) writes that *L. salicaria* is "detested for taking over areas it roots in, 'choking out' other plants."

A brochure published by the Ontario Federation of Anglers and Hunters (2003) mentions that this plant is "a beautiful but aggressive invader" which "can rapidly degrade wetlands, diminishing their value for wildlife habitat". This brochure also advises members of the public that if they have purple loosestrife in their gardens, they should remove the plants, including the roots, because otherwise the presence of this plant could contribute to the loss of suitable habitats for local fish and wildlife. Gardeners were advised not to buy this plant even if it is legally available for sale. An updated version of this brochure, containing pretty much the same information, including the statements quoted above, was published in 2009. Several other organizations contributed to the production of this brochure, including: the Canadian Wildlife Service, the Manitoba Purple Loosestrife Project, the Minnesota Department of Natural Resources, the Ontario Ministry of Natural Resources, the US Great Lakes Sea Grant Network, and the University of Guelph.

I even found a recent reference to this species in *University Affairs*, a publication freely distributed to university professors in Canada, which normally deals with issues related to academic life and contains various advertisements for academic jobs at Canadian universities. In this article, however, purple loosestrife was once again referred to as "an aggressive invader that crowds out native vegetation" (Lougheed, 2008).

Purple loosestrife is also featured on the cover of a book on invasive plants published by the Brooklyn Botanic Garden (and edited by Randall and Marinelli, 1996). I always thought of botanical gardens as areas where exotic plants can find refuge, but this particular publication makes it clear that quite a few introduced species are decidedly unwelcome, and, within the book, several authors discuss various methods of controlling non-native plants, including the use of certain herbicides. In the section on purple loosestrife, this species is once again accused of replacing native plants, threatening rare and endangered plant species, and reducing the availability of food and shelter for wildlife. As a result, we are informed that *L. salicaria* is considered a noxious weed in many US states, where the sale and growth of this plant are not allowed (Blossey, 1996).

Myers and Bazely (2003) discussed the fact that purple loosestrife, which is widely dispersed in Canada (from the Atlantic shores of Newfoundland to the Pacific coast of British Columbia), has been declared a noxious weed in some Canadian Provinces, such as Alberta and Saskatchewan, but not in others. The definition of the term "noxious weed" varies from one jurisdiction to another. According to Myers and Bazely (2003), in Canada, there is little to no connection between the density of purple loosestrife populations in particular jurisdictions and the inclusion or non-inclusion of this species on the noxious weed list.

In a textbook about wetland ecosystems, Mitsch et al. (2009) briefly mentioned that the presence of *L. salicaria* is causing great concern for wildlife managers working in North American marshes, and concluded that, as a result, "Many freshwater marsh managers have implemented programs designed to control purple loosestrife by chemical and mechanical means."

School children attending a nature day camp at the Kortright Centre for Conservation (just north of Toronto) approximately 25 years ago were also taught about the dangers of purple loosestrife and then asked to participate in activities which required them to remove these plants under the supervison of a local naturalist. After quite a few of the purple loosestrife plants were removed and placed in buckets, the naturalist told the children that native plants will be better off because of their efforts (Matthew Ronan, August 3rd, 2014 – personal communication). The Kortright Centre is currently described as "Ontario's premier environmental and renewable energy education and demonstration centre" (http://kortright.org, 2015).

So, with all this anti-purple loosestrife information being widely distributed and all these definitive statements made, and control programs undertaken, against this plant, one would think that the scientific evidence proving the negative effects of *L. salicaria* on North American wetlands and local native species is very clear, conclusive, and abundant. The reality, however, is quite different, and, once again, there seems to be quite a gap between what is often said or written against this species, and what we actually know, from a scientific standpoint, about its ecological impact in North America.

In two thoughtful and comprehensive review studies, Anderson (1995) and Hager and McCoy (1998) analyzed the available information, and found no solid evidence in support of the idea that purple loosestrife has a negative effect on wetlands and other species. In fact, according to these authors, it appears that the accusations made against *L. salicaria* often preceded the collection of any thorough data, and the few studies claiming to show the negative effects of this plant were rather inconclusive, and had important methodological problems.

First, Anderson (1995) reviewed 71 articles about purple loosestrife, as well as conducting his own fieldwork on the ecology of this plant species in a wildlife sanctuary in Massachusetts. In addition to identifying some clear problems and biases associated with the anti-purple loosestrife studies, Anderson (1995) also found "no correlations between species richness and loosestrife abundance" based on his own fieldwork, and was "unable to locate any documented examples in the literature of species extinctions and declines caused by loosestrife". The claims that purple loosestrife may endanger other species were found to be hypothetical and based on "speculative scenarios". For example, there was published speculation that a particular duck species might be threatened by purple loosestrife, even though this plant was not even present yet in the particular region where the duck species was known to breed (Anderson, 1995).

Next, Hager and McCoy (1998) looked at the history of purple loosestrife studies and control programs in North America, and found that the most commonly cited scientific report dealing with the effects of *L. salicaria* on native species was inconclusive and had certain problems, including, for example, difficulties with the experimental methods and design, a lack of relevant statistical comparisons, and discrepancies between the claims made by the authors and the actual data they presented. The analysis by Hager and McCoy (1998) also revealed that the anti-purple loosestrife rhetoric and control programs started well before any scientific quantitative data about the impact of this plant species were gathered. Such control programs continued, and the negative comments about *L. salicaria* were repeated, again and again, despite the continued shortage of relevant data. Overall, Hager and McCoy (1998) concluded: "The hypothesis that purple loosestrife has detrimental effects on North American wetlands has been widely accepted, despite a lack of reliable scientific evidence." The authors also found that, despite the momentum gained by biological control programs in both the USA and Canada, and despite the great support received from certain organizations and some members of the public for these control programs, "there is currently no scientific justification for the control of loosestrife".

Ironically, at the time the study was conducted, both Hager and McCoy were associated with the University of Guelph, one of the organizations which supported, and cooperated with, the production of the previously discussed anti-purple loosestrife brochure in 2003.

Treberg and Husband (1999) investigated the relationship between the presence or absence of *L. salicaria* and plant species richness in areas along the banks of the Bar River in Ontario, Canada. Purple loosestrife had been present in this region for at least 12 years before the study was conducted, and this period of time was considered sufficient for a proper assessment of the effects of *L. salicaria* on local ecosystems. The authors compared vascular plant species richness in plots containing purple loosestrife and plots of identical size where this species was absent, and found no significant difference in mean plant species richness between these two categories of plots. In other words, according to these findings, the presence of purple loosestrife did not reduce plant biodiversity in this area. In the plots where purple loosestrife was present, the abundance of this species did not have any significant effect on plant species richness either. Furthermore, this study also showed that no plant species were more likely to be found in plots without purple loosestrife than in the plots where *L. salicaria* was present. Thus, it appears that in this region, at least, purple loosestrife did not competitively exclude any other plant species. Treberg and Husband (1999) concluded that "these results provide no support for the idea that the spread of *L. salicaria* is associated with a reduction in plant diversity in North American wetlands."

More recently, Farnsworth and Ellis (2001) conducted a quantitative field study on the effects of purple loosestrife in Connecticut, and reported that "*Lythrum salicaria* does not appear to threaten the diversity or density of other wet meadow species". These authors also stated that: "Truly monospecific stands of *L. salicaria* may be relatively uncommon, as other species appear to coexist with even dense stands of small stems." In fact, Farnsworth and Ellis (2001) confirmed the earlier findings of Anderson (1995) and Treberg and Husband (1999) which had shown that purple loosestrife does not have a negative effect on the species richness of North American wetlands.

A similar conclusion was reached in a more general study on plant biodiversity conducted by Houlahan and Findlay (2004) in 58 Ontario wetlands. The relationship between the abundance of four common native wetland plants (including two *Typha* or cattail species), and four so-called invasive species (including *L. salicaria*), and native and rare native species richness was analyzed. The results showed that exotic species were not more likely to dominate a wetland than native species, and that the effects of common introduced and native plants on native species richness were the same. Overall, no evidence was found in support of the hypothesis that non-native plant species competitively exclude native plants in south-eastern Ontario wetlands. The authors concluded that "exotic plant species are, by and large, a minor threat to native wetland plant diversity", and, referring specifically to purple loosestrife, they commented that the majority of empirical studies showed "no effect of *L. salicaria* on native plant diversity."

Hager and Vinebrooke (2004) conducted a study in 10 wetlands located in Minnesota and found, much to their surprise, that plant species richness and diversity were actually significantly higher in wetland regions invaded by purple loosestrife than in other wetland areas where this species was not growing. Furthermore, *L. salicaria* did not form monocultures and only colonized some areas of these wetlands, despite the fact that viable seeds of this plant were found in the soil of wetland sections where purple loosestrife did not grow. In other words, purple loosestrife did not take over these wetlands and was clearly associated with higher, rather than the expected lower, plant species diversity.

Lavoie et al. (2003) studied the distribution and history of exotic plants in St. Lawrence River wetlands in Québec during a large-scale field survey, and found that purple loosestrife does not usually form monocultures at these locations. In fact, purple loosestrife covered more than 50% of the examined areas at only 9% of the sampling stations where this species was present. So, despite the fact that *L. salicaria* is thought of as a highly invasive species in Québec, where this plant has been present at least since 1865, Lavoie et al. (2003) stated that their data "do not support the hypothesis that purple loosestrife reduces wetland plant diversity."

Anderson (1995) and Hager and McCoy (1998) also found the overall evidence for the alleged competitive exclusion of cattails by purple loosestrife to be inconclusive.

However, before purple loosestrife became the latest high profile enemy species for North American wetland conservationists, native cattail species (*Typha* spp.) used to be referred to as "aggressive" plants which crowded out other presumably more valuable plant species, and interfered with human recreational activities such as boating or fishing. As a result, control programs were undertaken against cattails. More recently, without any new evidence about the ecological effects of cattails, the focus appears to have shifted from cattail control to the protection of these plants from the perceived purple loosestrife hostile takeover. In effect, ironically, the conservation of the cattail, a plant disliked in the past by wildlife managers, is now cited as a reason to fight against purple loosestrife, the new "menace" to wetlands (Anderson, 1995; Hager and McCoy, 1998).

Not that cattails have been entirely rehabilitated. In their comprehensive textbook on wetlands, Mitsch and Gosselink (2007) mentioned that, although cattails are considered perfectly acceptable plants for use in restored wetlands in other parts of the world, in the United States, *Typha* spp. are "championed by some and disdained by others". The detractors think that cattails can be rapid colonizers of limited value to wetland wildlife species, which can also slow the flow of water in canals (Royer and Dickinson, 2004; Mitsch and Gosselink, 2007). These accusations have been directed at purple loosestrife as well. It seems that whenever some people don't like a plant, they discover that it is of limited value to wildlife. I always found this to be a particularly subjective and difficult to substantiate accusation. Nature is complex, and it is hard to believe that abundant and widely dispersed species, such as cattails or purple loosestrife, would not be used as valuable resources by a variety of other organisms. For example, during my own fieldwork in southern Ontario, I observed that cattail stands seem to provide good habitat for the burrowing crayfish

Fallicambarus fodiens, a vulnerable species in Canada, as well as for several types of frogs and birds. These were just casual observations, but if I really wanted to quantify the number of species associated with cattails or purple loosestrife, in a thorough and systematic way, I am certain I could have come up with a fairly impressive list (particularly if little known invertebrate species, such as soil nematodes, would also be taken into account). I doubt that most of the individuals talking or writing about the lack of value for wildlife of introduced species ever took the time to determine whether that is actually true. Generally, when a species is declared of little value, what is really implied is that it is of little perceived value from the point of view of a particular human interest group (such as, for example, cattle ranchers or duck hunters). It is perfectly appropriate for representatives of such groups to make their concerns known, but, ideally, such concerns should be based on conclusive and objective scientific evidence. Also, in general, what is good for a particular human interest group is not necessarily the same as the much broader "good for the environment" argument that is frequently invoked when control programs are justified.

The discouraging part, from my perspective, is that many of the people advocating, or participating in, control programs targeting purple loosestrife do not seem to be aware of the flimsiness of the scientific evidence against *L. salicaria*, as discussed in the important studies by Anderson (1995), Hager and McCoy (1998) and others. As can be seen in this review of the relevant information, despite the fact that these thoughtful papers were published about two decades and a decade and a half ago, respectively, the anti-purple loosestrife propaganda continues unabated. It seems to me that before taking drastic steps against a particular species, the least one can do is keep well informed of the relevant scientific facts. Unfortunately, however, in the simplistic "good species" vs. "bad species" world of some invasion biologists, there is no room for nuance, subtleties, gray areas, doubt, and restraint.

In a *National Geographic* article, for example, purple loosestrife was included on a list of the world's 100 "least wanted", or worst, invasive species. The article also mentioned that purple loosestrife is "choking out native plants that are important food sources for birds and wildlife" (McGrath, 2005). And in a very recent field guide about the wetland plants of the Chicago region (Chadde, 2011), *L. salicaria* is, predictably, once again referred to as "a serious threat to native flora and of little value to wildlife." So, we can see that despite all the scientific studies cited above, which tell a very different and much more complex story, magazine and newspaper articles, as well as some field guides, making reference to this species continue to promote the same simplistic anti-purple loosestrife information, and are even using similar language about "aggressive" invaders which are "choking" native plants and/or wetlands. Knowing what I know about the research on the effects of purple loosestrife on North

FIGURE 13 *Purple loosestrife* (Lythrum salicaria) *growing together with Queen Anne's lace or wild carrot* (Daucus carota) *– another introduced plant species – in the Toronto region*
PHOTO BY DORIS WONG

American wetlands, and seeing that this plant species is still considered to be one of the worst 100 invasive species in the world, I cannot help wondering about the quality of the evidence against the other species on this list…

The assertion that purple loosestrife reduces local biodiversity, and that insects and birds, for instance, do not use this plant as a resource (as stated by Kershaw, 2002; Lockwood et al., 2007; and others) is contradicted by studies where actual relevant field data were collected. Diehl et al. (1997) collected insects associated with purple loosestrife in the Canadian province of Manitoba, and found that numerous insects, including many native ones, could feed on, or be otherwise associated with, this plant. Overall, 63 insect genera, belonging to 38 families and seven orders were collected on the *L. salicaria* plants. Forty-one of these insect genera were herbivorous, and members of 14 of these genera were directly observed feeding on purple loosestrife.

Anderson (1995) confirmed that purple loosestrife flowers are a rich source of nectar and pollen for a variety of beneficial insect species in North America. Honey bees, for example, commonly spread the pollen of *Lythrum salicaria*. In fact, Fletcher (2006) reported that beekeepers have planted this species recently, as a food source for their bees. Apparently, native North American bees like purple loosestrife as well (Slobodkin, 2001). Cox (2002) pointed out that *L. salicaria* is pollinated by insects and produces plenty of seeds, indicating that there is certainly no shortage of insect visitors attracted by the purple loosestrife flowers. I have personally observed several species of bees and bumblebees visiting purple loosestrife colonies in the Toronto area. At the same time, several species of butterflies feed on the nectar of this plant. Overall, based on these observations, it is probable that large populations of purple loosestrife may represent an important resource for many local insects in North America (Anderson, 1995; Slobodkin, 2001).

Control campaigns against purple loosestrife in North America were based to a large extent on the assumption that this plant is unsuitable for providing bird habitat and has detrimental effects on native wetland bird species in general, and waterfowl in particular (Anderson, 1995). Ironically, purple loosestrife was one of the wetland plants which managed to spread in parts of the US, due to attempts to create additional habitat for waterfowl by deliberately flooding the floor of hardwood forests and manipulating wetland water levels (Hager and McCoy, 1998).

However, several species of local waterfowl and songbirds have been observed to nest in *L. salicaria* stands. Some of these North American birds, as reported by Anderson (1995), included American coots, pied-billed grebes, black-crowned night-herons, and gray catbirds. In addition, purple loosestrife seeds are apparently eaten by ducks and red-winged blackbirds. Kiviat (1996) found that the American goldfinch (*Carduelis tristis*) also built nests in purple loosestrife plants in two counties in New York State. Mature purple loosestrife plants appeared to protect the American goldfinch nests from the sun, wind, and rain, as well as from predators and the brown-headed cowbird (*Molothrus ater*) – a well-known brood parasite, as discussed elsewhere in the book. Since there is broad overlap between the North American ranges of purple loosestrife and the American goldfinch, Kiviat (1996) concluded that this introduced plant can provide suitable nesting habitat for this native bird species over a good part of its breeding area.

The seeds of *L. salicaria* can also germinate in water, and may serve as a food source for some herbivorous fishes. The common carp (*Cyprinus carpio*), a fish introduced into North America from Europe, may feed on purple loosestrife seedlings. In addition, pigeons and pheasants, two other Old World species introduced into North America, have been observed to eat seedlings of this plant, as well (Anderson, 1995). Thus, these three non-native species can perhaps help in keeping the purple loosestrife populations in check naturally, to some extent.

Whitt et al. (1999) studied the use of purple loosestrife-dominated habitats by birds in the coastal wetlands of Saginaw Bay, Lake Huron. The authors observed ten breeding bird species in habitats dominated by purple loosestrife. Among these species were swamp sparrows, whose nests were most abundant in areas where purple loosestrife was predominant, mallards, Virginia rails, American bitterns, sedge wrens, yellow warblers, American goldfinches, and red-winged blackbirds. Species such as swamp sparrows and red-winged blackbirds seemed to actually prefer habitats dominated by purple loosestrife. Overall, Whitt et al. (1999) concluded that the avian use of purple loosestrife appears to be higher than was commonly believed. This conclusion clearly challenged the widely held view that *Lythrum salicaria* did not provide valuable habitat for birds.

In a recent detailed review, Lavoie (2010) could find no evidence in peer-reviewed studies that purple loosestrife has a negative impact on waterfowl or fishes. The author also mentioned that the so-called "large negative impacts" presumably caused by purple loosestrife on wetlands are likely exaggerated.

Furthermore, Anderson (1995) reported that several species of North American mammals, such as white-tailed deer, muskrats, and rabbits, also feed on purple loosestrife shoots. Thus, contrary to the anti-purple loosestrife propaganda, a variety of native and introduced animal species appear to use this plant as a valuable resource. The list of such species will surely grow, as the currently little known ecological interactions involving *L. salicaria* in North America will be studied in more detail. Unfortunately, to date, the efforts spent trying to demonize and control this species were much greater than the genuine attempts to gather careful, detailed, and objective scientific data about the actual impact of purple loosestrife on the North American ecosystems.

In my own experience, during some of my fieldwork looking for crayfish in a stream located reasonably close to the city of Toronto, an area containing many purple loosestrife plants was far from the "silent monoculture" referred to by Kershaw (2002), and seemed, in fact, to be teeming with life. A variety of insects, including bees, bumblebees, and various types of flies, butterflies, and beetles appeared to be actively feeding on, or seeking, the *L. salicaria* plants, while several small birds were using the area to either rest or hunt for insects. In addition, although the beautiful purple loosestrife flowers were particularly showy and visible from a distance, a closer look revealed a variety of plants growing in and around the stands of *L. salicaria*. So, this was not really a monoculture at all, and there were plenty of apparently complex ecological interactions all around, involving many species. Furthermore, the fish and crayfish populations in the nearby stream seemed to be thriving, and I could not find any evidence that the stream ecosystem had been degraded in any way. Thus, in response to Bright's (1998) bizarre "equation", which was quoted above, the wildlife "value" of this beautiful natural area seemed considerably higher than that of a parking lot, at least to me.

Interestingly, it appears that this plant can also act as a sort of "pioneer" species, and is capable of colonizing areas not already occupied by other plants. For instance, during the summer of 2011, after I had already completed the initial section of the book dealing with this species, I observed a thriving colony of *L. salicaria* in a formerly empty field, just north of the city of Toronto, and close to the large Keele campus of York University. This field had been plowed only a few months earlier, and I was not aware of the presence of any wetlands in the area. And yet, in the middle of this recently plowed field, where few other plant species managed to establish themselves, I found this healthy population of beautiful purple loosestrife plants, which seemed to attract most of the conspicuous insects in the area, such as butterflies and bees.

If purple loosestrife would have been considered a native species in North America, local conservation biologists would have undoubtedly praised this plant for its great beauty and its ability to support a variety of animal species and therefore contribute to the maintenance of local biodiversity. There may well have been guided tours for the public to impressive and spectacular purple loosestrife dominated wetlands each summer, to show just how glorious nature can be, and provide some wonderful picture taking opportunities. Any species perceived to threaten purple loosestrife would probably have been actively fought against. After all, we have to protect a beautiful local natural treasure. And *L. salicaria* would have been considered an asset instead of an enemy.

But, since this species is suspected to have been introduced from Europe, about 200 years ago, the beauty and positive contributions brought by this plant are minimized or ignored, and the focus of conservationists is solely on taking drastic action against *L. salicaria* and "educating" members of the public

FIGURE 14 *Purple loosestrife plants in a field just north of Toronto*
PHOTO BY ILEANA GUIAȘU

about this great purple plant menace in their midst. It is all a (very subjective) matter of perception. For instance, Bright's (1998) description of the beautiful purple loosestrife flowers during the summer comes with a warning: "It is a spectacular sight, and could easily be mistaken for a perfectly natural one." Why exactly is this group of flowers unnatural? After all, these are not plants made of plastic or cardboard, and, as we have seen, they are genuine contributors to their ecosystems. Just in case somebody may actually still enjoy looking at the purple loosestrife fields without feeling very, very guilty, or even scared, Bright's (1998) warnings for the uninitiated become darker: "…; if you do not know, you may just enjoy the view. It is possible to look right in the face of a monstrous change and not see it – if the monster does not fit into your way of seeing." In my opinion, this type of condescending and over-the-top comment speaks for itself, really. All I can say is that I carefully reviewed the relevant scientific literature, and, as both a professional biologist and a nature lover, when I look at a wetland which includes purple loosestrife flowers, I still very much enjoy the view. Sadly, for a lot of other people, the enjoyment of some types of otherwise perfectly beautiful and functional natural ecosystems may be spoiled or significantly diminished by dire anti-exotic species warnings and comments, such as the ones cited in this section.

The purple loosestrife stands are there either to be appreciated or resented, depending on one's subjective point of view, and science has very little to do with this, because, as mentioned, the evidence used to justify the persecution of, and discrimination against, this wetland plant species is at best insufficient and inconclusive.

Case Study: The Zebra Mussel

Earlier in this chapter I referred to the article by Ruesink et al. (1995), according to which all non-indigenous species should be regarded as "guilty until proven innocent". In the opening paragraph of that article a few species were cited as examples of presumably particularly bad non-native species. One of these species was purple loosestrife, and I have, of course, already discussed this species in detail in the previous section. Another one was the famous zebra mussel (*Dreissena polymorpha*). In fact, when I publicly tried to make the argument that perhaps not all introduced species are always bad, I was sometimes asked the question: "But what about the zebra mussel?"

In response, let me start by saying that we should not have guilt by association. In other words, just because some introduced species may cause problems at some locations, under certain circumstances, and during a particular period of time, this does not mean that all introduced species will cause problems everywhere at all times. So, using the same familiar examples (the brown tree snake in Guam, or the zebra mussels in North America, and so on) as an indication of what is likely to happen any time any new species arrives anywhere is not a particularly useful, or rational, or fact-based approach to such situations. This is, in fact, guilt by (very tenuous) association, which, just like the "guilty until proven innocent" slogan, has rather unsavoury, intolerant and repressive connotations we would expect to encounter in the propaganda literature of police states, as opposed to scientific papers or articles about some of our fellow species. If accuracy, fairness, and finding real solutions to real problems truly interest us, then we should treat each introduced plant and animal species at each location as an independent and unique case study, to be analyzed and explored on its own merits.

Secondly, and just as importantly, as we have seen with the purple loosestrife example, and are about to see with the zebra mussels, even the famous case studies of invasive species can be considerably more complex than some invasion biologists would suggest. So, having said all that, let us now really talk about the zebra mussel.

FIGURE 15 *Zebra mussels* (Dreissena polymorpha) *attached to a rock taken from an Ontario lake*
PHOTO BY DORIS WONG

At least a couple of books about invasive species have photographs of zebra mussels on their front covers. One of these books is a more recent reprint (from the year 2000) of the classic work on invasive species by Elton, which, as previously discussed, was originally published in 1958. Zebra mussels were not a species of concern in 1958 – about three decades before they were discovered in North America, and, as a result, they are not even listed in the index of Elton's original book. However, in a possible attempt to make this book more relevant to a present day North American audience, the publishers of this newer edition decided to have pictures of clusters of zebra mussels on the cover, as a dramatic visual reminder of what invasive species are all about. Furthermore, in the Foreword to this edition of Elton's book, Daniel Simberloff – the current Editor-in-Chief of the journal *Biological Invasions* – refers to some invasive species as "horrors", and clearly identifies the "European zebra mussels" as one of these terrible species, since he cites the zebra mussel's arrival in North America as a clear example of a "disastrous" invasion, immediately after the use of the word "horrors" (Simberloff, 2000). Can anyone imagine watching a Hollywood horror movie, or a thriller (*Revenge of the zebra mussels!*), about some peaceful little molluscs, quietly filter feeding, for two hours or so?

The other book, titled *Bioeconomics of invasive species* and edited by Keller et al. in 2009, is obviously quite recent, and, this time, there are about 20 entries in the index under the heading "zebra mussels". In fact, all of the index entries listed for the letter "Z" refer to this mussel species. Two conclusions can be drawn from this: first, the zebra mussel is now one of the most famous invasive species – a symbol for invasive species in general, certainly in North America, and, second, a lot more attention is paid to this species these days, than was the case several decades ago, in Elton's time.

A photograph of zebra mussels also appears on the cover of a recent book titled *Protecting the Great Lakes from invasive and nonindigenous species* (Wouters, 2010), together with pictures of three other species – the purple loosestrife (shown in a rather out of focus image), the Canada goose, and what appears to be the spiny water flea. The inclusion of the Canada goose (*Branta canadensis*) on the cover of this book is rather odd, since this large bird is clearly native to the Great Lakes region. While the Canada goose is quite common in the area now, it was almost hunted to extinction in southern Ontario by the beginning of the 20th century (McKeating, 1990; Bezener, 2000). In any case, the title of the book and the fact that the Canada goose is on the cover reinforce the previously made point that there is a great deal of confusion in the field, with respect to what exactly is an invasive species. In other words, the book title clearly suggests that the Great Lakes (of North America) need to be protected from both invasive and nonindigenous species (even though, as we know, most introduced species will never become invasive). Since there are numerous non-native species currently residing in the Great Lakes region, "protecting" the Great Lakes from such species would imply the removal of a significant part of the present day biodiversity of the region – a task which would come with unimaginably large ecological and financial costs. Furthermore, the removal of so many species (which is really impossible to seriously contemplate, anyway) would undoubtedly pave the way for fresh "invasions", as new species would simply move in to fill ecological niches made available by the radical control programs. The appearance of the Canada goose on the cover of the book may suggest that, to some, the "invasive species" label should be applied to pretty much any currently successful species (native or introduced), regardless of its geographic origin. So, according to this line of reasoning, we should fight against species that are becoming a bit too successful for some people's taste, and label all such species as "invasive", even if they evolved in the region, and, in addition, we should control or eliminate nonindigenous species, in general, regardless of their current relative impact or level of success, simply because of their "alien" origins. This way, we would make sure no species become too common, and no "foreign" species at all would be allowed. This is obviously extremely silly and impractical (to put it very, very politely), and would appear to be a monumental waste of time and resources.

But, anyway, back to the zebra mussel.

The zebra mussel has been much maligned since its introduction in the Great Lakes region of North America probably in the mid-1980s. I remember reading an article titled *Introduction of non-native freshwater fish can certainly be bad* (in general, the titles of invasion biology articles tend not to be overly subtle…), and in the abstract of this article, the authors wrote: "We demonstrate that many real-world examples of freshwater fish introductions have catastrophic ecological consequences. We detail a few noteworthy examples, such as those of the Nile perch, carp, tilapias, catfishes, and the zebra mussel." (Vitule et al., 2009). Well, perhaps we can blame the zebra mussel for a lot of things, but we cannot accuse it of being a freshwater fish… Therefore, it seems to me that this species should not be listed as one of the main culprits even in an article such as the one just mentioned. In fairness, the article was written as a response to an earlier one by Gozlan (2008), and later on, in a separate section, Vitule et al. (2009) eventually refer to *D. polymorpha* as an invertebrate. Nevertheless, one can still question the validity of the decision to include a separate section about zebra mussels in an article which deals with non-native fish. Why was this considered necessary, particularly since the Gozlan (2008) article that Vitule et al. (2009) responded to did not have a zebra mussel section? In fact, Gozlan (2008) only mentioned the zebra mussel very briefly in a more general discussion.

Although we are quite accustomed to all sorts of alarmist statements regarding invasive species by now, the opening paragraph of an article published in 1993 – only a few years after the discovery of the zebra mussels in North America – by Ludyanskiy et al. in the reputable journal *BioScience* still stands out, even in the context of this type of literature. According to this intense paragraph:

> Just when the threat of a Russian invasion of North America seemed to have disappeared with the end of the Cold War, an invasion has been found to be not only under way but proving to be successful. Rather than missiles, a naval force of hordes of zebra mussels has secured beachheads in many US and Canadian lakes and rivers. One could call it biological warfare, but not directed by any human admiral.

So much for the assertion that journalists are mostly to blame for the overblown rhetoric and militaristic metaphors associated with invasion biology… Apparently, zebra mussels are not just a successful introduced species, but also members of a sophisticated naval force, with a complex invasion plan and command structure. Very impressive indeed! We must rewrite the mollusk behaviour sections of invertebrate biology textbooks.

The "Russian invasion" analogy is obviously based on the fact that *Dreissena polymorpha* is believed to have been originally native to freshwater ecosystems in southern Russia. This species may have originated near the Caspian Sea. In the early 1800s, the zebra mussel started to expand its range westward, into central Europe, using some of the newly built local canal systems and benefitting from increased transportation and commerce on European rivers (Ludyanskiy et al., 1993; Barr, 1996). In a discussion of freshwater invasive species in Poland, Jazdzewski and Konopacka (2002) mentioned that the zebra mussels used to be widely distributed in Europe in previous interglacial periods, and then retreated to the Black Sea and Caspian Sea regions during the last glaciation. Therefore, from this longer-term perspective, the current ongoing colonization (or re-colonization) of Europe by this species can be seen as a part of a cycle of range expansions and reductions on this continent. In a sense, humans may have helped the zebra mussel reclaim its former interglacial European distribution.

This mussel species was first discovered in North America in 1988 in the Lake St. Clair and western Lake Erie regions, but the initial North American introduction is believed to have taken place two or three years earlier (in 1986 or 1985), probably in Lake St. Clair (Ludyanskiy et al., 1993; Barr, 1996).

Numerous other articles have documented the negative or potentially negative effects of this species in North America since its introduction about three decades ago, and I will not review all that information here, since, as previously stated, this is not the main purpose of this book. If one is interested in the claims made against the zebra mussels, there is certainly a wealth of information out there – not only many articles, but also book chapters, cautionary brochures, and so on. I do not intend to dispute the fact that the zebra mussel can, under certain circumstances, cause problems for some other species – certain native North American mussel species such as the pearly mussels, for example, and I am also well aware that control programs against zebra mussels can add to the costs of operating power plants. The zebra mussels are not causing problems maliciously, of course, but simply as a result of their ability to attach themselves to hard surfaces, which may include rocks, pipes, or the shells of larger mussels. It is clear that the zebra mussel can become very abundant under optimal conditions, at least initially, at some of the new sites this species has colonized. As a result of its success, the zebra mussel can be quite influential in the freshwater ecosystems it becomes established in and can have a multitude of effects – some positive and some negative – on the other species present there.

FIGURE 16 *A closer look at a few zebra mussels collected in Ontario and the now discontinued Canadian one cent coin. The small coin is shown to give a sense of scale*
PHOTO BY DORIS WONG

However, there is another side even to this story of one of the most widely and intensely despised introduced species in recent history. The first time I became aware that there may be more to this story was when I read an interesting article provocatively titled *Our friend the zebra mussel* (Barr, 1996). I was still a graduate student at the time – just a few months away from the defense of my Ph.D. thesis, and I had already begun to think about some of the discrepancies between the claims made about invasive species and the facts I was able to gather from the relevant scientific literature. Despite its title, the article presented a balanced view of both the negative and the positive aspects of zebra mussels in their new habitats, and seemed to offer a refreshing alternative to the shrill and one-sided writings I had been exposed to earlier on this subject. The article was written by Dr. David Barr, who was at the time both a Curator of Invertebrate Zoology at the Royal Ontario Museum and a Professor in the Zoology Department at the University of Toronto. As a disclaimer, I should mention that David Barr and I worked together on a couple of studies (Guiașu et al., 1996 a and b), including a fairly comprehensive paper on the changing distributions of some Ontario crayfish species. Years later, I decided to include David Barr's article about zebra mussels in a course kit for one of the more advanced environmental studies courses I am teaching at Glendon College, York University, in order to give the students in my class a broader perspective and stimulate discussions (according to David Barr, generating debates about this issue was one of the reasons behind the publication of the article). The course kit includes, among other things, several papers about introduced species. Some of these articles offer perhaps the more "traditional" invasion biology view, while others give a somewhat different perspective on non-native species.

I must confess that I was initially reluctant to include a discussion of the zebra mussel in this book, not because such a discussion would not be potentially informative, but mainly because I strongly suspect that some invasion biologists may simply focus only, or mostly, on this notorious invasive species, and ignore many of the other case studies mentioned in this volume. In my view, this is essentially what happened when I published a general article about invasive species (Guiașu, 2008a) a few years ago. I did not include the zebra mussel in the article at first, because I had plenty of other case studies to illustrate the main points. However, the editor of the magazine where the article was published – who was very supportive of the article and very helpful during the publication process – suggested that perhaps some information about this famous invasive species should be added, and I agreed, because this gave me an opportunity to show that, even in this seemingly very straightforward and well-known case, there is more to the story than many of the related headlines would have us believe. In the rebuttal to my article, however, Mandrak (2008) predictably focused mainly on the zebra mussel, and largely ignored my other case studies – again, the guilt by association argument I mentioned earlier. In this case, the argument seems to unfold something like this: look how bad the zebra mussel can be, and it is invasive, so we shouldn't trust any other introduced species, because they will probably turn out to be just as bad. Thus, by using such a well-known species as an example, I felt that I might just invite similar responses to this book. In the end, though, I obviously decided that including species such as *D. polymorpha* should not be avoided, because the more complete story of this species in North America is more nuanced than commonly assumed. Furthermore, this story is still unfolding and changing, of course. So, once again, yes, I am well aware that the zebra mussel can have some negative effects, but, in my opinion, this is not the only thing worth knowing about this resourceful species.

Inevitably, the extent of the costs associated with the invasion of the zebra mussels remains unclear and controversial. In their study of the costs of non-native species in Canada, Colautti et al. (2006) admitted that: "Despite the prominence given to these mussels in the scientific and popular press, obtaining accurate cost estimates proved challenging." Part of the problem, according to these authors, seemed to be that some possible invasive mussel-related costs, such as those linked to the potential changes in the extent and types of various fisheries, were impossible to quantify. It should be mentioned that, since the presence of zebra mussels can actually favour certain fish species, there may also be some benefits to particular fisheries, as a result of the introduction of these mussels.

Pimentel et al. (2005) mentioned that zebra mussels can clog the water intake pipes and have negative effects on the functioning of water filtration and electricity generating plants. The costs associated with zebra mussels cited by these authors added up to a staggering $1 billion (US) per year. However, other authors gave significantly lower estmates of such costs, and painted a less alarming picture of the overall impact of these invasive mussels on water treatment and power plants.

Connelly et al. (2007) conducted a thorough survey of various managers of drinking water treatment plants and electric power companies. The study covered the 1989 to 2004 time period, and the entire North American range of the zebra mussels. The estimated total economic costs incurred by these plants and companies due to invasive mussels for the entire 1989–2004 period was $267 million (US). This is still a very large sum, of course, but it is obviously much lower than the $1 billion per year cost estimate cited by Pimentel et al. (2005). Furthermore, Connelly et al. (2007) estimated that the annual costs reported by water treatment and power plants as a result of zebra mussels declined over time. These annual costs were $44,000 per facility in the early years of the zebra mussel invasion and $30,000 per facility in the more recent years included in the study. The relative decline in such annual costs is even greater if the costs are adjusted for inflation. It seems that costs associated with zebra mussels went down over time partly because managers at these facilities learned from their previous experiences and developed more effective and efficient methods of dealing with the problem. The respondents to the survey also mentioned that it was difficult to separate the costs linked exclusively to zebra mussels from other more general expenses. As an example, chlorine is used both as a disinfectant and to control zebra mussels in water intake pipes. So, even if zebra mussels are not present, chlorine would still be used. Therefore, it is perhaps unfair to add the entire cost of chlorine treatments exclusively to the list of zebra mussel-related expenses. Overall, Connelly et al. (2007) concluded that some of the predictions made in the early 1990s about the costs associated with invasive zebra mussels in North America – including an estimate of $4 billion over 10 years in the Great Lakes region – "may have been overblown".

Barr (1996) reported that the additional costs incurred by water purification and power plants in Ontario as a result of the control of zebra mussels were relatively small as a percentage of the total costs of operating such plants. It seems that the managers at the Ontario facilities, just like their counterparts elsewhere in North America, have, for the most part, learned to cope increasingly well with these introduced mussels.

Zebra mussels were first found in Lake Ontario in 1989, and only two years later, in 1991, these mussels were quite abundant in shallow waters in the western part of this lake. By 1993, zebra mussels were found in fairly large numbers in both the western and the eastern parts of Lake Ontario (Barr, 1996; Wilson et al., 2006). Since *D. polymorpha* was first introduced in North America in Lake St. Clair (as far as we know), and was first detected in Lake St. Clair and Lake Erie, the gradual colonization of Lake Ontario from the west to the east makes sense. However, the early success of the zebra mussel in Lake Ontario was not going to last, and the dominant role of this species in this ecosystem was about to be challenged.

Another invasive dreissenid mussel – the quagga mussel (*Dreissena bugensis*) – arrived in Lake Ontario in 1990. In 1993, this new arrival was still rare in Lake Ontario (Wilson et al., 2006). In fact, in 1996, David Barr reported that *D. bugensis* represented "only a minor component of invading mussel communities" in Ontario, and had somewhat different habitat requirements than the zebra mussel.

However, by the summer of 2003, when Wilson et al. (2006) conducted a thorough survey of dreissenid mussel populations along the Canadian shoreline of Lake Ontario, the situation had changed drastically, and the quagga mussels were now dominant and had replaced the zebra mussels almost completely in this region.

Karatayev et al. (2013) estimated that the quagga mussels replaced the formerly dominant zebra mussels in both Lake Erie and Lake Ontario in the mid-1990s, and that, ten years later, the quagga mussels represented more than 90% of the total population of dreissenid mussels in these lakes.

Although the zebra mussels remain fairly common, and quite widespread, in some parts of the Great Lakes of North America, the rapid replacement of *D. polymorpha* by *D. bugensis* clearly shows that zebra mussels are not invulnerable, and can be affected quite dramatically by changes in local ecosystems and environmental conditions.

Not surprisingly, the abundant zebra and quagga mussels can provide a valuable resource for a variety of local species, including native ones, in North America (Guiașu, 2008a). In the Inner Bay, at Long Point, on Lake Erie, native waterfowl species such as the lesser scaup (*Aythya affinis*), the greater scaup (*Aythya marila*), and the bufflehead (*Bucephala albeola*) eat large quantities of these introduced mussels, therefore keeping the mussel populations in check naturally. Zebra mussels first appeared at this location in 1990, and populations were monitored during the first half of the 1990s. Due in large part to the heavy predation on mussels by local waterfowl, the abundance, density, and distribution of the zebra mussels at this location declined consistently and substantially throughout this period (Petrie and Knapton, 1999). This also illustrates once again that invasive species such as the zebra mussel often reach the highest population numbers shortly after they arrive in a new area, and then the numbers drop, often significantly, in subsequent years. This also happened with zebra mussels after their range expansion into central and Western Europe – in the first few years after their arrival, the numbers of mussels grew rapidly, but then generally declined and stabilized at lower levels later on. In many ways, after being present for almost two centuries in parts of central Europe, the zebra mussels have become a natural part of this region's freshwater ecosystems (Barr, 1996).

Fish, such as the pumpkinseed (*Lepomis gibbosus*), the freshwater drum (*Aplodinotus grunniens*), and the round goby (*Neogobius melanostomus*), and crayfish, such as *Orconectes propinquus* and *O. rusticus*, were also observed feeding on zebra mussels in North America (French III, 1993; Martin and Corkum, 1994; Perry et al., 2000; Naddafi and Rudstam, 2014). In fact, crayfish and zebra mussels can have effects on each other – the crayfish can eat the mussels, as mentioned, and the mussels can attach themselves to the bodies of crayfish, presumably interfering with the normal activities of these crustaceans and requiring the crayfish to spend more energy to move about. Brazner and Jensen (2000) reported the presence of zebra mussels attached to the bodies of rusty crayfish in the Great Lakes region of North America, and I have seen photographs of crayfish almost completely covered by zebra mussels (although I have never seen zebra mussels on crayfish during my many years of fieldwork, collecting crayfish in various streams and lakes in Ontario). Presumably, the attached mussels can slow down the escape behaviours – such as tail flip – of the crayfish confronted with potential predators. However, as Barr (1996) also pointed out, crayfish shed their old exoskeletons at fairly regular intervals, during the moulting process which allows them to grow, and, as a result, crayfish can periodically get rid of attached zebra mussels as well.

Zebra mussels are prodigious filter feeders, of course, and their activities can lead to a dramatic increase in water transparency or clarity (Holland et al., 1995). This allows higher levels of light to penetrate deeper into the water, which promotes the growth of benthic filamentous green algae and associated changes in the benthic food webs (Lowe and Pillsbury, 1995).

Zebra mussel colonies can also add structure to lake environments – a bit like a freshwater version of coral reefs – and offer valuable shelter and habitat for several other types of small invertebrates, including snails, crustaceans, insect larvae, and the hydra. In the western region of Lake Erie, the zebra mussels assist the growth of young yellow perch (*Perca flavescens*) by providing enhanced habitat for the small crustaceans these fish feed on (Barr, 1996).

As an additional benefit, zebra and quagga mussels can be useful as biomonitors for various pollutants, including metals and organic compounds, in both European and North American freshwater ecosystems. Biomonitors can be helpful in allowing researchers to detect contaminants which are present in very low concentrations in the water and cannot be measured by other methods. Zebra mussels and quagga mussels were found to be potentially effective biological indicators for a number of toxic chemicals in the Niagara River in Ontario (Barr, 1996; Richman and Somers, 2005).

Case Study: The Impact of the Rusty Crayfish

Whereas the zebra mussel and the purple loosestrife were disliked in part because they were intercontinental travelers, which arrived in North America from Europe, there is no doubt that the rusty crayfish is a North American native. And yet, as discussed earlier in the book, this species is also very much disliked by many freshwater ecologists, mainly because it currently appears to be thriving and expanding its range on its home continent, without having received official approval to do so from the proper authorities. This type of audacity, success, and resourcefulness seems to offend certain biologists, and various methods have been explored to try to stop the rusty crayfish, and other such currently successful species, in their tracks.

The rusty crayfish (*Orconectes rusticus*) is accused of outcompeting other crayfish species of its genus, such as *O. virilis* and *O. propinquus*, in parts of the US and Ontario. However, it should be noted that, while *O. rusticus* does indeed seem to replace species such as *O. virilis* at certain locations, this phenomenon is highly variable and influenced by local conditions. Capelli (1982) studied the population dynamics of *O. propinquus*, *O. rusticus*, and *O. virilis* in several northern Wisconsin lakes and found that while *O. rusticus* did increase in abundance at the expense of one or both of the other crayfish species in some of these lakes, this did not take place everywhere. In certain lakes changes in the relative numbers of individuals belonging to these crayfish species did occur over several years, as might be expected, but, as the author noted, "Changes favoring *O. rusticus* were less dramatic or absent in other lakes," and in one lake, for example, "O. rusticus showed no change from its very low original density, and all three species remained in similar proportions." (Capelli, 1982).

In fact, as far as I can tell, the now familiar "rusty crayfish as a nasty invader" story is based largely on studies conducted in some northern Wisconsin lakes. These studies can be traced back to a handful of research laboratories, and a few professors and their graduate students and research associates.

Furthermore, both *O. virilis* and *O. propinquus* remain quite abundant and very widespread in North America, in general, and in Ontario, in particular. In fact, *O. virilis* is the most widely dispersed crayfish in Canada (Hamr, 1998; Phillips et al., 2009). Its impressive Canadian range stretches from the province of New Brunswick, in the east, to northern Alberta, in the west (McAlpine et al., 1999; Phillips et al., 2009). Not one of Ontario's lake and river crayfish species is currently threatened. The only vulnerable crayfishes in this province are the semi-terrestrial burrowing ones, but their plight has nothing to do with the rusty crayfish (Guiașu, 2002b, 2007a, 2008a).

According to the Natural Heritage Information Centre web site (2007, 2008, and 2011) of the Ontario Ministry of Natural Resources (OMNR), *O. virilis* is also "the most widespread and abundant crayfish in Ontario", while *O. propinquus* is considered to be the "second most widespread crayfish in Ontario, and most commonly encountered species in southern Ontario, occurring ubiquitously in suitable habitats". I can confirm that in my own fieldwork, I found these two crayfish more often than any other species, and in many local streams *O. propinquus* seemed to be the most common crayfish species. The ratio of *O. propinquus* to *Cambarus robustus* specimens collected was sometimes 10 to 1 or even higher. Clearly, these two *Orconectes* species are not in trouble in this province, at this time.

Furthermore, the ranges of species keep changing, of course, and we cannot necessarily expect *O. virilis* and *O. propinquus* to remain the most abundant crayfishes in Ontario forever. Sooner or later, the crayfish species composition and relative abundance at various locations will inevitably change. While *O. virilis* may perhaps be displaced, or become less common, at some Ontario or Wisconsin locations, this species has recently expanded its range in other parts of North America, such as North Carolina and Washington State in the United States and the Canadian provinces of New Brunswick and Alberta (McAlpine et al., 1999; Cooper and Armstrong, 2007; Phillips et al., 2009; Larson et al., 2010).

However, when species such as *O. virilis* and *O. rusticus* expand their ranges into new areas, they tend to be regarded as "invasive" – a label which makes these range expansions appear illegitimate somehow. Conservationists and invasion biologists seem to feel regret for the apparent decline of *O. virilis* in parts of its native range, and hold the rusty crayfish responsible for this. However, when *O. virilis* spreads outside its former range, this phenomenon is also sometimes labelled as an "invasion" and frowned upon. Again, this reinforces the perception of conservation and invasion biology as fields favouring a static view of the natural world. Species are expected not only to stay put in their so-called "native" ranges (as we define them), but also to preserve about the same relative abundance there year after year. I suppose that, according to some invasion biologists, the only sensible things a newly arrived species can do in its new home would be either to remain as inconspicuous as possible or simply disappear.

When it comes to the impact of the rusty crayfish, or other non-native species, at various locations, we frequently encounter what Davis (2009) identified as an issue of authority related to citation practices. In such cases, a preliminary or tentative set of findings from a particular location is used as proof of the negative impact of an introduced species at all locations (everywhere and under all circumstances), even though the evidence in support of these sweeping generalizations is either controversial or does not exist. As an example, let us look at this statement by Williams and Mason (1999) taken from their discussion of the impact of the presumably introduced rusty crayfish in the aquatic ecosystems of the Toronto region:

> In other parts of Ontario, this species has replaced native species such as *O. virilis* by virtue of its higher aggression (due to its larger claw size), its lower susceptibility to predation, and its interference with the reproductive success of residents.
>
> GARVEY AND STEIN 1993

As can be seen, the study by Garvey and Stein (1993) is cited by Williams and Mason (1999) as the only reference in support of this statement. After seeing this particular statement, a reader unfamiliar with the original reference cited may reasonably conclude that Garvey and Stein (1993) actually studied the interactions between *O. rusticus* and native crayfish species such as *O. virilis* in Ontario, and found that the rusty crayfish is replacing native species in this Canadian province for all the reasons mentioned above. However, the use of the Garvey and Stein (1993) reference is somewhat misleading in this context, since these authors actually worked with crayfish collected in some northern Wisconsin lakes and not in Ontario (as we have seen, northern Wisconsin is the area where most research on *O. rusticus* as an invasive species was conducted). So, in fact, the Garvey and Stein (1993) study does not tell us anything about the impact of the rusty crayfish in Ontario or the possible reasons for the perceived success of this species in this part of the world. Thus, Williams and Mason (1999) are evidently just repeating some of the conclusions taken from the Garvey and Stein (1993) article, and are applying these findings automatically to Ontario, without presenting any direct evidence of their own, or citing any Ontario-based studies. In my experience, this sort of thing happens a lot in invasion biology. If an introduced animal or plant is found to cause some problems at one location, the species will frequently be assumed to be a problem everywhere it is encountered outside its so-called native range, even if the evidence for such generalizations is not there. Why bother to collect new data, if you can just use someone else's study, even though it may have been conducted in very different conditions at far away locations, to demonize introduced species in your own backyard?

There are other problems with the use of the Garvey and Stein (1993) reference by Williams and Mason (1999), as well. I re-read the former study carefully, and it turns out that Garvey and Stein (1993) measured and compared the claw lengths and widths, as well as the carapace lengths, of *O. rusticus*, *O. propinquus*, and *O. virilis* crayfish from two Wisconsin lakes, but did not actually study the interactions among these crayfish directly. In other words, these authors did not collect data on possible mechanisms, such as reproductive interference, which may account for the relative success of *O. rusticus* in northern Wisconsin. Instead, in their study, Garvey and Stein speculated that reproductively active *O. rusticus* males use their relatively larger claws and body sizes to "likely outcompete" Form I (or reproductive form) *O. propinquus* males in the possible struggle for access to *O. propinquus* mature females (since interbreeding between these two species has been reported). In fact, although interspecific reproductive interference was mentioned as a potential factor in the replacement of *O. propinquus* and *O. virilis* by *O. rusticus* in Wisconsin, the authors were simply discussing this as a possibility, while also mentioning that "Mechanisms explaining this replacement remain elusive."

Furthermore, while interspecific reproductive interference may be a factor in the replacement of some crayfish species by *O. rusticus* at some locations, under certain circumstances, this is not always the case. Kuhlmann (2008) analyzed populations of both *O. rusticus* and *O. propinquus* in the upper Susquehanna River watershed in New York State, where the rusty crayfish is expanding its range. *O. propinquus* is considered native to this river system, while *O. rusticus*, as usual, is perceived as invasive. Variables such as the fecundity of the females and the frequency of occurrence of non-viable eggs were studied and compared for both species, in allopatric and sympatric populations. Kuhlmann (2008) concluded that the results of his study "do not indicate that reproductive interference is currently contributing to the success of *O. rusticus* as an invader in the upper Susquehanna River watershed." Thus, in this case, the presence of the rusty crayfish did not have a negative effect on the reproduction of *O. propinquus*, which, at the very least, shows that the reproductive interference mechanism does not explain the success of *O. rusticus* at all locations. This reinforces the point that one should not cavalierly apply findings, or tentative findings, from one area to other regions with different environmental and ecological conditions.

Coming back to the Garvey and Stein (1993) study, a couple of additional points should perhaps be made. These authors essentially attribute the success of invasive crayfish species, such as *O. rusticus*, in displacing other crayfish to the relatively bigger claws and body sizes of the "invaders" in comparison to those of the native species. However, in this same study, we learn that "chelae of equal-sized life stages of *O. rusticus* and *O. propinquus* differed little". So those differences may not explain everything after all. In order to overcome this apparent dilemma, the authors then inform us that the larger overall body sizes of *O. rusticus* may give these crayfish the edge over the smaller *O. propinquus*. Perhaps, but then how do we explain the fact that both *O. rusticus* and *O. propinquus* are dominant over, and apparently displace, *O. virilis* in some Wisconsin lakes, despite the fact that "*O. virilis* grow larger than *O. propinquus* and *O. rusticus*" (Garvey and Stein, 1993)? We are all free to come up with various hypotheses about such issues, of course, but it seems that we still have a lot to learn about these interspecific interactions and the potential mechanisms behind them. In the meantime, it may be helpful to keep an open mind, and perhaps ease up on the control programs and the nasty anti-rusty crayfish rhetoric.

Since the relatively larger claw size of the rusty crayfish, at least in comparison to some other species of the same genus, is often emphasized in studies that tend to blame this species for all sorts of problems, it should perhaps be mentioned that crayfish with larger claws may be overrepresented in minnow traps. As Somers and Green (1993) wrote, in a study about patterns in trap catches for some Ontario crayfish: "catches with baited traps generally attract proportionately larger males of the most aggressive species".

So, perhaps *O. rusticus* may be caught more often, in some lake surveys, because of their relatively larger claws and possibly greater aggression. This, however, does not necessarily mean that rusty crayfish are always actually more common than other crayfish – maybe *O. rusticus* are simply better at getting into those minnow traps and staying there, or maybe their larger claws make it more difficult for them to get out of these traps. When it comes to interpreting the relative abundance of crayfish of various species caught in baited minnow traps, other variables, such as bait type, water temperature, and even lunar phase may also be significant, and different species can respond differently to changes in some of these, and other, variables, depending on their respective life histories (Somers and Stechey, 1986).

In addition, different crayfish species may have different habitat preferences, of course, and this should be taken into account when we interpret the results of various crayfish collecting efforts (Somers and Green, 1993). In their study on the spread of rusty crayfish and the apparent decline of native crayfishes in Wisconsin, Olden et al. (2006) mentioned that the sampling locations were found mainly in close proximity of boat launches and road crossings – in other words, areas where introduced aquatic species can be expected to be more common. At the same time, as Olden et al. (2006) admitted, the choice of these sampling locations, while undoubtedly more convenient for the researchers, "could underestimate the occurrences of native species." This type of admission is useful, but also usually buried in the middle of the Methods section, and rarely noticed by readers who tend to focus mainly on the title and the abstract of a scientific article. It is possible that the rusty crayfish is just better at surviving in areas disturbed by humans, and therefore more likely to be found there.

On the rusty crayfish watch, or identification, card, which will be discussed in more detail later on in another chapter, the information about this species is predictably brief and unequivocal. For instance, we are informed that rusty crayfish "are more aggressive than other native crayfish". Well, that depends on which native crayfish species we may be referring to. When I conducted some studies on interspecific aggression and shelter competition between evenly size-matched *Cambarus robustus* (a native species in Ontario) and *Orconectes rusticus* (presumably an introduced species in this province) crayfish, I found that *C. robustus* were much more aggressive than, and overwhelmingly dominant over, the rusty crayfish, in most cases (Guiașu, unpublished data). In fact, even though I had explored other types of intra- and interspecific agonistic contests in intermoult, intact crayfish in several previous studies (Guiașu et al., 1996b, 2005; Guiașu and Dunham, 1997a, b, 1998, 1999a, b, and c), I was surprised by the very high level of aggression displayed by *C. robustus* crayfish towards *O. rusticus* of the same size. In one such interspecific contest, for example, a large *C. robustus* individual actually put its claw right through the side of the carapace of an equally large *O. rusticus* opponent. None of the specimens used in these contests had recently moulted or were about to go through the moulting process, and, therefore, the exoskeletons of all these crayfish were suitably hard. Furthermore, all crayfish had been intact and in good condition prior to the contests. I had analyzed a variety of agonistic contests in *Cambarus* and *Fallicambarus* crayfish species in previous years, and I had never seen anything like this before. In the hundreds of intra- and interspecific agonistic contests between paired, evenly size-matched crayfish belonging to the species *Cambarus bartonii*, *C. robustus*, and *Fallicambarus fodiens* I saw over the years, I never witnessed any injuries to any of the combatants. For the most part, *C. robustus* were clearly dominant over *C. bartonii* of similar sizes, but no injuries to the subordinate crayfish were ever recorded (Guiașu and Dunham, 1999a and 1999b). And yet, in this contest between *C. robustus* and *O. rusticus*, to my surprise and discomfort, the rusty crayfish was injured, possibly due to a thinner carapace than that of the *Cambarus* crayfish, and I had to stop the contest right away in order to prevent additional injuries. This may have been an isolated incident, but, overall, *C. robustus* seemed dominant over *O. rusticus* in the majority of the contests analyzed. As a result, *C. robustus* gained control of valuable disputed shelters in these laboratory studies (Guiașu, unpublished data). During my fieldwork, I also noticed that big *C. robustus* individuals tended to be found under large, flat rocks on the bottom of streams – possibly the best crayfish shelters available (Guiașu, 2002a), while some *Orconectes* crayfish present in the same bodies of water were often restricted to apparently less optimal areas with fewer and smaller rocks or a sandy substrate.

The outcome of the interspecific contests between the native *C. robustus* and the possibly introduced *O. rusticus* should not really be all that surprising, given that, even though the crayfish were size-matched as closely as possible for each contest, *C. robustus* individuals appear to be much more powerfully built and possess relatively bigger and heavier-looking claws than *O. rusticus* specimens of similar body sizes. So, in this instance, the native crayfish species usually has the bigger claws in comparison to the notorious "invasive" *O. rusticus*. Therefore, the generalization that the larger claws and greater aggression of the rusty crayfish always give it an advantage over native crayfish obviously does not apply in this case.

However, just because *C. robustus* is capable of dominating *O. rusticus* in interspecific agonistic contests, this does not mean that *C. robustus* is, or will be, actually displacing the rusty crayfish at various locations. While sometimes more aggressive crayfish species can indeed replace less aggressive ones, at some locations, this is not always the case (Guiașu, 1997). For example, Rorer and Capelli (1978) reported that *Orconectes spinosus* crayfish appeared to replace *Cambarus bartonii* in a Virginia lake between 1933 and 1957, even though *C. bartonii* consistently dominated *O. spinosus* of similar sizes during interspecific agonistic contests, in the laboratory, both in the presence and in the absence of shelters. Thus, in this case, competitive exclusion mediated by interspecific aggression does not seem to apply, since the relatively more aggressive species was replaced by the less aggressive one. The reasons for the replacement of *C. bartonii* by *O. spinosus* in this lake were unknown, but, since *C. bartonii* remained present in nearby streams (Rorer and Capelli, 1978), it is quite possible that environmental changes to the lake itself may have been responsible. In such a case, and in the absence of any clear evidence, perhaps even stating that one crayfish species displaced another at a particular location can be misleading. Maybe one species did better, while the other one gradually disappeared, because changing environmental conditions – for example changes in average water temperatures or water chemistry – favoured one species over the other. Perhaps there was little or no direct competition at all between the two species in this instance.

These examples also suggest that caution should be exercised when applying the results of laboratory studies on interspecific agonistic contests to the observed changes in the distributional patterns and relative abundance of crayfish species in the field. A more aggressive and dominant species in interspecific contests staged in artificial laboratory conditions may not necessarily take over and displace other species in the field (Guiașu, 1997).

On the other hand, laboratory studies can answer some questions about possible mechanisms of displacement and the interactions among species, for example by eliminating some possibilities. The crayfish species *Orconectes hylas* has been introduced in the St. Francis River drainage system, within the state of Missouri, in the United States. This crayfish has been expanding its range, while at the same time two local endemic species of the same genus – *O. peruncus* and *O. quadruncus* – have been declining in this region. Both endemic species are listed as imperiled in Missouri. The assumption was that the introduced species may be more aggressive than the native ones, and this may be the reason, or a reason, for the decline of the native crayfishes. Several types of interspecific contests between the non-native and the native crayfishes were analyzed in a series of laboratory experiments. The results showed that the introduced crayfish, *O. hylas*, was not more aggressive than, or dominant over, the native species *O. peruncus* and *O. quadruncus*. Therefore, higher aggression of the introduced species, leading to the possible displacement of native ones, was not the explanation for the decline of the imperiled Missouri crayfishes, in this case. Other factors, such as habitat degradation, or changes in certain environmental variables, may be responsible (Rahm et al., 2005).

Coming back to the previously mentioned problems with citation practices, when it comes to generalizations about the negative impact of invasive species, I should also mention that in the Introduction of their article, Rahm et al. (2005) cited one of my own studies (Guiașu and Dunham, 1999a) in a somewhat misleading way – no doubt inadvertently. In their article, Rahm et al. (2005) referred to the negative effects of invasive crayfishes on native crayfish species with similar habitat requirements and life histories and cited my previously mentioned co-authored 1999 study as a reference in support of that statement. In the Guiașu and Dunham (1999a) article, interspecific agonistic contests between *Cambarus bartonii* and *Cambarus robustus* were analyzed and discussed, and, as already mentioned, *C. robustus* was found to be dominant over *C. bartonii*. Furthermore, these two Ontario crayfish species do indeed have similar habitat requirements and life cycles (Guiașu, 2002a) and *C. robustus* may be expanding its range in this province, possibly at the expense of *C. bartonii*, at least at a few locations (Guiașu et al., 1996b). The problem, however, is that *C. robustus* is not considered to be an invasive species in Ontario. In fact, both *C. robustus* and *C. bartonii* are considered native to Ontario, which is probably why the apparent range expansion of *C. robustus* in this province did not cause any concern (Guiașu, 2008a). Therefore, we obviously cannot use this as an example of the negative effects of invasive species on native ones.

In a major study, Edwards et al. (2009) found significant and geographically widespread declines in all crayfish species (native and non-native) sampled in 100 lakes in south-central Ontario, within the last twenty years or so. However, this decline could not be attributed to *Orconectes rusticus*, since this species was absent from almost all of the lakes sampled for the study. Therefore, the authors stated that the rusty crayfish "was not a factor in the declines." So, here we have an important recent study, documenting major declines in all crayfish species in 100 lakes, distributed over an area of more than 20,000 square km, and the rusty crayfish is not involved in this at all. Other factors, such as some changes in water chemistry, were cited in the study as possible causes for the observed reductions in the crayfish populations since the early 1990s.

Rusty crayfish are also frequently blamed for negative effects on macrophytes or aquatic plants (Carpenter et al., 2007). However, all crayfish species seem capable of eating and damaging macrophytes, if given the opportunity. Studies have found that crayfish such as *Orconectes virilis*, *Orconectes propinquus*, and *Procambarus clarkii* are also feeding on macrophytes (Chambers et al., 1990; Olsen et al., 1991; Carreira et al., 2014). In a comparative study, Olsen et al. (1991) looked at the effect of three *Orconectes* species (*O. propinquus*, *O. rusticus*, and *O. virilis*) on macrophytes and found that "*O. rusticus* did not destroy (consumption plus fragmentation) more macrophyte biomass than its congeners."

Dorn and Wojdak (2004) reported that *O. virilis* consumed large quantities of fish eggs, filamentous green algae, and vascular plants in experimental ponds set up in Michigan.

The effect of crayfish on aquatic plants can be complex. While crayfish eat some of these plants, as we have seen, they also feed on snails, and the reduction of the snail populations benefits the growth of plants (Chambers et al., 1990). Olsen et al. (1991) showed that *O. rusticus* ate more snails than either *O. propinquus* or *O. virilis*, which, presumably, may benefit the growth of macrophytes.

Since most, if not all, crayfish species have a broad diet, we shouldn't be surprised that they can feed on a variety of plant and small animal species. It seems that the crayfish effects on macrophytes are only considered problematic when a non-native species, such as *O. rusticus*, is involved. This reinforces my earlier point that any normal activity of introduced species is viewed as a potential problem, whereas the very same activities by similar native species are generally ignored or even encouraged. It is a matter of perception.

And, as discussed in the previous section, rusty crayfish can also prey on invasive zebra mussels, and reduce mussel populations to a significant extent (Perry et al., 2000).

Sometimes, *O. rusticus* can be accused of causing problems even at locations where it is not even present. In a new book titled *Field guide to invasive plants and animals in Britain*, the rusty crayfish is listed as one of the invasive species in that country, and receives a full-page description. Here, we are informed that *O. rusticus* "may have an impact on native species diversity", which is a bit puzzling, considering that, according to information found on the same page, the rusty crayfish is "not yet present" in Great Britain (Booy et al., 2015).

Fish Case Studies: The Round Goby and the Common Carp in North America

When it comes to perceived "invasive" species, it seems that sometimes any reason, no matter how flimsy, can be used against them. For example, I read a brief interview about invasive species with Hugh MacIsaac, the director of the Natural Sciences and Engineering Research Council's Canadian Aquatic Invasive Species Network II, in a local newspaper. During the interview, MacIsaac listed some familiar invasive species in the Great Lakes region of North America, and included the round goby fish among the five most damaging invasive species in Lake Ontario. When asked about the ecological and economic cost of invasive species, MacIsaac said that "Round gobies are a complete nuisance for anglers" (Menon, 2011). Well, frankly, I would imagine that anglers are a much bigger nuisance, not to mention a potentially lethal danger, to the fish (native or introduced) unfortunate enough to be caught in the hooks and the nets set up to capture them.

A few years ago, I was walking by the shores of a local lake, when I saw a fisherman who was deliberately smashing the head of a freshly caught round goby

FIGURE 17 *The round goby* (Neogobius melanostomus)
PHOTO BY STAN YAVNO

against a large rock. When he saw me staring at him in disbelief, the fisherman greeted me with a friendly smile and told me not to worry, because this was, in his words, "a garbage fish – not from around here." I imagine that many invasion biologists may have approved of his actions and words. I was going to ask where he thought the fish he had just pulled from the lake was actually from, but I suddenly felt very tired, and simply kept on walking.

In any case, the fact that some anglers would rather catch other fish than the round gobies is hardly a valid reason, in my view, to demonize the fish. After all, the round gobies do not want to be caught, and they do not waste the anglers' time on purpose. Perhaps the anglers should be the ones adjusting to the changing fish species composition of the lake and coming up with inventive ways not to catch unwanted fish species, particularly since most, if not all, of the anglers I have seen in these parts are recreational fishermen, who catch fish as a hobby, and not for any other more vital reasons, such as subsistence. Besides, with all the chemical pollution of the Great Lakes, I certainly would try to avoid eating any of the local fish. Perhaps if we paid more attention to improving the water quality of the Great Lakes, and less time going after round gobies, we would all enjoy healthier aquatic ecosystems in this part of the world.

The round goby (*Neogobius melanostomus*) came to North America from the Black Sea and Caspian Sea regions – the same general area where the zebra mussel is supposed to have originated. These two introduced aquatic species arrived in North America almost at the same time, in roughly the same place (the round goby was first discovered by anglers in the St. Clair River in 1990, and the zebra mussel only two years earlier in Lake St. Clair), and by similar means (most likely in the ballast water of ships) (Barr, 1996; Ray and Corkum, 1997; Winterton, 2005).

Furthermore, the introduced round goby is a major predator of zebra mussels, and can therefore be considered as a natural control agent for this invasive mollusk. Ray and Corkum (1997) showed that zebra mussels represented a majority (58%) of the prey items eaten by round gobies collected from the Detroit River, within the North American Great Lakes region. Another non-native fish species in North America, the common carp, is capable of feeding on zebra mussels as well (French III, 1993).

The common carp (*Cyprinus carpio*) is a species native to parts of Europe and Asia, which was successfully introduced into the United States in 1877 by the US Fish Commission, after a number of unsuccessful introduction attempts. This species was brought into Ontario in 1880 (Scott and Crossman, 1979). *C. carpio* is now widely distributed in North America, but it is still not accepted by many as a truly local fish, even though no human being alive today was born before the introduction of the carp on this continent. Therefore, there are no North Americans today who could possibly say that they remember the way things were before the carp arrived. As a result, when some people talk about the restoration of former ecosystems to a state like the one which existed before the introduction of species such as carp, all we are really dealing with is some sort of vague nostalgia for a time we never actually knew or experienced. Just remove the carp, or the purple loosestrife, and we will all be happily back to the early 19th century, in an idyllic setting we remember from an old painting in the National Art Gallery… Wouldn't that be nice?

But, anyway, coming back to reality, I remember an ecology lecture from a course I took as a graduate student, at the University of Toronto, long ago. I had just started my Ph.D. studies at the time, and the professor teaching the course was a very nice man, as well as a very reputable and well-respected aquatic ecologist, with many years of research experience. At one point, he mentioned something about the common carp not being an edible fish in North America, and being a bit of a nuisance species here, as opposed to its native range in Europe. The clear message was that carp were fine as long as they stayed home in Europe, but here in North America, they suddenly and inexplicably turned into bad fish (I could just visualize the professor carrying a placard with a sign that read: "Carp go home!", while marching in protest around a local pond; it's a bit silly, I know, but the mind can drift sometimes, particularly during longer lectures). Normally, I took notes quietly and did not ask questions during the lectures, but this comment intrigued me, since I used to eat carp, once in a while, during my European childhood (these fish were commonly available in local markets), and the carp I saw in North America looked identical to the ones I remembered seeing in Europe. So, I politely asked the kind, elderly professor (who was also of European descent, by the way), why he thought that local carp were not edible, unlike the European ones, and why these fish were considered to be a problem here. To my surprise, and slight discomfort, the otherwise very knowledgeable professor seemed somewhat taken aback by my question, and he hesitated before answering vaguely that the carp in North America are simply "different" than the European ones. No other explanation was given. I didn't pursue this discussion at the time, but I was left with the distinct impression that the reason why the carp in North America were considered to be "different" was simply because they were not "native" to this place. Otherwise, they obviously look and behave the same way in both Europe and North America. What some really object to, deep down, is the fact that the carp on this side of the Atlantic have been brought over here by people, long ago. And that is when I started to get really interested in introduced species...

Winterton (2005) acknowledged that carp in North America are indeed edible, if they come from clean waters, but added that: "Unfortunately, many Great Lakes carp contain relatively high levels of PCBs and toxic chemicals and cannot be marketed as food fish." Fair enough, but then it seems to me that the problem does not lie with the carp, but with the appalling pollution of our lakes with all kinds of nasty chemicals. Besides, if the waters are polluted, then surely native fish are contaminated as well, particularly favoured local salmon species which, as top predators, tend to bioaccumulate large quantities of toxins in their bodies.

Along the same lines, in a book about the freshwater fishes of Manitoba, we can find the following paragraph about the introduced common carp (*Cyprinus carpio*): "Although carp are valued as food in Eurasia, their flesh does not adapt well to the 'fillet and fry' approach to fish cooking used so commonly in North America" (Stewart and Watkinson, 2004). So, in this case, a perfectly good to eat introduced fish, which could be considered a valuable new resource, is rejected, because it doesn't lend itself easily to a particular local cooking method. Bring on the control programs...

Fortunately, not everyone is that fussy, and people in some other parts of the world where this fish species was introduced seem to find it valuable. For example, in an article about the development of aquaculture in Africa's most populous country, Anadu (1996) reported that the common carp (*Cyprinus carpio*) is the "most successfully, widely cultured exotic fish in Nigeria."

FIGURE 18 *The common carp* (Cyprinus carpio) *in North America*
PHOTO BY LYNDA CORKUM

The common carp is also considered "one of the most popular angling fish" in Britain (Booy et al., 2015) and some other parts of Europe. Despite this fact, and although this species has been introduced in Britain before 1496 as a valuable food fish, *C. carpio* is still considered as an invasive species there (Simmons, 2001; Booy et al., 2015). Being present in a particular country for more than 500 years is clearly still not enough to allow a species to be considered a "natural" part of local ecosystems, at least in this case.

Coming back to the carp's presence in Manitoba, it seems that the people in charge of the management of this species cannot quite decide which approach should prevail. Thus, on the one hand, there were repeated, and unsuccessful, attempts to remove and exclude carp from certain areas of Lake Winnipeg and Lake Manitoba, and, apparently, local anglers and commercial fishermen often catch these fish, and then usually discard them as unwanted bycatch. On the other hand, there is a large local commercial fishery specifically targeting carp, and most of the catch is then exported by the Freshwater Fish Marketing Commission. Furthermore, while Manitoba-based anglers do not appear to hold the carp in high regard, it seems that a valuable recreational fishery for this species has been established in the province's Red River, because of the arrival of European tourists with a keen interest in fishing for carp (Stewart and Watkinson, 2004).

There is also confusion about the nature of the perceived problems we accuse such invasive freshwater aquatic species of causing. For example, carp are blamed for increasing water turbidity while feeding in the muddy substrate on the bottom of a lake or pond (Stewart and Watkinson, 2004). On the other hand, zebra mussels are disliked for being wonderful filter feeders, which remove floating particles, and therefore make the water very clear in the lakes they inhabit – the exact opposite effect from that caused by carp. So, perhaps we should make up our minds. If we prefer clear waters, then zebra mussels should be useful, and if increased turbidity is what we want, then carp should be able to do the job. And if we want water that is partly clear and partly turbid, then maybe we should place both carp and zebra mussels together in a lake, and see what happens… In any case, it seems strange that we can blame both carp and zebra mussels for changing water clarity through their respective feeding actions, even though, as we have seen, the two species have opposite effects from this point of view. This reinforces the notion that introduced species just can't catch a break – whatever they do will always be perceived as bad and harmful by the ever vigilant invasion biologists, and any activity of a non-native species can be used against it.

In a fairly recent discussion, found within the *Encyclopedia of biological invasions*, Sorensen and Bajer (2011) acknowledge that when it comes to poor water quality and the loss of native fish and waterfowl, "the precise role played by carp is not well understood." Despite that fact, these authors also mention that "there has been a recent surge in interest in using integrated pest management strategies to suppress carp." So, at this time, we don't know exactly if carp is really to blame for some of the environmental problems we hold it responsible for, but we want to find new and better ways to suppress this species anyway… Therefore, we can conclude that this "recent surge in interest" in finding better ways to control carp is not really related to new and conclusive evidence that carp is harmful, but is mainly linked to the growing popularity of invasion biology and the control programs some invasion biologists advocate as the solution to many perceived ecological problems.

I was just reading a recently published field guide to hiking opportunities in some conservation areas in southern Ontario (Perrett, 2012), when I came across some information on the Rattray Marsh. This marsh is located within the Greater Toronto Area, so, not surprisingly, as the author notes, the conservation area is "surrounded by urban development and paved surfaces". While discussing some of the species present in this marsh, Perrett states that: "Non-native species also affect ecosystems and Rattray Marsh is no exception. Common carp and purple loosestrife are two alien species that adversely affect the marsh." Well, first of all, any species present at a particular location will have some sort of an effect on the ecosystems, by virtue of its very existence in the area. Whether these effects are considered "good" or "bad" is largely a subjective matter, which depends on exactly which type of nature some people may prefer, often for their own ideological reasons. For example, a version of nature which eliminates or minimizes the impact of so-called "alien" species happens to be in fashion in certain circles these days, but that does not mean that this version of nature is the best or most desirable in absolute and universal terms, particularly when we are talking about a marsh located right within the boundaries of the largest urban area in Canada. Essentially, the region from Toronto to Hamilton (another major Canadian city) is one uninterrupted urban area, and the roughly one hour drive between the downtown cores of these two cities takes place exclusively through a thoroughly urban landscape. Secondly, this is a guide to some of the hiking opportunities in southern Ontario, and these types of general anti-introduced species blanket statements strike me as misplaced in a book which is mainly about describing the beauty of local landscapes for potential hikers and nature lovers. This is perhaps not the best forum for discussing the ecological impact of particular species on the ecosystem, particularly since the scientific literature is not as clear on these subjects as the statements found in such guides or more general articles. Nevertheless, Perrett continues the discussion, by mentioning that: "There are also plans to prevent, or minimize, the impacts of carp on the ecosystem."

Drastic actions against carp species in North America have been taken before, of course. In the 1970s, the writer, painter and fishing enthusiast Russell Chatham described the effects of a decision taken by the Department of Fish and Game to deliberately poison the waters of a river in the US with rotenone, in order to get rid of carp for good. This type of poison kills fish indiscriminately, however, and, in this case, it turned out that unintended targets, such as the smallmouth bass, were more vulnerable to rotenone than the carp. As a result, the smallmouth bass died and many carp recovered and became even more abundant in the river than before (Chatham, 1988 – the story is part of a book initially published in 1978).

Although, according to Sorensen and Bajer (2011), the life history of the common carp is poorly understood, we do know that this species is very resourceful and resilient. For example, although *C. carpio* can thrive in warm waters with temperatures of 25 to 30 degrees Celsius, these fish can also survive harsh winters and live under the ice. In fact, during my early childhood, well before I even thought of becoming a biologist, I became aware of the remarkable abilities of the carp to survive in extreme conditions. As mentioned, the carp is a popular food fish in parts of Europe, and my parents would often purchase live fish of this species from a local market. The fish would be brought home wrapped tightly in a plastic bag, without any water, and then placed in the freezer for a few hours. One day, when I was about five or six years old, I went to the freezer and took out what seemed to be a completely frozen carp. The fish had been in the freezer for quite some time, and appeared dead. However, with the hopeful enthusiasm of the young child I was at that time, I removed the large fish from the plastic bag that was covering it tightly, and then I filled the bathtub with cool water and placed the frozen fish in the water. To the adults in the house, this seemed like a silly and futile undertaking, but, only a few minutes later, as I recall, I was vindicated, when the carp not only showed signs of life, but actually started swimming actively in the bathtub. That day, I not only gained new respect for this fish, but I also learned a valuable early lesson about the remarkable abilities of some other species to survive in very challenging circumstances.

Furthermore, the common carp is also known to be very tolerant of low dissolved oxygen concentrations in the water, as well as high turbidity, and even somewhat increased salt content in its mainly freshwater habitats (Sorensen and Bajer, 2011). Wichert (1995) looked at the effects of urbanization on the fish associations of Toronto streams, and found that the common carp and the related goldfish (*Carassius auratus* – the popular aquarium fish and another introduced species in North America) showed the greatest tolerance, overall, among the 39 fish species analyzed, to water quality-related features such as high chlorine, low dissolved oxygen, high turbidity, and habitat destruction. Habitat destruction included the loss of riparian vegetation, channelization, high sedimentation, and loss of substrate variety.

The fact that the common carp can tolerate a wide range of temperatures as well as low oxygen conditions, habitat destruction associated with urbanization, and some increased salinity should make it an ideal fish for the streams, marshes, lakes, and ponds located in the Greater Toronto Area (known locally as the GTA). Here, in this region, we certainly get extreme temperature fluctuations. Air temperatures can have an annual range of over 50 degrees Celsius – from more than +30 degrees Celsius in the summer to less than –20 degrees Celsius in the winter – with a corresponding effect on water temperatures. During my fieldwork in Toronto area streams, I collected crayfish and fish in waters with temperatures ranging from 4 degrees Celsius in early November to 23 degrees Celsius in July and August (Guiașu, 2002a), and, of course, local bodies of water can freeze over during the winter. In addition to the wild seasonal temperature swings, GTA streams and lakes are also affected by urban pollution, associated increased turbidity, and quite a bit of road salt (used to minimize the effects of ice and make winter driving somewhat less dangerous) that eventually ends up in local waterways. Thus, we should be grateful to have resourceful fish such as carp around, which can tolerate these difficult conditions few other fish species could put up with. The common carp has been in Ontario for more than 130 years, and seems well adapted to some of our aquatic habitats. Perhaps it is time we accept these facts, leave these fish alone, and move on.

CHAPTER 5

Positive Contributions of Introduced Species

> So we should venture on the study of every kind of animal without distaste; for each and all will reveal to us something natural and something beautiful.
>
> ARISTOTLE

FIGURE 19 *Native Canada geese feeding on introduced dandelions in the Toronto region*
PHOTO BY DORIS WONG

Several years ago, a tiny white mulberry tree sapling sprouted in my front yard. It came seemingly out of nowhere, as there are no other white mulberry trees on our street and nobody in my family had anything to do with this (in other words, we most definitely did not plant this tree). It is almost certain that the seed which eventually developed into this tree was inadvertently dropped into our yard by a bird, which had eaten a white mulberry somewhere else in the city. The white mulberry tree (*Morus alba*) is native to eastern Asia, and is well-known for its importance for the silk industry, since the leaves of this tree are used as the main food source for silkworms (which are actually the caterpillars of the mulberry silk moth – *Bombyx mori*), especially in countries such as

 | DOI 10.1163/9789047426134_007

China. In fact, it seems that this relatively small tree species was first brought to North America in an attempt to start a local silk industry on this continent. This tree is now capable of growing and spreading quite well on its own, with the help of birds, in southern regions of Ontario and in some other parts of eastern North America (Kershaw, 2001). Clearly, the arrival of the white mulberry tree in my front yard was not assisted by humans in any way, and was therefore an entirely natural process, even by the dubious standards of invasion biology.

The fragile-looking sapling grew into a beautiful mulberry tree, and now, this introduced tree species provides a reliable food supply every late spring and early summer for a variety of native and introduced insect, bird, and mammalian species. The mulberry crop is produced mainly in late May and throughout June, but some fruits can still be found on the branches in July. I was aware that cedar waxwings (*Bombycilla cedrorum*) are found in the Greater Toronto Area, because I read this in relevant field guides (for example, McKeating, 1990). However, I had never actually seen any cedar waxwings, until the young mulberry tree in my front yard started producing its first fruits. Then, all of a sudden, each year, throughout the month of June, flocks of up to 20 cedar waxwings would descend upon my mulberry tree several times a day to feed eagerly on this rich fruit supply. After the mulberries were gone, the flocks vanished as suddenly and mysteriously as they had first appeared.

In his book *Birds of Ontario*, Bezener (2000) writes that: "Native berry-producing trees and shrubs planted in your backyard can attract Cedar Waxwings, often encouraging them to nest in your area." The word "native" seems unnecessary in this brief statement, since it is clear, as the example with my white mulberry tree also illustrates, that these birds will be attracted by a variety of fruits, produced by either native or introduced species of trees and shrubs. Thus, I conclude that the inclusion of the word "native", in this context, was meant simply to promote only the planting of native species in private Ontario yards, possibly for ideological reasons. However, cedar waxwings evidently do not care whether the berries they enjoy eating are produced by native or introduced trees and shrubs, and these birds, unlike some of the members of our own species, are not interested in discriminating among various types of useful berry-producing trees, based on their initial geographic origins.

The cedar waxwings were not the only native species to benefit from this new and valuable resource provided by my exotic white mulberry tree. Other native species seen feeding regularly on the mulberries available in my yard included American goldfinches, American robins, northern cardinals, gray squirrels, and raccoons. A few striped skunks sometimes arrived at night attracted by the fruits which fell to the ground. House sparrows (an introduced species in North America) were also seen feeding on mulberries, but, for some reason, other non-native bird species present in my neighbourhood, such as pigeons (rock doves) and starlings never visited the tree (at least, not when I was watching). In any case, the mulberry tree provided a bountiful feast for many birds and quite a few mammals for several years now, and the fruit supply was so vast, that I rarely saw any overt competition or aggression among the many animals feeding in the tree at the same time. Occasionally, a robin would attempt to monopolize the tree, and chase away smaller birds, but this seemed futile, since there were simply too many other birds around, and enough mulberries for everyone, so, for the most part, birds were feeding peacefully, busily, and without rancour, within the protective dense foliage of the tree. To some extent, this reminded me of the famous salmon runs of north-western North America, when so many large fish would travel upstream, that the normally solitary and aggressive grizzly bears would simply feed on the fish, quite peacefully, without bothering to chase other bears away, or engage in much intraspecific aggression. When there is more than enough food for everybody, there is no reason to waste precious time and energy in pointless, and potentially dangerous, fighting.

In addition to the birds and the mammals, large numbers of insects of many species were attracted by the mulberry fruits, both on the branches of the tree and on the ground. In fact, I suspect that some birds may have come to the tree not only for the fruits, but also to hunt some of the many insects which seemed to hover around during the mulberry season. On the other hand, the cedar waxwings were clearly there exclusively because of the availability of fruits. Cedar waxwings are avid fruit eaters, with a reputation for sometimes getting drunk after eating too many overripe berries. However, I did not notice any intoxicated birds on my tree, possibly also because the mulberries were eaten well before they had a chance to become overly ripe.

The multitude of birds feeding on mulberries even attracted predators. For example, some neighbourhood domestic cats seemed to take an unusual interest in the mulberry tree at this time of the year, and often sat on the ground, near the tree, closely following the movements of the small birds among the branches. Once, I even saw a hawk charge the tree at top speed, although this happened so fast, that I really can't be entirely sure if the hawk actually caught anything, other than air, or not.

All of this activity started with a seed dropped into my yard, most likely by a bird, only a few years ago. As a result, I gained a beautiful and hardy tree, without spending any money or having to visit a garden centre, and I had the opportunity to witness many interesting interactions among individuals of various local species from the comfortable vantage point of my front window. Just a few days ago, for instance, as another mulberry crop has appeared in early June, I watched a pair of northern cardinals strengthen their bond by using the fruits of the tree. The brightly coloured red male picked up a mulberry from the tip of a remote branch, and then, instead of eating the fruit, the male offered the mulberry to the more discreetly adorned female. This nuptial gift was accepted readily, and then the two birds flew away together, possibly to a nesting site. The mulberry tree sustains a lot of other species, and makes bird watching particularly easy and rewarding throughout the spring and summer.

Actually, over the years, I noticed that the trees which grow and survive best are the ones that sprout on their own in various yards or city parks. Every year I witnessed some tree-planting programs, which these days tend to often promote native species only, for ideological reasons, and a few weeks or months later, a high percentage of these young trees just wilt and die in the relatively harsh city conditions. On the other hand, trees that spread and grow on their own, which include a variety of introduced and native species, are often pulled out or cut, even though it seems quite obvious that these are the hardiest and best adjusted trees in the region, and simply letting them grow in parks and yards would be the best and least expensive way to have a healthy and naturally regenerating urban forest.

I have two brief examples, to illustrate these points. One of these examples is contained within this paragraph, and the second one will open the next section of the book. A couple of years ago, on my university campus in central Toronto, I noticed a young horse chestnut sapling growing among some planted shrubs. The sapling of this introduced tree (initially native to Eurasia) originated from a chestnut which had most likely been buried there by a native gray squirrel – a perfectly natural process. However, even though this was a "semi-wild" area of the campus, where a variety of shrubs and trees were growing, the healthy and rapidly growing horse chestnut sapling was eventually removed by the persons responsible for the campus green spaces, presumably because it was growing in the "wrong place". In other words, because they did not plant this tree, and the tree grew on its own (with some initial assistance from a squirrel), they could not allow it to survive. Had the sapling been left alone, it most likely would have grown into a spectacular horse chestnut tree, and, over time, would have provided a natural replacement for some of the beautiful, but aging, trees on our campus, at no cost to the university.

Norway Maples and Other Useful Non-native Plants in the Toronto Region

In my own backyard, I set aside a certain area where no lawn-mowers were used. As a result, a variety of mostly introduced flowering plants, whose seeds were brought by the wind, or birds, or small mammals over the years, could

FIGURE 20 *The seeds of the Norway maple* (Acer platanoides)
PHOTO BY DORIS WONG

grow naturally there. Dozens of such plants, including some with very beautiful flowers, temporarily established themselves in this section of the garden, before being replaced by other plant species. After a few years, several particularly vigorous young Norway maples started to emerge in the portion of my backyard set aside for nature to take its course without human interference, and these trees are an obviously healthy and useful addition to the local canopy. If we can get over our recent fixation with native and non-native species, and simply enjoy the plants growing naturally all around us here and now, we will most likely have much healthier urban forests and gardens for a considerably lower cost.

In fact, in a study of the benefits provided by the tree species present in Allan Gardens, in downtown Toronto, Millward and Sabir (2011) found that the Norway maple (*Acer platanoides*) was the most valuable tree, overall. This species brought the largest contribution to air quality improvement, through the removal of pollutants, as well as providing the most energy savings as a result of shading, the greatest carbon dioxide reduction, and the highest stormwater reduction and aesthetic benefits. Due to its impressive canopy, *A. platanoides* has a greater leaf area coverage than any other tree species not only in Allan Gardens, but in the entire City of Toronto. The local importance of this tree species, which was introduced into North America from Europe in the 1700s and was planted widely in Toronto especially since the 1960s, is emphasized in the following passage from the Abstract of the study by Millward and Sabir (2011):

> The flow of benefits from Allan Gardens' urban forest is heavily dependent upon Norway Maple (*Acer platanoides*), a finding mirrored across much of Toronto's urban forest. Norway Maple provides the greatest overall annual benefits ($4846 total; $113 per tree) and as a species contributes 17.5% of the environmental and 20% of aesthetic value provided by trees in the park.

Despite its usefulness and its tolerance of air pollution and soil compaction – which make it an ideal tree for urban areas – the Norway maple is currently out of favour with organizations such as Toronto's Urban Forestry Department and is "rarely planted anymore by the City of Toronto", because this tree is non-native and perceived by some as invasive (Millward and Sabir, 2011).

At the end of a book chapter in which he makes a compelling case for finally granting "Canadian citizenship" to the Norway maple and accepting this tree as a legitimate local species, Larson (2015) mentions being asked to cut down a perfectly healthy and mature tree of this species growing on his property, in southern Ontario, by "a local plant nativist". The author resisted and did not remove the tree, but this incident shows just how far some people would go to convince their neighbours to get rid of non-native plants. The pressure is on to minimize or eliminate the use of introduced species even in private gardens and to get rid of established non-native plants growing there.

These days, as previously discussed, it seems hard to find a recently published field guide that does not mention non-native species, usually in a negative way, while at the same time uncritically promoting native species at all costs. In the Introduction section to his book on some Ontario conservation areas, Perrett (2012) suggests, quite sensibly, that we should become more tolerant of certain much maligned species, such as the introduced dandelions, and should rely less on pesticides and other such chemicals to maintain largely artificial lawns. So far, so good. However, in the very next paragraph, the author suggests planting and promoting native plants only in our yards, supposedly in order to create a more "natural" environment. This is clearly a mixed message. If we are supposed to create more "wild places", as Perrett advises, and allow nature to follow its own course, then surely we should not interfere by promoting only certain "pre-approved" native species, while excluding non-native species which have been around for decades or even centuries, and are obviously perfectly well adapted to current local conditions. If we interfere in this way, we would not be promoting truly wild places or the best adapted species to current local conditions, choosing instead to prop up yet another artificial and unsustainable version of nature that satisfies the fashionable nostalgia some may feel for a time long gone. In defense of his position, Perrett also states that "native plants require less water than non-native species, and they are more adapted to the area." I dealt with the dubious notion that native species are better adapted than non-native ones before, but this claim is so pervasive these days that it may require a bit more attention. Again, if native species are so well adapted, then why do they need our help? Secondly, environmental conditions are constantly changing, of course, and so does the human impact on the environment. Therefore, it should be fairly obvious even to some conservation biologists that the conditions in a region such as the Greater Toronto Area are quite different in 2016, in comparison to, say, 1850 (when the GTA as we now know it obviously did not exist). As for the idea that native plants need less water than non-native ones – it just seems very unlikely that such overly general statements have much validity. Are we expected to simply accept uncritically the proposition that all native plants require less water than all the non-native ones? Do we really know this to be true for each native and non-native plant species in this area? Has anybody actually studied the water requirements of all these species, under various circumstances, in a systematic and scientific way? We experienced somewhat of a drought here in southern Ontario a few years ago, and the month of July 2012 was a particularly dry one. Predictably, lawns turned a depressing shade of yellow, and many native plants seemed to be under severe stress. On the other hand, just as an anecdotal observation, many of the common introduced wildflowers, such as chicory, bird's-foot trefoil, Queen Anne's lace, several types of thistles, and other so-called weeds, appeared to be thriving. Frequently, these were the only green spots on an otherwise dismal brownish-yellow field landscape. As the climate and other environmental conditions change, the numbers and types of species found in southern Ontario, and elsewhere in the world, will clearly change as well. We can either come to terms with this straightforward fact, or continue to spend vast resources desperately trying to hang on to, or recreate, the presumed species compositions of 200 or 300 years ago.

Perrett (2012) also mentions that it is important to plant native species only, because, presumably, native insects do not feed on non-native plants. This is the type of information I also found in some fairly recent newspaper articles (for instance, Scrivener, 2012 a and b) as a justification for promoting only native plants while disparaging the introduced ones. As we have seen during the discussion of the purple loosestrife case study, this is not actually true for at least some exotic plant species, and it is probably untrue for most of them. Purple loosestrife, as you may recall, is a valuable resource for a variety of native and non-native insects, and I strongly suspect that the same applies to many

FIGURE 21 *Motherwort* (Leonurus cardiaca) *plants*
PHOTO BY DORIS WONG

other non-native plant species, particularly those that have been around for quite some time. Furthermore, there are quite a few non-native insect species here, in Ontario, as well, including very useful ones, such as honey bees, and these insects certainly appreciate having non-native flowering plants as valuable local resources as well. Finally, given the great variety of both exotic and native plant and insect species found in our region, I very much doubt that the usefulness of numerous introduced plant species for native insects was ever properly assessed. Certainly, when I look at some introduced flowering plants even for a few minutes, during the spring and summer months, I see a fairly wide variety of insects visiting these flowers. If someone would systematically study just how many insects choose to visit the flowers of common introduced plants, I am quite confident that the findings of such formal studies would not support the general statements suggesting that native insects do not use exotic plant species. The addition of non-native plants actually brings new and more diverse resources to our local ecosystems, particularly in urban areas, and this, in turn, may benefit a variety of animal species in these ecosystems.

As an example, five or six years ago, a few introduced motherwort plants (*Leonurus cardiaca*), which are members of the mint family originally native to Central Asia and Europe, appeared on my front lawn, by unknown natural means – just like some dandelions and the white mulberry sapling previously did. I let some of these motherwort plants develop, and they kept resurfacing for several years afterwards. Despite the much talked about recent decline in the populations of local honey bees and bumblebees, the motherworts attracted impressive numbers of both types of insects, as well as some other insects, each summer. In fact, these plants seem to be particularly favoured by bumblebees. I don't recall ever seeing as many bumblebees in our garden before the motherwort plants appeared, although we always had several other types of flowering plants on our property.

Native Butterflies and Exotic Plants in California

Ecologist Arthur Shapiro (2002) found that native butterfly species in parts of northern California "breed mostly or entirely on alien plants, especially naturalized weeds." This conclusion is based on more than 30 years of data and observations collected by the author on the butterflies and flowering plants found in and around the city of Davis. The vast majority of these butterflies are considered to be native and the vast majority of the flowering plants are alien. There are 32 butterfly species in the Davis region, and more than 90% (29) of them breed at least partially on non-native plants, while over 40% of the butterflies (13 of 32 species) breed exclusively on the introduced plants. Without the presence of these exotic plants, the author of this study predicts that native butterflies could well vanish from this region, particularly in urban and suburban areas. This state of affairs is not unique to Davis, California. According to the author, native butterflies also breed on alien plants in the eastern United States, as well as in much more remote regions, such as Patagonia. In addition, a number of butterfly species native to Ontario – including the eyed brown (*Lethe eurydice*), the broad-winged skipper (*Poanes viator*), and the Dion skipper (*Euphyes dion*) – appear to feed on the presumably invasive lesser pond sedge (*Carex acutiformis*) (Hall et al., 2014).

Shapiro (2002) mentioned that many people seem to like butterflies but hate flowering plants perceived as weeds. As a result, and also because of the rise of invasion biology, there is now more of an emphasis on eliminating introduced plant species in urban areas. These initiatives, in Shapiro's words, "appear driven more by ideology than ecology", and can be very damaging to the butterfly biodiversity in and around cities. So, providing suitable habitat for the popular urban butterflies actually means leaving the unpopular so-called alien weeds alone, as much as possible.

Convincing people that these weeds should be tolerated on scientific grounds can be a tall order these days, as Shapiro (2002) also described, since many gardeners and other members of the public have come to believe that "native" plants have to be the only suitable ones for native butterflies, while introduced plants must be nasty and generally bad for the environment in every conceivable way. As also discussed in the previous chapter, this is a difficult mentality to overcome, because it seems to make sense, at a superficial level, even though there is often no actual scientific evidence in support of the often repeated claims that native plants are somehow better for local animal species and ecosystems than introduced wildflowers and trees.

I have experienced such attitudes myself, when I allowed the motherworts and other such mostly introduced plants to grow freely for a while on what used to be a dreadfully dull and uninspiring conventional lawn. The presence of such plants on lawns, or instead of lawns, apparently offends conformist suburban sensitivities. Although some people may be more open-minded and receptive to ecological explanations about the potential virtues of lowly exotic weeds, by and large, this is a still a hard case to make, and bringing scientific studies into the discussion tends to make certain individuals even more suspicious. However, despite such entrenched biases, introduced plants can be very helpful for a variety of native and exotic insects.

Beneficial Non-native Bumblebees

Exotic insects can also bring useful contributions, for example by successfully pollinating vulnerable native plants. In a study conducted in the temperate forests of the Southern Andes region of Argentina, Sanguinetti and Singer (2014) found that introduced *Apis mellifera* bees, as well as the exotic bumblebees *Bombus ruderatus* and *Bombus terrestris*, pollinated the native orchids *Chloraea virescens* and *Brachystele unilateralis*. In fact, these novel pollinator-flowering plant associations led to an unusually high reproductive success for the orchids, indicating that the non-native pollinators were particularly good at performing this task. The non-native bees and bumblebees were deliberately introduced for other purposes – *B. ruderatus* in 1984 in order to pollinate red clover, and *B. terrestris* in 1998 as a pollinator for greenhouse tomatoes. However, these versatile insects clearly discovered, and adjusted to, other plant species, including the native orchids, fairly quickly and helped these flowering plants form fruits and seeds at remarkably high rates. Although the conservation status of these orchids – both of which are endemic to the southern Andes – is yet to be assessed, the authors estimate that at least one of the species (*B. unilateralis*) should be added to the list of threatened local plants, due to its restricted distribution in Argentina. Sanguinetti and Singer (2014) warn that future plans to limit the populations of the introduced bumblebees should take into account the fact that these newly arrived pollinators are largely responsible for the current reproductive success of native orchids in this region.

Interestingly, even though this study shows that hardy and adaptable exotic insects were clearly beneficial for vulnerable native plants, the authors were very careful to start and end their article with the familiar warnings about the negative effects of invasive species, in general. This is a trend I encountered quite frequently in such articles. It seems that many authors are particularly cautious, and feel almost guilty, when their research reveals positive features of alien species and mutually beneficial interactions between introduced species and native ones. Perhaps this may also be a strategy designed to make the articles a bit more acceptable to the readers (and potential reviewers) whose profound dislike and suspicion of non-native species are not easily concealed.

Since *B. terrestris* is a particularly effective pollinator of a variety of crops, including the previously mentioned greenhouse tomatoes, this economically valuable bumblebee species was reared in laboratories in Europe. This species was also deliberately introduced to countries located far away from the European continent, such as New Zealand and Japan. Ono et al. (1994) found that the addition of newly emerged *B. terrestris* workers greatly assists the queen of the native Japanese bumblebee *Bombus hypocrita* with nest building and brood rearing activities. The exotic bumblebee workers and the native bumblebee queen work together, and this results in the production of many more workers of the native Japanese bumblebee species in the early colony formation stages. Therefore, it appears that, in this case, introduced *B. terrestris* workers can be used to help the reproduction and population increase of the native *B. hypocrita*.

Mute Swans and Trumpeter Swans

Trumpeter swans (*Cygnus buccinator*) are considered to be a native species in North America, while mute swans (*Cygnus olor*), as previously discussed, have been introduced to this continent from Eurasia in the 19th century (Sandilands, 2005). Intriguingly, Hughes (2001) mentioned that fossils of birds similar to modern day mute swans were discovered at several North American locations in sediment layers deposited during the Pleistocene. The implication appears to be that mute swans, or very close ancestors of theirs, may have been native to North America as well, in the more distant past.

Trumpeter swans were ruthlessly hunted for their meat, skin, and feathers, and became nearly extinct, as a result. In Canada, most of these swans were killed between 1772 and 1903 (Lumsden and Drever, 2002; Shea et al., 2002).

FIGURE 22 *A tagged trumpeter swan* (Cygnus buccinator) *in the Stong Pond, on the Keele campus of York University in north Toronto. The yellow tag probably indicates that this swan was raised in captivity and then released*
PHOTO BY RADU GUIAȘU

Sandilands (2005) stated that trumpeter swans "did not occur naturally in Ontario for approximately two centuries," since the extirpation of this species in this province took place, for all intents and purposes, at the beginning of the 19th century. The last *C. buccinator* record in Ontario, prior to recent reintroductions, was a bird which migrated from western North America and was shot in 1886 (Lumsden and Drever, 2002).

However, trumpeter swans were reintroduced in Ontario, starting in 1982 (Sandilands, 2005; Lumsden, 2007). Eggs and mature birds were brought from western North America, and some breeding pairs of adult swans were fed and cared for by people in fenced ponds, which were kept free of ice in winter. Some of the eggs were placed in artificial incubators. The young swans (or cygnets) were housed in covered enclosures at locations such as the Metro Toronto Zoo and the Wye Marsh Wildlife Center (Lumsden and Drever, 2002).

In Ontario, mute swans have been enlisted in the effort to assist and enhance the breeding success of trumpeter swans. The eggs of trumpeter swans were sometimes placed in the nests of mute swans, and the mute swans incubated these eggs and raised the cygnets of trumpeter swans. While the mute swans were not quite as successful as the trumpeter swans at taking care of these eggs and cygnets, several young trumpeter swans raised by mute swans survived to adulthood and at least one of these trumpeter swans bred successfully with another member of its own species. Although these trumpeter swans lived in a fairly large group of mute swans, they apparently did not attempt to breed with the mute swans (Lumsden and Drever, 2002; Sandilands, 2005). This also indicates that trumpeter swans can be tolerated by the mute swans in populations containing members of both species. In other words, the two species may be able to coexist. So, the mute swans assisted trumpeter swan conservation efforts and brought a contribution to the survival and reproduction of trumpeter swans in this province.

And yet, despite this, Lumsden and Drever (2002) stated that: "The Canadian Wildlife Service and Ontario Ministry of Natural Resources deemed it desirable to try to replace Mute Swans with native Trumpeter Swans." As far as I can tell, the reason or reasons why such a change in swan species may be desirable was not explained in that particular study.

It is perhaps worth pondering the fact that the extirpation of the trumpeter swans in Ontario was due to hunting, as mentioned, and had absolutely nothing to do with the arrival of the mute swans. In fact, trumpeter swans disappeared from Ontario before the mute swans arrived in this province. In addition, since trumpeter swans were absent from Ontario for almost two centuries, the current local mute swan populations obviously have a longer continuous presence here than the reintroduced trumpeter swans. Mute swans were first observed breeding and building nests in the wild in Ontario in 1958 (Sandilands, 2005; Badzinski, 2007), and they did this entirely on their own, and survived without having the benefit of fenced ponds, covered enclosures, and artificial incubators. Thus, the supposedly "non-native" mute swans were born and lived their entire lives in Ontario for several generations.

On the other hand, the presumably "native" trumpeter swans were recently introduced to Ontario from parts of western North America. In fact, most of the trumpeter swans in Ontario are descended from a Greater Yellowstone population in the United States. These swans have low genetic variation due to inbreeding and this may possibly explain why many trumpeter swan eggs do not hatch (Lumsden and Drever, 2002; Lumsden, 2007).

It is not immediately obvious, at least to me, why a successfully established mute swan population needs to be replaced with a recently introduced trumpeter swan population in the name of conservation. If some wildlife managers insist in bringing trumpeter swans back to Ontario, then perhaps these birds should be allowed to fit in among the existing waterfowl species, rather than having other swan species artificially removed. The survival of trumpeter swans in North America is no longer in doubt, since healthy and growing populations of this species are now found in Alaska and parts of western North America (Shea et al., 2002; Kerasote, 2005). Thus, bringing trumpeter swans to Ontario cannot be justified on the basis of rescuing the trumpeter swan from extinction.

Also, Ontario has clearly changed quite a bit in the couple of centuries or so since the trumpeter swans were extirpated from this province by hunters. More than 75% of the original wetlands of southern Ontario were either completely destroyed or severely altered, and wetland habitats are considered endangered in this region (Findlay and Houlahan, 1997; McAllister, 2000; Guiașu, 2002b and 2007a). As a result, there is considerably less suitable waterfowl habitat available here now, than there was before the original trumpeter swans were eliminated. The situation is quite similar in the United States, where more than half of the wetlands have been destroyed during the last 200 years (Primack, 2008). Some experts worry that the artificial reintroduction of trumpeter swans in eastern North America may just create more problems for the management of the remaining wetlands left in this part of the world and the species currently occupying these wetlands. Furthermore, while there is evidence showing that trumpeter swans did live in eastern North America centuries ago, some prominent ornithologists do not think that these birds actually bred there. In other words, there is a debate on this topic with clear conservation implications. Some wildlife managers are introducing, or hoping to introduce, trumpeter swans to the eastern regions of this continent, in an attempt to establish breeding populations in these areas. However, if certain ornithologists are correct, and the trumpeter swans migrated to eastern North America but never actually bred there, prior to their extirpation, then the present day establishment of breeding *C. buccinator* colonies in the East cannot really be perceived as a genuine restoration effort, since these breeding populations never existed before (Kerasote, 2005). The controversy also highlights, yet again, just how little reliable historic information is available in many of these situations. Researchers are looking at old maps or vague journal entries by early explorers to determine which species might have existed in a given region centuries ago. Sometimes, conservationists may attempt to recreate a mythical past that never was, while perhaps ignoring present day realities and possibly undervaluing the fauna and flora we currently have in fascinating, self-sustaining, and vibrant novel ecosystems.

Perhaps it may be wiser to leave the trumpeter swans alone in the West, and let the mute swans survive in the East, and, if the two species eventually meet naturally in some Ontario, or other North American, wetlands, to allow them both to coexist and let nature take its course, so to speak.

Furthermore, since some wildlife managers and conservationists had expressed concerns about the presence of the big and presumably aggressive mute swan in local wetlands, does it then make sense to deliberately introduce the trumpeter swan – another big and aggressive waterfowl species – in this region as well? Why do we apply different standards to these two rather similar swan species? The answer to these questions seems obvious. The trumpeter swan receives preferential treatment because it has been labelled as native, even if its direct or most recent ancestors have never lived in Ontario, while the mute swan is disliked by wildlife managers because it was introduced in North America several generations ago, and, therefore, it will never be allowed to receive the privileged "native" status on this continent, even though this bird is generally well-liked by many people.

Perhaps I may be a bit too sensitive about such issues, but it seems to me that there is something quite opportunistic, and perhaps even somewhat unseemly, in using mute swans to assist the growth of the trumpeter swan population, while at the same time planning to replace the mute swans with the trumpeter swans in the longer run. If the mute swans are good enough to help out the trumpeter swans, then they should also be acceptable as a permanent resident of Ontario's wetlands.

Introduced Species and Biodiversity

In the Greater Toronto Area (or the GTA), the vascular plant biodiversity has actually increased within the last century or two because of the addition of exotic species. According to Eckenwalder (1999), approximately 750 of the 1,000 or so native plants initially found in the GTA before rapid urbanization began are still present in the city, and about 500 exotic plants coexist with native ones in the city's ravines, creating a unique urban ecosystem. Non-native plant species bring positive contributions to our local environment, in the Toronto region, in different ways – by generating oxygen and storing carbon dioxide, and providing food and shelter for a variety of native and non-native animal species. The seeds of the introduced bull thistle are a favourite food of the native American goldfinch, for example. Local non-native animals can offer helpful services and resources to native species, as well. Introduced honey bees pollinate a variety of flowers, and non-native city pigeons are the main prey of the endangered peregrine falcon, to list just a few of the many cases of vital native-exotic interactions. Trying to pull apart the complex ecological strands that currently tie all these species together and removing plants and animals simply because they may not have been here at some arbitrary point in the

FIGURE 23 *Canada thistles in the Toronto region*
PHOTO BY DORIS WONG

past could actually make our fragile urban ecosystems less functional and less diverse.

At the Don Valley Brick Works, in midtown Toronto, an abandoned old brickyard was recently converted into a valued green space which includes wetlands and ponds. During the restoration of this area, volunteers initially spent approximately 80% of their time trying to eradicate invasive thistles. The task proved to be not only cumbersome and futile – since the thistles were difficult to remove and always came back – but also counterproductive. It turned out that the native wildflowers planted by the volunteers could not survive in the thin, unstable soil without the help of the "invasive" Canada thistles (which, as we have seen earlier, are actually of European origin, despite their common name). The roots of the resourceful thistles kept the vital loose soil in place and prevented erosion. When the persecution of the thistles stopped, the native wildflowers started to become established and thrive, together with, and because of, the Canada thistles. So, in this case, the invasive plants were actually vital for the survival of the native ones. This example directly contradicts the commonly held belief that alien species outcompete and eliminate native vegetation. Fortunately, at the Toronto Brick Works site, after the failed eradication attempts, common sense finally prevailed, and the non-native plants were eventually allowed to bring their useful contributions to the restored ecosystem (Foster and Sandberg, 2004).

Heard and Sax (2013) reported that native and non-native plant species coexisted for more than 200 years along beaches in Rhode Island and Massachusetts, and during this long period of time, by and large, the exotic plants did not appear to displace the native ones. Such long-term coexistence between native and non-native plants seems common and was observed at other locations, in various parts of the world.

During some of my own research, conducted in collaboration with several students at York University in Toronto, I found that the numbers of both native and introduced local wildflower species around campus woodlots went up (or down) together. In other words, when the location was favourable, the biodiversity of both native and exotic wildflowers increased. If non-native plants would indeed suppress the growth of native ones, then we would expect to have fewer native wildflowers, not more, when the number of introduced wildflowers is high. However, as mentioned, that was not the case at any of the locations examined in and around the Greater Toronto Area.

Similarly, an analysis of the influence of several environmental factors on the distribution of plant species on 29 islands in the Galápagos Archipelago showed that in locations where there were a lot of endemic species there were also plenty of so-called immigrant species. Habitats which can sustain large numbers of immigrant species can also support, at the same time, high numbers of endemic species. The endemic species were considered unique to the Galápagos Islands. The immigrant species category included both plants which had colonized the islands long ago and had arrived probably by natural means (dispersed by winds, ocean currents, birds, etc.) and the more recently arrived species brought to the islands possibly as a result of human activities. In any case, in the Galápagos Archipelago, island elevation was the most important factor in this analysis for explaining plant diversity. The higher the elevation, the greater the number of vegetation zones on the island, and, therefore, the greater the diversity of habitats available for various plant species. When an island contained sufficient relevant resources, a great plant diversity, consisting of higher numbers of both endemic and immigrant species, was found there (Guiașu and Guiașu, 2010c).

In a previously mentioned new study, Thomas and Palmer (2015a) analyzed the distribution and abundance of plants in Britain between 1990 and 2007, and discovered that at locations where the diversity of non-native species increased, the diversity of native species increased as well. The authors concluded that: "For Britain, at least, the non-native species have supplemented, rather than excluded, the native flora." In other words, the plant diversity or species richness of Britain was enhanced, overall, by the addition of non-native species.

Sax et al. (2002) found that the number of non-native vascular plant species added to the flora of oceanic islands since the arrival of humans was far greater than the number of plants which became extinct on the islands during the same periods of time. As a result, due to the addition of introduced plant species, the net number of plant species successfully established on these islands doubled. In this study, the non-native species included those introduced by people and those which may have arrived without human help, after people had reached the islands. As discussed elsewhere in this book, it is often difficult to be sure exactly how species travel to various islands and other remote regions. Therefore, from a practical standpoint, separating species possibly brought somewhere by people from species which may have arrived by other means is sometimes simply not possible. Presumably, even species which may have arrived on islands recently, perhaps without our help, can benefit from human-made habitat transformations, which allow these species to become established on the islands.

Sax et al. (2002) also cited evidence of substantial increases in the number of plant species on continents. For example, only approximately 28 plant species became extinct in California, but more than 1,000 introduced plant species became established there.

Theoretically, a higher number of plants should provide a wider selection of resources for both native and introduced animal species on both continents and islands.

The addition of non-native species has led to substantial increases in the species richness of other groups, as well. For example, the diversity of freshwater fishes on oceanic islands was greatly enhanced by such introductions (Sax et al., 2002). There are only five presumably native freshwater fish species in Hawaii – four of these are true gobies, belonging to the family Gobiidae, and the other one is the Hawaiian sleeper goby of the family Eleotridae. However, 44 freshwater fish species, belonging to 15 different fish families – including Gobiidae, but not Eleotridae – became established in Hawaii (Ziegler, 2002). About 90% of all Hawaiian freshwater fishes are non-native. Therefore, there has obviously been a huge increase in the biodiversity of this group in Hawaii – in fact, the Hawaiian freshwater fish species richness went up by almost 900% (880%, to be exact) due to the addition of exotic species, and, at a higher taxonomic level, the number of freshwater fish families went up from only 2 to 16.

It seems that groups of organisms with poor natural dispersal abilities are particularly likely to benefit from human assistance, since individuals of such groups would rarely be able to reach distant oceanic islands otherwise (Sax et al., 2002).

According to Pedreschi et al. (2014), the number of freshwater fish species found in Ireland has more than doubled due to the addition of non-native species – there are only 11 freshwater fish species native to Ireland, and 13 non-native ones. Pedreschi et al. (2014) also point out that "almost half of the introduced fish species now present in Ireland have been here for many hundreds of years", before adding that these species "have no known date nor source of introduction."

So, as we have seen, the addition of introduced species can often increase local biodiversity in various regions, and, quite likely, the complexity of the ecological interactions among these higher numbers of species. And yet, even when the number of species in a particular ecosystem goes up, overall, due to the addition of non-native species, exotics are still blamed for lowering biodiversity (despite the frequent lack of direct proof implicating alien species in the demise of native ones, and the previously mentioned positive interactions between these two categories of species).

Let us consider the following statement from Zimmer (2001): "Some islands risk losing most of their native life. Native species on the island of Mauritius have declined from 765 to 685, while 730 alien species have made it their home." Well, first of all, island plants and animals are usually especially vulnerable to changes in their environments, and we cannot easily extrapolate conclusions drawn from the studies of islands (particularly small and isolated islands) to continents. Secondly, while the extinction of any species is a sad event, of course, and Mauritius, like other islands, has lost its share of unique species, including the dodo bird (a poster species for extinction in historical times), just looking at the numbers given above, we can see that the number of new species added to Mauritius far exceeds the number of species lost. Thus, if we use species richness as the key indicator, the biodiversity of Mauritius has clearly increased. Furthermore, according to this information, the vast majority of the native species are still present on the island, and it is not at all clear that all, or even most, of the native species which have become extinct on Mauritius disappeared because of the arrival of exotic species (other than humans). Perhaps some native species may be doing better than before, or were saved from extinction, because of positive interactions with newly introduced species, and the new resources such species may have provided. Also, over long periods of time, some of the exotic species on Mauritius, which are now isolated from the rest of the world, may change into unique local populations and subspecies, and even represent the starting point of new speciation events.

As an additional example, let us take the following paragraph from Holdich (1999) about the crayfish in his country: "Prior to the 1970s Britain only had its single native species, *Austropotamobius pallipes*, and now in the 1990s there are five alien species in the wild and one kept by aquarists." The author of this article considers the introduction of "alien crayfish" into Britain to be a bad thing, but, clearly, the crayfish biodiversity of the British Isles has increased considerably due to the addition of these exotics. After all, who says that Britain should always have only one species of crayfish, and that this species should always be *Austropotamobius pallipes*? Why is this the only permissible state of affairs? Undoubtedly, the addition of the other crayfish species brought some changes to certain local freshwater ecosystems, but these changes do not have to be necessarily considered as negative ones. Invasion biologists often regard any changes in species assemblages over time as bad, and potentially disastrous, ones. But, as often discussed throughout the book, changes in the numbers and types of species in regions all over our planet have always occurred, since the beginnings of life on Earth, and will always occur. If it weren't for such changes, mammals would never have evolved, dinosaurs wouldn't have become extinct, and we wouldn't be here today. Trying to stand in the way of such changes at all costs is often not only just the result of misguided nostalgia, and possibly misplaced guilt for past human misdeeds towards other species, but also yet another way in which humans interfere with the progression of life on Earth. In other words, by preventing or slowing down certain inevitable changes in the species composition of a given ecosystem, we are also causing a change in the way in which nature functions (or would function, if we didn't meddle). If we try very hard to save a doomed species, which may otherwise have become extinct naturally, we may prevent the formation of new species in the long run, or inhibit the development of some new ecosystems and biological communities, better suited for the conditions of today.

Other Benefits Provided by Non-native Species

The snail kite (*Rostrhamus sociabilis plumbeus*) is an endangered bird of prey found in Florida's Everglades region. The recent increase in the population of this rare native bird can be attributed to a significant extent to the arrival of the introduced herbivorous island apple snail (*Pomacea maculata*), which may have reached Florida through the aquarium trade. The invasive snail species forms a substantial part of the highly specialized diet of the kite. Since this snail uses a non-native aquatic plant (*Hydrilla verticillata*) in order to reproduce, wildlife managers in the Everglades found it desirable in this case to preserve both the exotic plant and the invasive snail in order to help the endangered kite (Gewin, 2014). This example also shows that native species, including rare and vulnerable ones whose populations were declining, can adjust to, and take advantage of, new food sources and other novel resources and opportunities provided by introduced species. Increasingly, ecologists and wildlife managers will have to take such matters into account, when assessing the complex effects of species introductions into various ecosystems.

Even the dreaded invasive ship rat (*Rattus rattus*), also known as the black rat, can help out native species, in some cases. Pattemore and Wilcove (2012) demonstrated that this rat, as well as the introduced silvereye (*Zosterops lateralis*) – a type of bird – are valuable pollinators for three endemic forest plant species in New Zealand. It appears that the rat and the silvereye – two species which arrived in New Zealand since the middle of the 19th century – manage to compensate for the loss of native vertebrate pollinators by providing vital pollination services to these plants. This is an example of mutually beneficial relationships between native and invasive species. The native New Zealand plants provide nectar and in return the invasive vertebrates assist the plant reproduction process. According to the authors, this study shows that invaders can also play important roles in maintaining ecosystem functions, in some cases.

Invasive species can also assist the recovery of ecosystems. In the last several decades, overfishing has severely reduced the number of native predators available in salt marshes located in the Cape Cod region, in Massachusetts, USA. In the absence of these predators, the local populations of the native herbivorous crab *Sesarma reticulatum* increased dramatically, and these crabs consumed large quantities of cordgrass, which led to the degradation or destruction of the salt marsh ecosystems. However, the introduction of another predator – the European green crab, *Carcinus maenas* – seems to offer hope for the recovery of these marshes. The European green crab can prey on the herbivorous *S. reticulatum*, as well as drive this herbivorous crab away from its burrows. As a result, the "invasion" of the area by European green crabs actually had a positive effect on the local ecosystems, since it led to a reduction in the number of herbivorous crabs, which, in turn, allowed the cordgrass and the salt marshes to recover (Bertness and Coverdale, 2013).

Rodriguez (2006) analyzed the relevant literature and found 61 studies which showed that invasive species can be helpful to native species. Despite this, Rodriguez concluded that the benefits provided by invasive species to native species are often ignored by invasion biologists. More recently, Schlaepfer et al. (2011) offered a review of some of the positive roles played by non-native species in ecosystems, perhaps indicating an increasing, and long overdue, interest in the conservation value of these species.

And, as we have seen in previous chapters, even notorious "invaders", such as purple loosestrife and zebra mussels can provide tangible direct and indirect benefits for various native species. In the next chapter I will also discuss some positive impacts on native species generated by another famous invasive species – the cane toad.

CHAPTER 6

Changing Ecosystems and Impacts of Introduced Species over Time

> You cannot step twice into the same river.
>
> HERACLITUS

The profound statement cited above is attributed to the Ancient Greek philosopher Heraclitus of Ephesus, although the words of Heraclitus are only known to us from the quotations by other famous philosophers such as Plato. Heraclitus, whose work was done about 2,500 years ago, was known as a "philosopher of change", because, as his famous statement indicates, he believed that all things are always changing, or, in another version of the statement attributed to him: "all things are flowing" (Kahn, 1979; Russell, 1984; McClellan III and Dorn, 1999). Clearly, what Heraclitus meant was that we cannot step into the same river twice, because the river always changes, and the waters that flow over our feet the second time are not the same waters we were exposed to the first time we stepped into the river. Furthermore, other features of the river also change over time, and we are always changing as well. I do not wish to take the river metaphor too literally, but I can illustrate it with an example from my own experience, since I have done quite a bit of fieldwork for several years in various streams scattered all over southern and south-central Ontario. For instance, during a camping trip with friends, several years after my last field trip to the area for research purposes, I decided to show my companions a particularly scenic spot – just upstream of the Elliott Falls on the Gull River – where I had conducted some of my fieldwork. It seemed like a great place for a picnic, as well. Although the area was still scenic, the water volume that day was much lower than before. The river I had stepped into a few years earlier was moderately deep and had a strong current, which made it challenging at times for me to keep my balance while looking for crayfish. In fact, because of the potentially dangerous rushing waters, I had advised one of my field assistants at the time not to attempt to cross the unpredictable stream. However, several years later, the comparatively puny and tame river my friends and I encountered seemed like a shadow (or a trickle) of its former self, and we could have crossed easily to the other side, without even getting very wet, by stepping on the now exposed boulders which used to be entirely covered by the rushing water. Furthermore, given the number of years which had passed

 | DOI 10.1163/9789047426134_008

since my last visit, I was well aware that it was highly unlikely that any of the particular crayfish I had seen during my earlier trips would still be alive now. A new generation of crayfish inhabited the river now, and, over longer periods of time, it would be reasonable to expect that the types of species of crayfish, and fish, found in the stream, and their relative abundance, would change as well, reflecting the effects of the changing environmental conditions and the passage of time. The riparian vegetation had changed as well, at least in part, and, therefore, the seeds and leaves which fell into the water from time to time belonged to some different species of shrubs and trees than before. So, although, according to the map, this was still the Gull River, the stream we encountered on my last visit was in many ways quite different from the one I remembered from my previous trip.

Observations and Unplanned Natural Experiments in a Home Aquarium and the Garden

Several years ago, a few duckweed plants appeared in one of my home aquaria, seemingly out of nowhere. I have had aquaria in my house ever since I was about ten years old, and this had never happened before. Within a few months, the entire surface of the water in my aquarium was covered with the little green duckweed plants, and, since this did not seem to have any obvious negative effects on the other residents of the aquarium (the freshwater fish, the snails and the other plants), I did not intervene. The duckweed "invasion" persisted for a few more months, as I recall, and, during this time, some tiny white insects I had also never seen before started to appear and walk along the duckweed leaves. I could be wrong, and I did not study this formally, but it seemed fairly clear at the time that the arrival of the small white insects was somehow connected with the thriving duckweed population. In any event, after several months during which I had done absolutely nothing to change anything in my aquarium (the water temperature and the pH, the lighting, the composition of the fish species assemblage, and so on, all remained basically the same), the duckweed population started to decline, and, eventually, no duckweed plants at all were left in my aquarium. The disappearance of these surface water plants was just as sudden and unexplained as their arrival had been, some months earlier. The entire episode took about one year, overall, and the duckweed plants have never returned to my aquarium since then. So, in a way, this can be considered as an example of a miniature "invasion" – a new species appeared in my aquarium seemingly out of the blue, and became quite successful for a while. A few tiny duckweed plants may have arrived, undetected at first, in a bag with new fish from the pet store, but the exact way in which the new plants reached my aquarium is unknown. After all, green algae naturally appeared on the walls of every aquarium I ever had, and I never put any algae in there. Upon seeing the duckweed plants "take over" the entire water surface in my aquarium, I could have overreacted and tried to remove all the plants, or, even worse, add poisons to the water, or toss out everything in my aquarium (innocent fish, snails, plants, as well as the gravel), in order to get rid of the uninvited "invader" and start fresh, with a clean and freshly scrubbed (and lifeless) aquarium. However, this seemed, frankly, ridiculous, and so I waited, and, over time, nothing drastic happened, and life in the aquarium continued unabated even after the addition of the duckweed plants. A new species had been added to my aquarium – actually, two species, if we include the small white insects – but all the other aquarium residents seemed to adjust to this change, and go on with their lives. Eventually, as mentioned, the duckweed plants disappeared on their own, without my interference, and without the need for any control programs or misguided knee-jerk reactions. This is a small example of the way in which nature always changes – species come and go, living communities always transform themselves, and new arrivals can live for a while in, and adjust to, their new surroundings, or die out. Furthermore, the impact and the fortunes of introduced species, and even the individuals of these species, can clearly change over time without human interference.

Such examples can be observed all the time, in a variety of ecosystems, if we care to notice. In the sizeable section of my backyard left untouched by lawnmowers for several years, a few species of plants became established for some time, before being replaced by others. At first, dandelions appeared, but later on the supposedly invasive garlic mustard (*Alliaria petiolata*) took over. However, the garlic mustard "invasion" did not last, and after only a couple of years, these plants were completely replaced by the native yellow avens (*Geum aleppicum*), which currently dominate the area. Who knows which species will be the next to establish itself, or when the new change will come? And the non-native motherwort plants (*Leonurus cardiaca*) which appeared on my front lawn grew bigger and more abundant each summer, for the first few years, but then declined gradually, and became smaller and less dense and widespread, before virtually disappearing from most of the area they once occupied. Other wildflowers took their place. The entire process – the rise and the fall of the motherworts – took about five years or so. Perhaps the motherworts were victims of their own success, and exhausted some vital nutrients they relied on in the soil, or perhaps some other minute environmental change or series of subtle changes led to their demise. I did nothing to either encourage or discourage these plants from growing there. The species composition changed in front of my eyes over several years, without my interference. And, clearly, the changes in the plant species present must have been associated with corresponding changes in the composition of various animal species – mostly insects and other small invertebrates – in the area.

Lady Bird Johnson and Native Plants

Lady Bird Johnson, the wife of the former US President Lyndon Johnson, once said, apparently while promoting the use of native plants: "I want Texas to look like Texas, and Vermont to look like Vermont. I just hate to see the land homogenized." She also mentioned that native plants "give us a sense of where we are in this great land of ours." In the early 1980s, the former first lady founded a wildflower research center in Austin, Texas, which showcases native plants, as well as serving as an education center (Russell, 2004; www.wildflower.org/, 2014). The now renamed Lady Bird Johnson Wildflower Center has been associated with the University of Texas at Austin since 2006. The conservation of native plants is a central part of the mission of this center. On the center's web site, in 2013, right under the heading *Keeping Texas native*, there was an announcement about training offered at the center for the survey of invasive plants by local volunteers. In a section on invasive species, on this web site, we are informed that more than 1,000 citizen scientists have been trained at the center since 2005 through the *Invaders of Texas* program (described as "our signature program") to report on the presence of various invasive species. And once again, in bold letters, invasive species are described in this section as "the second greatest threat to biodiversity after habitat destruction." This is repeated again, in another part of the web site, as the first item on a list of problems caused by invasive species (http://www.wildflower.org/, 2013). As discussed earlier in the book, this type of statement remains an often repeated and prominent part of the anti-invasive species propaganda, despite the fact that it is highly questionable and based on a very shaky foundation.

On the web site of this education and conservation center, under the title *Why native plants*, presumably native wildflowers are promoted because "they help conserve water, reduce mowing costs, provide habitat for birds, butterflies and other wildlife, protect the soil and save money on fertilizer and pesticides." That is all very nice, but these attributes are not unique to native plants. Surely, introduced wildflowers can bring similar benefits as well. We are also told that "unlike many non-native plants, native plants introduced into landscape plantings are hardy, less susceptible to pests and diseases and unlikely to escape and become invasive" (www.wildflower.org/, 2014). Again, we run into the conundrum analyzed before – if native plants are so hardy, then why are they considered to be in trouble, and why do they need to be protected by centers such as this one? Also, since native plants are defined on this very same web site as "those that existed here without human introduction" (assuming we can ever determine such things with certainty), wouldn't the prescribed introduction of such species into the landscape make them automatically introduced, as well? After all, without human assistance, these plants wouldn't exist at the locations where we decide to introduce them, often simply for ideological or aesthetic reasons. Furthermore, according to some invasion biologists, invasive species are successful partly because they do not have any pathogens and parasites in their new surroundings (having presumably left them behind back home in their initial "native" ranges). That would clearly contradict the notion that native plants are less vulnerable to pests and diseases. As for native plants being less likely to "escape" (presumably, this term describes plants which would grow wherever they can, as opposed to only where we want them to) and become invasive, well, it is quite possible for native plants to become very successful, and some have even been labelled as "weedy" (Dickinson et al., 2004).

Although it is nice to see prominent individuals taking an interest in the natural world, the logic of this particular initiative can certainly be questioned. First of all, Texas and Vermont are defined by lines drawn on a map – in other words, the boundaries of these two American states are arbitrary human concepts. Nature throughout the North American continent is a continuum, which does not change abruptly at the boundaries between states. The plant and animal species assemblages are obviously defined by such fundamental factors as climate, soil, topography, intra- and interspecific interactions, and so on. Secondly, given the distance and climatic differences between the modern day areas currently identified as Texas and Vermont, there is little danger that the floras of these two regions will end up looking identical, or even very similar, in the near future. I grew up in Bucharest, Romania, and moved to Canada when I was 18 years old, and I can tell you based on personal experience that, despite the fact that dandelions and a few other such common urban plants grow in both cities, Bucharest does not look like Toronto. Thirdly, the plant life of Texas or Vermont undoubtedly changes over time, as well, and it is not uniform across these states. So, when we talk about the area we now refer to as Vermont, a good question may be, which Vermont, or which version of Vermont's plant life would we like to preserve and why? After all, about 20,000 years ago, during the last Ice Age, this north-eastern American state was covered by huge sheets of ice, so recreating the way this area used to look like then may pose a bit of a problem. Since that time, the plant life of the region has undoubtedly changed continuously, and there is no particular scientific reason to assume that these changes will stop and that the wildflowers of Vermont or Texas witnessed or favoured by Mrs. Johnson, or anybody else, should be the ones receiving preferential conservation treatment. I suspect that at least some of the pretty flowers seen by Mrs. Johnson in Texas may well have been initially introduced, quite some time ago. And native plants are not always wonderful, simply because of their "correct" origins. In fact, some of them can have nasty side effects, as well. For instance, in North America, the allergy or hay fever inducing common ragweed (*Ambrosia artemisiifolia*) is considered to be native.

In 2010, freelance writer Emma Marris wrote an article in *Conservation* magazine, in which she suggested that human-modified novel ecosystems, containing a mixture of native and introduced species, also have intrinsic value and should perhaps be considered worthy of protection and open-minded study. The article was based in part on discussions with, and the scientific work of, ecologists such as Joe Mascaro, Ariel Lugo, and Peter Kareiva. One reaction to this article came a few months later (spring 2011), in the form of a letter to the editor published in the same magazine. The letter was written by Travis Gallo, an ecologist at the Lady Bird Johnson Wildflower Center, and Steve Windhager, the president and CEO of the Santa Barbara Botanic Garden. The authors seemed to represent the wildflower center, and took issue with the main arguments presented in the Marris article. They acknowledged that novel ecosystems, such as "an exotic forest", may be stable, but concluded that "stability alone is not a desirable trait" and gave as an example a corn field in Iowa, which, in their view, "is quite stable yet would not be considered a success in ecological restoration." I found this particular example rather strange and inappropriate. First of all, a corn field in Iowa is clearly not stable, from an ecological standpoint. Without constant human interference and maintenance, the corn field would be quickly overrun by other plants, and would then likely get replaced altogether by different forms of wildflowers, and eventually by trees and shrubs. Due to the modifications which took place during the process of domestication (which applies to plants as well), crop plants such as wheat, barley, and peas are notoriously unable to spread their seeds well on their own these days, and therefore require our continuous assistance in order to survive and reproduce (Diamond, 1999). In addition, the corn field is largely a monoculture, whereas novel ecosystems can shelter a remarkable variety of species – sometimes more species than were found before in the same area. Forests full of exotic trees can be self-sustaining, and can provide vital ecosystem services on a regular and reliable basis, without our further involvement. All we have to do is leave these forests alone. It is also somewhat surprising to see the desirability of stable ecosystems questioned in this letter to the editor, since, generally speaking, many conservation and invasion biologists are suspicious of changes and would prefer to keep species associations within ecosystems in a stable and stagnant, or at least carefully managed, state. I suppose modifications in ecosystems are only considered bad if they favour species we don't approve of. As long as ecologists themselves get to make the changes, and decide which species should be eliminated and which ones should be introduced or "restored", then everything is fine...

Gallo and Windhager (2011) also questioned the need to protect novel ecosystems, and seemed to advocate control programs against the introduced species found there. They also cited Lady Bird Johnson's statement, expressing her wish for "Texas to look like Texas and Georgia to look like Georgia" – in the version of the quote given by Gallo and Windhager, Georgia became a substitute for Vermont, but I suppose the main point is essentially the same – before concluding with the familiar argument that invasive species are making the world's ecosystems more and more similar and homogenized. This is a common, and rather dubious, claim, which ignores the fact that, given sufficient time, ecosystems and populations constantly change on their own, and frequently follow divergent evolutionary paths, influenced by local environmental conditions. In fact, so-called novel ecosystems are just the current versions or stages particular ecosystems have reached, and there is usually nothing "unnatural" or "damaged" about them.

Cardinals in Ontario

As far as I know, the northern cardinal (a species with the relatively easy to remember scientific name *Cardinalis cardinalis*) is considered to be a native species in Ontario. In fact, this is now one of the best known and most often photographed bird species in this part of the world. Artists often paint this bird, in various poses, and it is difficult not to find some souvenirs with northern cardinal images in local gift shops, or various Christmas cards showing these very colourful birds. The males of this species, in particular, have a very distinctive bright red plumage, which makes them stand out in almost any setting. The useful and nicely illustrated *Birds of Toronto and vicinity* guide book, written by Gerald McKeating (1990), has northern cardinals on the front cover, as well as the back cover. And yet, in this book, we find out that the northern cardinal is a recent arrival in Ontario, from the United States. It seems that the first nest built by a northern cardinal in Ontario was found as recently as 1901, at Point Pelee, in the extreme south of the province. Since then, this bird species has gradually expanded its range northward, and has become common all year round in most of southern Ontario, including the Toronto region. More recently, *C. cardinalis* has also become a resident of the Ottawa region, further north. This recent range expansion of the northern cardinal into Canada, throughout the 20th century, has apparently been assisted by urbanization and related phenomena, such as the planting of many shrubs and small trees (many, no doubt, exotic) in urban and suburban gardens, yards, and parks. Apparently, this type of vegetation provides ideal nesting conditions for this species. The northward range expansion of this species in Ontario seems to be continuing, assisted by the increased presence of bird feeders and suitable habitat conditions particularly in urban and suburban areas, as well as the warming temperatures in this province over the last few decades (McKeating, 1990; McLaren, 2007). So, while it is great to have the beautiful northern cardinal added to our native bird fauna in Ontario, one has to ask why do we have one set of standards for northern cardinals, and another for, say, brown-headed cowbirds? After all, both species have expanded their North American ranges, taking advantage of human modifications to the local landscapes, and, as we have seen earlier in this book, cowbirds reached Ontario in the 1870s, about 30 years before the first northern cardinals were found nesting here. So, again, why are cowbirds considered "invasive" and northern cardinals are "native"? And why is one species despised and controlled, while the other is celebrated?

Changes in Aquatic Ecosystems

When it comes to the numbers and types of species found in various regions, things can change quite a bit even in a couple of decades or less. For instance, in the Preface of the new edition of their *Peterson field guide to freshwater fishes of North America north of Mexico*, published in 2011, Page and Burr explained that in the 20 years that passed since the publication of the first edition of their guide, 114 newly recognized native species and 16 newly established non-native species were added to the ichthyofauna of this region, while, at the same time, a few exotic species once thought to have been established have disappeared since 1991.

As mentioned in a previous chapter, when the impact of some well-known invasive species was discussed, the zebra mussel (*Dreissena polymorpha*) was almost completely replaced by another "invader" – the related quagga mussel (*Dreissena bugensis*) – along the Canadian shoreline of Lake Ontario in a little over a decade (between 1990 and 2003) (Wilson et al., 2006). Shortly after its arrival in North America, the zebra mussel seemed unstoppable, and both its range and its numbers grew at an impressive rate. However, after only a relatively few years, the zebra mussel populations started to drop at some locations due to predation by local species, such as some native waterfowl, the spread of the newly arrived quagga mussels, and other causes, including possible environmental changes.

Interestingly, in that fairly recent newspaper interview I had mentioned earlier in the book, Hugh MacIsaac – a prominent invasion biologist working in the Great Lakes region – included the quagga mussel, but not the much more famous zebra mussel, among the five most damaging invasive species in Lake Ontario (Menon, 2011). So, it seems that the zebra mussel may already have become yesterday's news, at least for some, as restless invasion biologists are moving on to new target species. The quagga mussel has become the new fashionable enemy now. However, I would suspect that, due to its current notoriety, the zebra mussel will continue to be used for the foreseeable future, whenever there is a new campaign to "educate" the public about the danger posed by exotic species in this part of the world. I can just imagine the new posters now: "Listen to your parents kids, or the zebra mussel will get you!"

Case Study: The Cane Toad in Australia

Although the exact geographic origin of the marine, or cane, toad (*Bufo marinus*) is apparently unknown, the native, or "natural", range of this species is assumed to include most of Mexico, a tiny part of southern Texas, in the USA, pretty much all of Central America, and roughly the northern (or tropical) half of South America (Lever, 2001). This toad was introduced to a variety of other areas – particularly to many Caribbean and Pacific islands, as well as the continent of Australia – during the past two centuries or so. As the widely used common name for this species, cane toad, indicates, *B. marinus* was introduced to many of these places, including Australia, because of its perceived usefulness as a biological control agent against insects that affected the, also introduced, sugar cane crops (Easteal, 1981). Some of the introductions of this toad species, to countries such as Egypt, Thailand, or Cuba, did not succeed, and, as a result, *B. marinus* did not survive there in the long run (Lever, 2001).

However, the cane toad's introduction to Australia was quite successful, at least from the toad's point of view. The toads were brought to Queensland, in north-eastern Australia, in 1935, not from their native range, but from Hawaii – another place where they had been introduced (Easteal, 1981). Since then, the toads have expanded their Australian range dramatically, and now occupy an area of perhaps about one million square kilometers, mainly in tropical and subtropical regions found along the northern and eastern coasts of the

FIGURE 24 *The cane toad* (Bufo marinus) *in Australia*
PHOTO BY MATTHEW GREENLEES

continent (Phillips et al., 2006; Letnic et al., 2008). Additional range expansions in these parts of Australia have been predicted, although some of these predictions are also dependent on expected changes in climate associated with global warming (Phillips et al., 2003). However, given the fact that they require water to complete their life cycle, the toads are unlikely to be able to find suitable habitats in the extremely dry and hot regions of central Australia.

Bufo marinus is resented mainly for being very good at defending itself against potentially deadly predators. This toad species possesses two large poison glands on its back, and the milky poison secreted by these glands can kill a predator within as little as 15 minutes or so. Australian predators which can be killed by the toad's poison include snakes, monitor lizards, and even large, 2 meters long freshwater crocodiles (*Crocodylus johnstoni*) (Phillips et al., 2003; Aldhous, 2004; Letnic et al., 2008). Letnic et al. (2008) reported that when it comes to the dead freshwater crocodiles, at least, numerous autopsies were performed, and cane toads were found only occasionally within the stomachs of these reptiles. Thus, direct evidence for the poisoning of crocodiles by the toads may well be insufficient, although Letnic et al. (2008) have little doubt that the circumstantial evidence implicating the toads in the deaths of these crocodiles is strong enough. However, the impact of the cane toads on these crocodiles can be very different at some locations in comparison to others.

While some Australian freshwater crocodile populations have declined after the arrival of the toads, other such populations were not affected in any significant way. Somaweera and Shine (2012) found that there was no decline in the number of freshwater crocodiles present in Australia's Lake Argyle, even though the crocodiles coexisted and interacted with the toads at this location for two years, and both species were quite abundant. The authors did not find any evidence of crocodile deaths caused by cane toads at Lake Argyle, and concluded that, given these observations, there may not be a pressing need to protect the crocodiles from the toads in this case.

When we consider the negative impact of the toad on some predators, at some locations, we should remember, however, that the cane toad is simply protecting itself. Apparently, just holding a toad is usually not enough to trigger the release of the poison. Only when the toad is violently attacked and/or held in the jaws of a potential predator (toad enemies can also include domestic dogs) is the poison secreted, as a desperate last-minute form of self-defense.

The concern is that some vulnerable Australian species, such as the northern quoll (*Dasyurus hallucatus*) – a small marsupial carnivore that tries to prey on any small animal it can catch, including cane toads – may become casualties of the toad's formidable toxic defenses. However, it should be noted that these quolls, and many other small mammals of northern Australia, are declining even in areas where the cane toads are not present. This may have to do with completely different environmental variables such as changes in fire management patterns. In addition, the northern quolls tend to have small populations and the type of unusual life history that makes them especially sensitive to environmental changes (Aldhous, 2004).

Webb et al. (2008) studied the feeding behaviours and prey selection of another small marsupial predator of the family Dasyuridae – the common planigale (*Planigale maculata*) – and determined that this Australian carnivore can survive exposure to the toads and learn to avoid them. The common planigale lives in some of the same tropical areas of northern Australia where the cane toad "invasion" is under way. These marsupials are generalist predators, and therefore feed on a variety of small animals, including insects and frogs. The vast majority (90%) of the so-called naïve common planigales attacked and tried to eat the cane toads. The naïve planigales are those without prior experience with these toads. However, 83% of these planigales survived their encounter with the toads, because they had limited exposure to the toad poison – the predators were repelled by the toad's taste and either released the prey relatively unharmed or stopped eating it before reaching the main poison glands. The encounters with cane toads obviously must have been quite unpleasant for these surviving planigales, since afterwards they quickly learned to avoid the toads for up to 28 days following the initial exposure. In most cases, a single encounter with a cane toad was sufficient to teach the planigales to avoid these toxic, and potentially deadly, invasive amphibians. When presented with cane toads again, these experienced planigales refused to attack them. This was clearly a win-win situation for both the planigales and the cane toads – they both walked away from such subsequent interspecific encounters unharmed – and the study demonstrated that native Australian marsupial predators can quickly learn to avoid the cane toads, and therefore survive this new potential danger by making the appropriate behavioural adjustments. Some Australian fish predators learned to avoid cane toad tadpoles, as well, so these types of adjustments are clearly not restricted only to mammalian species.

Interestingly, after the initial exposure to the toads, the planigales stayed away not only from these introduced amphibians, but also from native Australian frogs. The planigales did not attack native frog species for up to nine days after an encounter with the invasive toad, even though the native frogs were palatable and somewhat different in size and body colouration from the toads. After some time, the planigales learned to distinguish the native frogs from the cane toads by using chemical cues. Despite this, however, these marsupial predators remained more cautious and hesitant while approaching the native frogs, and waited longer before attacking. Presumably, in the field, this type of hesitation on the part of the predators would give the potential prey more time to escape. Therefore, the presence of the cane toads in this part of Australia actually benefits local native frog species, by minimizing predation on them. And, as we have seen, the toads did not seem to cause much long-term harm to the common planigales. In fact, common planigales and cane toads appear to coexist in northern Australia without significant problems for either species (Webb et al., 2008).

This type of research led to the realization that predators such as the northern quolls can be trained to avoid the toads. The training enhances the ability of these predators to coexist with cane toads in the long run (Shine and Doody, 2011).

The effects of the cane toads on native species are difficult to determine and hard to place in a broader context, because of the absence of data about the impact of the toads in the first few decades after their introduction into Australia. It is hard, if not impossible, to compare the present day situation with the way things were decades ago, if we don't have any data about the way things used to be right after the arrival of *B. marinus* (Crossland, 2000; Phillips et al., 2003; Aldhous, 2004). Although, as we know, the cane toad was introduced in

Australia in 1935, it appears that the first anecdotal information about the possible effects of *B. marinus* on native species was not published until the 1960s and quantitative data on interactions between these native species and the cane toads did not appear until the 1990s. Some of these studies showed that "the ecological impact of toads may be less extreme than might be supposed from intuition or public concern" (Phillips et al., 2003). In addition, Crossland (2000) mentioned that the effects of *B. marinus* cannot be separated from those of other anthropogenic actions such as habitat destruction or modifications. In other words, there were many other important human-caused environmental changes on the Australian continent since 1935, and analyzing the effects of any one of these changes in isolation could be misleading.

Overall in Australia, according to Aldhous (2004), "it's impossible to blame the toads definitively for the decline of any particular species." Aldhous (2004) also quotes Tony Griffiths, an ecologist at the Key Centre for Tropical Wildlife Management, associated with Australia's Charles Darwin University, who said, referring to the proof of the impact of the cane toads: "Most of it is anecdotal. There has been very little scientific evidence." And, in a study which will be discussed in more detail shortly, Greenlees et al. (2007) wrote that "we still have little understanding of the ecological impact of cane toads."

In some cases, *B. marinus* appears to have no significant negative effects, overall, on potentially vulnerable species. For example, although cane toads can eat the eggs and the nestlings of ground-nesting birds, they rarely actually do so, and the impact of *B. marinus* on these types of birds has been described as "minimal" in Australia. Therefore, at this time, the toads are not considered to be a serious threat to the 80 ground-nesting Australian bird species they coexist with (Beckmann and Shine, 2012).

And, while the arrival of cane toads in a new region of Australia may have a negative effect on some predators – at least before the predators learn to avoid the toads – there can also be related indirect positive effects for other species. For instance, shortly after the toads had reached the Daly River region of the Northern Territory, there was a decline in the numbers of yellow-spotted monitor lizards (*Varanus panoptes*). Presumably, this occurred because the monitor lizards were poisoned while attempting to eat the toads. However, the monitor lizards are generalist predators and the reduction in their numbers benefitted at least one of the other species they preyed on. The monitor lizards used to eat about 20% of the eggs of pig-nosed turtles (*Carettochelys insculpta*) per year, but after the arrival of the cane toads, the predation on the turtle eggs by these lizards stopped. This led to an increase in the population of the *C. insculpta* turtles – a species of conservation concern. Since *V. panoptes* also eats about 50% of the eggs of the flatback sea turtle (*Natator depressus*) in northern Australia, the death of these predators – possibly because of exposure to the cane toad poison – can be expected to benefit the sea turtle species as well (Doody et al., 2006). Thus, the types of effects the cane toad may have on Australian species can be mixed. While some predators may suffer, the conservation status and future prospects of species such as the pig-nosed turtle and the flatback sea turtle should improve as a result of the cane toad invasion.

Similarly, the arrival of *B. marinus* may actually help some native Australian frogs, while having a negative impact on other such species – or at least on certain populations of other frog species.

The ornate burrowing frog (*Limnodynastes ornatus*) and the desert tree frog (*Litoria rubella*) are native to Australia. Both species can sometimes reproduce in the same ponds in northern Australia, and they can also share breeding areas with introduced cane toads. *L. ornatus* tadpoles are major predators of *L. rubella* eggs and hatchlings. The *L. ornatus* tadpoles also try to eat the eggs and larvae of cane toads. However, the toxicity of the cane toad eggs and hatchlings is lethal to the *L. ornatus* tadpoles. As a result, by reducing the number of *L. ornatus* predatory tadpoles, the presence of *B. marinus* actually increases the survival of *L. rubella* eggs and larvae. It should be noted, however, that, even after many decades of coexistence with the invasive cane toads, the native *L. ornatus* frogs remain very common overall in parts of northern Australia where these amphibian populations were monitored. Cane toads and ornate burrowing frogs do not always share breeding sites and may reproduce at different times of the year. As a result, *L. ornatus* tadpoles may not encounter cane toad eggs or hatchlings at all at some locations, while at other sites such encounters may be minimal. Furthermore, since *L. ornatus* tadpoles can also eat the eggs of conspecifics, the death of some of these tadpoles as a result of exposure to the toxic *B. marinus* may actually help the survival of ornate burrowing frogs born later in the breeding season. Thus, although some *L. ornatus* populations may be negatively affected by cane toads, the survival of this native frog species is not at risk at this time. In fact, the arrival of the cane toads did not cause the extinction of any native frog species in Australia. In addition, cane toad tadpoles do not appear to eat the eggs and larvae of native Australian frogs to any significant extent (Crossland, 2000).

Greenlees et al. (2007) analyzed the interactions between invasive cane toads and native giant burrowing frogs (*Cyclorana australis*) in large outdoor enclosures in Australia's Northern Territory, and found that the two species had largely different diets and, as a result, competition for prey between them was minimal. The cane toads did not have a negative impact on the feeding rates, the food intake, and the composition of the diet of these native frogs. The authors also noted that, while the possible competition between the invasive toads and native frogs in Australia was much discussed, in general terms, there had been no prior experimental studies on this subject. Greenlees et al. (2007)

concluded that the arrival of the cane toads did not reduce the native frog populations, because of possible adaptive behavioural adjustments by the native species coupled with the apparent lack of significant competition between the invasive toads and the native frogs.

In a detailed review of the impact of cane toads in Australia, Shine (2010) reported that there was no decline overall in native frog populations after the arrival of the cane toads, and in some areas of Kakadu National Park there was actually an increase in frog species richness and abundance after the toads came in.

Shine (2014) focused specifically on the interactions between cane toads and native Australian frogs in an even more recent review and concluded that "field surveys show that cane toad invasion has negligible overall impacts on frog abundance." The author also stated that: "The most clear-cut effect of toads on frogs is a positive one: reducing predator pressure by fatally poisoning anuran-eating varanid lizards."

Furthermore, some of the negative impacts on predators may become gradually less significant, because it turns out that the cane toads have changed from a morphological standpoint in response to their new environment, since their introduction into Australia. Phillips and Shine (2005) measured *B. marinus* specimens from the collections of the Queensland Museum and determined that both the overall body size of the cane toads and the relative size of their main poison glands (with respect to the rest of the body) decreased over time since the introduction of this species to Australia. Therefore, in the seven decades that passed between the introduction of *B. marinus* into Australia and the publication of this study by Phillips and Shine, the amount of poison stored by the toads, or the relative toxicity of the toads, has been declining, and this should result in a reduced cane toad impact on native predators. Often, the impact of such successful non-native species may be at its peak shortly after the introduction, but then the extent and intensity of the negative effects of the invasive species on native ones may well decline naturally over time.

Phillips and Shine (2005) suggested that these morphological changes in Australian cane toads may have taken place either because of toad-induced changes to local environments, resulting perhaps in lower food supplies (which would lead to reduced size), or due to adaptive changes to new environments. Such adaptive changes could presumably arise in response to reduced predation pressure, after predators of cane toads are either killed or have learned to avoid the toads. With fewer predators around, having a big body and energetically expensive large poison glands may not offer a selective advantage anymore. In other words, with few deadly enemies left, the investment in huge defensive weapons may be unnecessarily costly from an evolutionary point of view. According to Phillips and Shine (2005), these two main reasons (toad-induced environmental changes and possible adaptive changes by the toads in response to their new environment) for the reduction in cane toad size and toxicity are not mutually exclusive.

Evolutionary Changes

In his discussion of the 1958 book which supposedly started the field of invasion biology, Daniel Simberloff (2000) mentioned that "Elton did not concern himself with evolution of introduced species in their new homes. He was an ecologist, not an evolutionist." Richardson and Pyšek (2008) agreed that topics such as "post-introduction evolution" were not addressed at all in Elton's book about invasive species.

However, as the well-known geneticist Theodosius Dobzhansky (1900–1975) stated, "Nothing in biology makes sense except in the light of evolution" (Beatty, 2009). This famous statement would also clearly apply to dynamic fields such as ecology and biogeography, in general, and to the dispersal and interactions of various species all over the world, in particular. Thus, given that Elton combined elements of these fields, and explored the impact of some invasive species in their new surroundings, in a book published almost one hundred years after Darwin's *On the origin of species*, the lack of an evolutionary perspective in *The ecology of invasions by animals and plants*, the book that is the very foundation of invasion biology, is a pretty big oversight indeed. This may have set the stage for a preoccupation almost exclusively with the short-term, and the immediate aftermath of an "invasion", without proper concern for long-term evolutionary implications of the introductions of species, among many of Elton's followers. In the longer run, with a bit of patience and foresight, we may find that a lot of ecological "problems" that seem urgent today can solve themselves, gradually and naturally, as long as we get out of the way and allow this to happen.

Evolutionary changes – in both native and non-native species – associated with ecological "invasions" have already been discussed earlier in the book, to some extent – for instance in Chapter 3, which dealt in part with processes such as hybridization and speciation, and in the previous section, where possible adaptations of the cane toads to new environmental conditions on the Australian continent were explored. However, given the significance of such evolutionary aspects for a thorough understanding of the long-term impacts of non-native species, some additional examples will be analyzed below.

So, returning for a moment to the cane toad case study, it should be noted that, in the decades since this species was introduced to Australia, the toads actually triggered evolutionary changes in their snake predators. Phillips and Shine (2004) measured several features indicative of the head and mouth size, as well as the overall body size, for four different species of snakes. The snake species are native to Australia. The data were gathered from specimens available in the collections of the Queensland Museum. The four snake species were distantly related frog predators with wide distributions in Queensland. All four snake species coexisted with cane toads since 1935. Two of these snake species – the red-bellied blacksnake (*Pseudechis porphyriacus*) and the green tree snake (*Dendrelaphis punctulatus*) – can be killed if they eat even fairly small cane toads. The other two snake species chosen for this study – the swamp snake (*Hemiaspis signata*) and the keelback (*Tropidonophis mairii*) – are generally not affected as much by the cane toads. Swamp snakes are small and their mouths are not large enough to allow them to eat a toad big enough to be lethal, and keelbacks are quite resistant to the cane toad poison. Therefore, the cane toads would be expected to exert a powerful selective pressure on the two snake species which are both extremely vulnerable to the toad poison and capable of eating larger, and therefore more dangerous, toads. Clearly, larger toads will have more poison than smaller conspecifics, and the size of the toad ingested may determine whether a vulnerable snake predator will live or die.

As expected, the red-bellied blacksnakes and the green tree snakes – the two species most likely to be killed by the ingestion of cane toads – developed gradually smaller mouths and larger bodies during the decades of exposure to the toads. In particular, there was a gradual and consistent reduction in the gape size of these vulnerable snake species. This reduction in gape size minimizes the risk of swallowing the larger and more dangerous toads. Therefore, having a smaller mouth has clear adaptive value for these snakes, since individuals with smaller mouths are likely to survive encounters with cane toads much better. At the same time, vulnerable snakes with larger mouths will gradually be eliminated from the population due to poisoning by the cane toads. Significantly, these types of morphological changes were not observed in the two snake species which were not likely to be killed by the cane toad poison. The swamp snakes and the keelbacks do not face strong selective pressures to change in response to the presence of cane toads, and, as a result, no such adaptations evolved in these snake species (Phillips and Shine, 2004).

Thus, during the eight decades since the introduction of the cane toads to Australia, evolutionary morphological changes occurred in both the invasive species and some of its native predators.

FIGURE 25 *House sparrows* (Passer domesticus) *in North America*
PHOTO BY DORIS WONG

Since their introduction into North America from England and Germany, in the mid-19th century, house sparrows (*Passer domesticus*) managed to distribute themselves over most of the USA and Mexico, and a good part of southern and central Canada. Birds from the initial breeding population established in New York in the early 1850s expanded their range dramatically in the next several decades, and reached Vancouver, on the West Coast, by 1900, and Mexico City, to the south, by 1933 (Johnston and Selander, 1964; Ridley, 1996).

Johnston and Selander (1964) recorded and measured several features related to the colour and size of a large number of house sparrows collected at various locations throughout their introduced range in North America and Hawaii, as well as in their native range in England and Germany. The data revealed that the house sparrows started to change and evolve in just a few decades in their introduced range. As a result, house sparrows from various regions of North America could be recognized reliably based on their newly evolved distinctive features. For example, house sparrows from Vancouver and more northern locations had dark plumage, while southern conspecifics from areas such as Texas and California – and especially California's Death Valley region – were considerably lighter in colour. The Hawaiian house sparrows had particularly distinctive colour patterns – so different, in fact, that even experienced bird watchers had a tough time identifying the birds at first. The house sparrows were introduced to Hawaii in the early 1870s (either 1870 or 1871) from New Zealand, where they had been brought from England in the late 1860s.

At the same time, in North America, house sparrows from northern locations are generally larger, while those from the south-western areas are smaller. Canadian house sparrows are larger than Central American ones. The larger body size in northern regions is presumably an adaptation which allows the birds to cope better with the colder conditions. The differences in colour and size among the house sparrows from different parts of North America, which were first identified in the 1940s and then further explored in the 1960s, showed that evolution in this introduced bird species can take place quite rapidly – perhaps in about 100 generations or so. In just a few decades, the introduced house sparrows became distinct from their European ancestors, and also different enough from one region of North America to another to justify perhaps a separation into distinct subspecies (Johnston and Selander, 1964; Ridley, 1996). Over much longer periods of time, and many more generations, it is certainly conceivable that the house sparrows descended from the initial birds brought over from Europe could perhaps evolve into several new species in the New World. If that happens, the avifauna of both North America and the world would clearly become richer and more diverse as a result of this introduction. At the very least, this example shows that the individuals of non-native species can change, often quite rapidly, in response to the new environmental conditions encountered in their introduced range.

Furthermore, according to Johnston and Selander (1964), the changes recorded in introduced house sparrows were similar to the types of changes observed in native bird species in response to conditions such as climate, for example.

Starlings and Crested Mynahs in Vancouver

In his essay *Requiem for a rare bird*, the award-winning science writer Wayne Grady described the rise and fall of an exotic species – the crested mynah (*Acridotheres cristatellus*) – in the Vancouver area, in the southern region of the Canadian province of British Columbia (also commonly known as B.C.). This gregarious bird species, with a talent for mimicking various sounds, was introduced in Vancouver from south-east Asia in the late 1890s. By 1904, the crested mynah was breeding in its new home, and by 1920, the Vancouver population of this species was estimated at approximately 20,000 individuals. The crested mynah appeared to be thriving. At this point, anybody studying the crested mynahs in Vancouver would have concluded that this is a very successful non-native species, with a potential for greatly expanding its North American range in the near future. The United States Department of Agriculture became concerned by the presence of this new species in south-western Canada, so close to the US border, and, in 1935, this department produced a report which advised that "every precaution should be taken to check the spread of this species and prevent its spread into the United States." However, as both Kaufman (1996) and Grady (2006) pointed out, the crested mynah did not move far beyond the Vancouver area, and, therefore, the report from the US Department of Agriculture referred only to a perceived hypothetical threat from a new species, and not to any real negative consequences of the presence of this species on US territory. It may be interesting to note, as well, that the first reaction of the US Department of Agriculture to this new bird species, which had not even reached the United States yet, was immediately negative. Even before a single bird crossed the border from Canada, there was instant apprehension and worry about this potential exotic species, as well as a determination to make sure that the crested mynah will not be allowed to enter the US How did the people who wrote the report know in advance that the crested mynah would not be a good, interesting, and useful addition to the avian fauna of the United States?

As is often the case, the initial healthy population growth trends for this recently introduced species did not continue beyond the first 20 to 25 years after its arrival on the southern B.C. coast. As it turned out, 1920 was the year when the crested mynah population of Vancouver reached its peak. After that, a gradual decline started, and this decline intensified particularly after 1950. The estimated total numbers of crested mynahs in Vancouver dropped steadily, to 906 individuals in 1971, then 630 in 1980, 98 in 1985, and possibly only 5 or 7 birds by 2002. The last crested mynah known to exist in the wild in Vancouver died shortly afterwards, leading to the local extinction of this species in this area, where it had been introduced a little more than a century earlier (Grady, 2006). Thus, the crested mynah's presence in Western Canada lasted about one hundred years, showing, once again, that sometimes taking a longer view of non-native species may lead to very different interpretations of their status and long-term impact on a particular region.

The introduction of the crested mynah into North America was also mentioned by Elton in his book on invasive species. On this subject, Elton wrote that the crested mynah "spread from Vancouver since about 1897 into parts of British Columbia." However, by the time Elton's book was published in 1958, the decline of the mynah in the Vancouver area was already under way, and, as discussed, the species would vanish from the wild in this part of the world just a few decades later.

Grady (2006) focuses on another exotic species, the European starling (*Sturnus vulgaris*), as the main likely culprit for the disappearance of the crested mynah from Vancouver, stating that: "The decline of Vancouver's Crested mynah population began with the arrival of starlings in 1950." As we have seen earlier in the book, however, just because one species declines after the arrival of another species, does not mean that the newly arrived species is necessarily responsible, or solely responsible, for the decline or the eventual demise of the first species. This could be a mere coincidence, or there may likely be several other factors behind the local extinction (also known as extirpation) of a particular population. It is worth noting that, according to Grady (2006), the crested mynah population of Vancouver reached its peak in 1920, which suggests that the actual decline started about 30 years before the arrival of the starlings. Kaufman (1996) wrote that the numbers of crested mynahs in the Vancouver area "have been declining gradually for more than half a century", which, given the year when Kaufman's book was published, would put the beginning of this decline well before 1950. Furthermore, Grady also mentioned that crested mynahs were observed together with starlings in Vancouver, within mixed species flocks, which is most likely a useful anti-predator strategy beneficial to individuals of both species. Nevertheless, Grady suggested that starlings may have pushed out the crested mynahs because of superior numbers.

It is worth remembering, however, that crested mynahs arrived from regions of south-east Asia, which have a considerably warmer year-round climate than Vancouver. Thus, even in southern British Columbia, which is known to have the mildest climate in Canada, the relatively low winter temperatures in particular should be fairly challenging for birds such as crested mynahs. On the other hand, the starling is a European immigrant, accustomed to a temperate climate, and, therefore, with colder temperatures and cool seasons. Thus, it is logical to expect the starling to adjust better to the Vancouver climate than the crested mynah. The higher numbers of this species in British Columbia may well be simply a reflection of that fact. In addition to possible superior physiological adaptations to a temperate climate, starlings also have behavioural adaptations to colder conditions, as well. Grady (2006) mentioned that starlings spend 24 hours per day incubating their eggs (the male and the female take turns doing this, of course), which is very useful in the cooler conditions of Vancouver, whereas the crested mynahs, which are accustomed to relatively tropical conditions, and temperatures which rarely dip below 25 degrees Celsius, spend only about half of the day incubating their eggs. Not surprisingly, these different strategies result in higher hatching success rates for the eggs of the starlings, in comparison to those of the crested mynahs, in a place such as

Vancouver. So, in a nutshell, the starlings were simply better suited for life in this part of the world than the crested mynahs. Perhaps the starlings may have contributed to the disappearance of the mynahs from southern B.C., but, even without the starlings, this region was quite marginal for the crested mynahs, from a climatic standpoint, and it is likely that, sooner or later, they would probably have declined there. It is possible, for example, that the local climate in Vancouver may have been milder than usual overall between the late 1890s and 1920, thereby facilitating the early success of the mynahs. However, that was bound to change eventually, and a few consecutive tougher winters may well have precipitated the decline of this species in the Vancouver area. To make things easier for ourselves, we often like to simplify things, and focus only on two species at a time (one predator and one prey, or two possible competitors, etc.), but, almost always, interactions among species, and among species and their respective environments, are considerably more complex and multidimensional than that.

In any event, the crested mynah (*Acridotheres cristatellus*) provides another example of an exotic species which seemed quite successful at first, and whose potential expansion was feared by some, before declining and probably disappearing altogether from the region where it had been introduced.

Starlings and mynahs are closely related species which belong to the same bird family (family Sturnidae) (Harrison and Greensmith, 1993). This may account for the fact that, just like the mynahs, the starlings also have a gift for mimicking a variety of other sounds, including the calls and sounds of other birds, and other types of animals. For example, starlings can convincingly reproduce not only human sounds like throat clearing, coughing, laughing, and sighing, but also fragments of human speech, such as: "you're kidding", "see you soon baboon", and "basic research" (as well as many others). In addition, these birds can accurately reproduce household sounds, such as those made by alarm clocks, telephones, pets (cats, dogs, etc.), and pieces of music they may hear from people. However, starlings often modify the original music they hear in creative ways. For example, one bird sang the first line of the song "I've been working on the railroad" with reasonable accuracy, but then modified the second line, and another bird regularly combined fragments from classical music with those from songs such as "Rockaby baby". Wild starlings have rich and very diverse repertoires of songs made up of a variety of whistles, clicks, rattles, snarls, and screeches. These birds combine all of these components, as well as some of the many sounds they hear and copy from their surroundings, and put them together into inventive and often unpredictable song patterns. Not surprisingly, it is apparently very difficult for researchers of bird songs to even come up with an accurate and comprehensive catalogue of starling songs, because of the great diversity of sounds and sound combinations encountered (West and King, 1990).

Overall, the songs of the European starlings are considered to be among the most complex of all birdsongs, and starlings are open-ended learners which modify their rich vocal repertoires from year to year, while also regularly incorporating new motifs and fragments overheard from the surrounding environment into their songs (Chaiken, 2004).

Furthermore, starlings are also capable of social learning through imitation of the actions of conspecifics. When starlings watched a "demonstrator" bird of the same species remove a plug from a hole in a platform, to get access to the food hidden underneath, they copied the type of plug removal method they saw. Thus, if the "demonstrator" bird pulled the plug upwards, the "spectator" starlings copied this method of plug removal, whereas, if the "demonstrator" bird (acting as a sort of role model or instructor) pushed the plug downwards, the "spectator" birds copied that specific method, instead. Interestingly, if the plug was removed using invisible wires, without a "demonstrator" bird performing the necessary actions, the "spectator" starlings did not imitate the particular direction of plug removal, indicating that social learning, from another bird of the same species, was required for imitation to take place (Boysen, 2009).

Somehow, I think that, once we become aware of these types of information, it may be harder for us to think of such talented, intelligent, and resourceful social birds as an invasive threat. Besides, starlings are successful in large part because they are a highly adaptable omnivorous species, with a diverse diet, very well developed social skills, complex vocal communication systems, and a capacity to learn from other members of their large communities of conspecifics. How can we really hate a species that is, in many ways, so much like our own?

However, in North America, starlings have often been referred to as aggressive invaders, which have a detrimental effect on native birds. For instance, in the detailed recent book *Birds of Canada* (edited by David Bird, and published in 2010), the European starling is described as "perhaps the most successful bird in North America – and probably the most maligned". In the same book, we are informed that "this adaptable and aggressive bird competes with native species for nest sites, and usually wins." The *Atlas of the breeding birds of Ontario*, a very comprehensive (containing more than 700 large format pages) and well respected publication for serious bird watchers and both amateur and professional ornithologists, considers the starling as "one of the most familiar, and most disliked, birds in Ontario." Why is the starling so disliked? The answer comes in the next sentence in this atlas: "Its unpopularity is in part related to

FIGURE 26 *Starling* (Sturnus vulgaris) *on the Keele campus of York University in Toronto*
PHOTO BY DORIS WONG

its being a non-native species that often out-competes native birds for nesting cavities in trees" (Cadman, 2007). In his book *Birds of Ontario*, Bezener (2000) stated that, in North America, the starling "spread quickly across the continent, often at the expense of many native, cavity nesting birds." In her own field guide on the birds of Ontario, Hughes (2001) delivers a similarly blunt message about the starling in North America: "Its phenomenal success on this continent has had a detrimental effect on many native cavity-nesting species." Kaufman (1996) is also very clear on this subject and reports that *S. vulgaris* "Undoubtedly has had a negative impact on some native hole-nesting birds." At least Kaufman wrote "some" native bird species, as opposed to "many", but, like the information in the other field guides, and some introductory biology textbooks (for example, the one written by Audesirk et al., 2003), his verdict on the negative impact of the starlings was unequivocal. I realize that the quotations given above are quite similar and often repetitive, but that is precisely the point I am trying to make. When it comes to well-known "invasive" species such as purple loosestrife and European starlings, most field guides (at least the ones I saw) offer almost identical negative assessments.

Despite these often repeated types of statements, a comprehensive study by Koenig (2003) largely exonerated the European starling. In his study, Koenig looked at the possible effects of starlings on 27 cavity-nesting bird species native to North America. These species were identified as among the most likely ones to be negatively affected by starlings, due to the types of nesting sites they usually selected. The population densities of these bird species were compared before and after the arrival of the starlings at particular locations, using data from several types of detailed bird counts and surveys. The results of the study, which were surprising even to the author, did not support the hypothesis that starlings have a severe negative impact on populations of native birds, despite common claims to the contrary in the general literature. It turns out that, of the 27 native bird species examined, only 10 showed some significant changes in mean population density after the arrival of starlings. However, at least half of these 10 species increased in population density, rather than decreased, after the arrival of the starlings. Among the species whose population density actually increased in the presence of starlings was the red-bellied woodpecker, which, in the study author's words, was "one of the only species for which there exists good evidence of nest usurpation by starlings". For the just 5 (out of the total 27 species examined) native bird species which showed some declines after exposure to the starlings, the results were inconclusive. Some of these declines showed up in certain data sets and analyses but not in others, and there were factors other than the presence of the starlings which may have accounted for the somewhat lower population densities, in these few cases. Therefore, as Koenig (2003) pointed out: "Taken as a whole, these data failed to support the hypothesis that the North American starling invasion has severely affected native cavity-nesting birds."

So, just as we have seen already in the case of the purple loosestrife, and as we can see now with the European starling example, it seems that when relevant field data are analyzed, the impact of these famous so-called "invaders" on native species is far less clear and far less negative than previously and commonly assumed. One problem is that most members of the public interested in birds or flowering plants, or other animals and plants, get their information about these species only from field guides and introductory textbooks, such as the ones cited above, or newspapers, television shows, and other forms of easily accessible media. Relatively few people – and almost nobody from outside the scientific community – read first hand scientific articles published in peer-reviewed journals. The average specialized scientific article is usually full of graphs, data tables, the results of various statistical tests, and many specialized terms. The authors of such articles are scientists who normally write very cautiously about their findings, taking into account statistical confidence intervals, several variables, and so on. Sometimes, after years of research, as the author of such a study, you may be able to state, timidly, that a certain number of individuals of a particular local population, belonging to a certain species, may sometimes have an effect of some kind on members of another species, under a given set of environmental variables and conditions, during a particular time of the year, and keeping in mind such things as sampling errors, observer biases, the types of data analysis methods selected, and so on. This is a far cry from a screaming headline, such as: "Invasive species takes over! End of the world expected to be near!"

The process of scientific discovery can often proceed, in fits and starts, along complicated winding pathways, which may appear like an impenetrable maze to the uninitiated. Therefore, when it comes to presenting invasion biology to the public, it is often much easier for journalists and textbook and field guide writers to just "stick to the main message", so to speak, and offer simple stories of "good" and "bad" species, the heroes and the villains of the natural world. In such simplistic and moralistic accounts, everything about "invasive" species has to be bad. Why clutter the main message with nuance and context? For example, instead of discussing the complexity of the starling's songs, such authors will offer unflattering assessments of the vocal capabilities of this bird species. And so, in her section on European starlings, Hughes (2001) writes, rather unkindly, that: "Roosting birds produce jumbled cacophony of harsh calls that can be annoying at close range." Well, perhaps that may sound like a jumbled cacophony to Hughes and the uninitiated, but to other starlings, and the researchers trying to decipher the complexity of bird songs, these may be challenging symphonies, full of meaning and beauty. Bezener (2000) states that "starlings are infamous for their ability to reproduce the sounds of other birds". Infamous? Why infamous? The word doesn't quite seem to fit in this context. This ability is usually considered as a sign of intelligence and resourcefulness by animal behaviourists. Let us be kind, and assume that the intended word had been "famous".

Since the standard anti-introduced species information provided in such second hand general sources is read by a lot more people than the more subtle, thorough, and nuanced scientific studies, which are based on direct research and published in specialized scientific journals, the public perception of these species will inevitably be shaped, almost exclusively, by the second hand reports. Therefore, despite valuable studies such as Koenig's, the information provided about European starlings (and other similarly successful non-native species) in the relevant field guides and even some introductory textbooks is unlikely to change anytime soon. It is easier to perpetuate a stereotype of an "invasive" species, than to come up with new and original insights, based on first hand research, or keep up with the scientific literature in relevant fields, and convey a more complex story, where shades of gray replace some of the ironclad black and white certainties.

The success story of the European starling (*Sturnus vulgaris*) in North America is a very familiar one to many ecologists and invasion biologists. A few birds (perhaps about 100) were introduced into New York City's famous Central Park, around 1890, or so, presumably by someone who wanted to bring the birds mentioned in Shakespeare's plays to North America (West and King, 1990; Cadman, 2007). At least, that is the most often cited reason for the initial introduction. Elton (1958) mentioned that the first few attempts to introduce this species into the United States were not successful. But, eventually, some of the birds released in Central Park evidently survived and reproduced, and the many descendants of these initial avian colonists spread throughout most of the North American continent during the 20th century. A map of the distribution of the starling in North America from 1891 to 1926 is given in Elton's (1958) book on invasive species. This map shows that by 1926, only three and a half decades after their introduction, starlings were already found over most of the eastern half of the United States and in south-eastern regions of Canada. The first record of a European starling from southern Ontario goes back to 1914, apparently (Cadman, 2007). An updated map of the North American distribution of starlings is shown in Stiling's 1999 ecology textbook. This much more recent map shows that the starlings reached the Rocky Mountains by about 1950, and the West Coast of the US by 1980 or so. This resourceful exotic bird species can obviously tolerate a fairly wide variety of climatic conditions, and can be found as far north as central Alberta and Saskatchewan, in Canada, and as far south as northern Mexico. There may now be as many as approximately 200 million starlings in North America (West and King, 1990; Cadman, 2007), although it is, of course, very difficult to be sure of the exact number. Some estimates are more vague, and put the number of North American starlings at "many millions" (Bird, 2010), and that seems like a reasonably safe assumption.

However, no species can expand its range and population forever, and, in Canada, for example, the starling population has declined steadily by 3.2% per year since 1968 (Cadman, 2007). Thus, despite common perceptions, the starling has actually been declining gradually in Canada, and possibly in other parts of its North American range, for several decades. The decline of this species in parts of Ontario is generally associated with the loss of open and agricultural habitats and older buildings, which may have offered more favourable nesting sites for birds such as starlings. It is clear that in Ontario, and most likely elsewhere in North America as well, the presence of starlings depends on the availability of human settlements and the development of agriculture (Cadman, 2007). Furthermore, it seems that the number of starlings is also decreasing in Europe, the original "home" continent of this bird, and particularly in the UK. As in Ontario, some of the possible reasons for this decline may have to do with the overall loss of farms and old wooden buildings, as the forested areas are gradually expanding in parts of the UK, and old buildings are replaced by modern ones, which are less suitable as starling nesting sites (Langton, 2006).

As we have seen, the European starling is widely disliked in North America, and, according to Bezener (2000), there were "many concerted efforts to control or even eradicate this species" on this continent. Langton (2006) described an anti-starling control program undertaken in the 1960s, which killed as many as 9 million birds. Sodhi (2011) mentioned that *Sturnus vulgaris* is one of the three bird species included among the world's 100 worst invasive species. The Appendix of the recently published voluminous *Encyclopedia of biological invasions* (Simberloff and Rejmánek, 2011) contains a list of "100 of the world's worst invasive alien species – a selection from the *Global Invasive Species Database*", and, indeed, the European starling is one of the three bird species found on this list, while, as previously mentioned, the purple loosestrife (*Lythrum salicaria*) makes the list in the land plants category. Some of the other species mentioned in my book that can be found on this list include the zebra mussel (*Dreissena polymorpha*), the common carp (*Cyprinus carpio*), the cane toad (*Bufo marinus*), the house mouse (*Mus musculus*), and the European rabbit (*Oryctolagus cuniculus*). According to the opening note of this Appendix: "Species were selected for the list using two criteria: their serious impact on biological diversity or human activities, and their illustration of important issues of biological invasion."

This list of the world's 100 worst invasive species can sometimes be used to stifle debate and silence critics of invasion biology. In his critical letter written in response to my *Specious claims?* article (Guiașu, 2008a), Mandrak (2008) mentioned this list and gave the web page address for it, before concluding that: "This short document leaves no doubt as to the devastating impacts that invasive species can cause." Well, perhaps, but then again, a short document is unlikely to provide all the relevant details, or include some of the references to controversial or less than clear issues regarding these species. Furthermore, in my article I did not question the fact that some invasive species can cause problems. That was not the point of my article – I was attempting to broaden the discussion by adding another perspective. Similarly, in their criticism of the previously mentioned insightful article by Thomas (2013), Simberloff and Genovesi (2013) brought up the IUCN's "100 of the world's worst invasive alien species" list as well. Apparently, these authors did not appreciate Chris Thomas' logical suggestion that novel ecosystems resulting from human activities may actually lead to increased biodiversity, through processes such as hybridization and speciation. What some invasion biologists may perhaps fail to appreciate is that invoking "the list", all the time, in response to any article that deviates from the standard invasion biology dogma, is not really the best way to address thoughtful concerns or build a compelling counterargument. When they are confronted with a legitimate criticism or an idea they may not like, such invasion biologists tend to slam the door shut, so to speak, on any real debate, by bringing up the same examples of terrible invasive species or the cost of fighting alien species in general, even if these much repeated topics are not directly applicable to the articles and ideas they are meant to put down.

These types of lists ("100 worst this" or "100 best that") also seem somewhat superficial to me – just a way of capturing the public's attention, a bit like "the 100 worst dressed celebrities" list, in a way. As scientists, perhaps we should just focus on the specific issues surrounding each of these species, and the relevant sets of data, and leave the lists to those working in the field of entertainment.

Starlings and Pigeons

In a reasonably balanced newspaper article about starlings, Langton (2006) quotes an amateur birdwatcher from southern Ontario who was disappointed to find that starlings, house sparrows, and pigeons were the common birds arriving at his bird feeder. The amateur birdwatcher is quoted to have said: "Of course, I was disappointed. It was even worse when I found out none of those birds are even native to North America." He then went on to express a strong dislike for starlings, and their large numbers in this region. Different people will have different reactions to such abundant species, naturally, but I will add that when I had a bird feeder for several years in my backyard, within the Greater Toronto Area, several bird species came to feed, but starlings, house sparrows, and pigeons were never among them, even though these introduced birds were commonly seen in the neighbourhood. Furthermore, those complaining about the noise generated by large flocks of starlings, or the droppings of city pigeons, would have probably been overwhelmed had they lived in the 19th century beside a massive breeding colony of the (sadly, now extinct) native North American passenger pigeons. Apparently, the branches of trees would break under the weight of the many passenger pigeons and their nests, and the layers of droppings in such areas could be knee-high (Ponting, 1993). And, at their peak, the numbers of native passenger pigeons dwarfed those of all the non-native birds put together.

As recently as the early to mid-1800s, there may have been as many as 5 billion (no, this is not a spelling error, I did write billion) passenger pigeons (*Ectopistes migratorius*) in North America, making this species the most abundant bird on this continent, and possibly in the entire world, at that time. According to some estimates, at one point in time, passenger pigeons represented about 40% of all North American birds. Although it is obviously not easy to accurately compile such numbers, apparently none of the living bird species anywhere in the world even come close to a total population of 1 billion these days. The passenger pigeons fed on acorns, nuts, and various caterpillars, among other things, and given the enormous numbers of these birds, they must have had a huge ecological impact on many plant and animal species, including the many predators that gathered around their gigantic breeding areas. And then, just a few decades later, due mainly to incredibly reckless hunting and habitat destruction, the passenger pigeons were gone forever. The truly massive slaughter of these birds started around 1850. The passenger pigeons were killed mainly for the palatable meat particularly the young birds could provide as well as for the long blue tail feathers, which made much sought after fashion accessories for ladies' hats. In addition, many pigeons were killed apparently just for "sport" in hunting competitions. It seems that the winner of one of these competitions had to shoot and kill as many as 30,000 birds. Not even a species as abundant as the passenger pigeon once was could sustain these types of losses for long, and, in 1914, the last known passenger pigeon on the planet – a female named Martha – died at the Cincinnati Zoo. This was followed shortly after by the extinction of the Carolina parakeet (*Conuropsis carolinensis*) – another formerly very common North American bird, and the only parrot species considered native to the United States at the time. These birds were also unusual because, unlike other parrots, they could survive as far north as New York and the southern Great Lakes, and could be found in places such as Albany, New York even during the winter. Since these parrots liked to eat various fruits and seeds, they were considered an agricultural pest, and were therefore killed in large numbers. By 1918, only four years after the extinction of the passenger pigeon, the last Carolina parakeet died, also, as it happens, at the Cincinnati Zoo (Ponting, 1993; Fuller, 2001; Flannery and Schouten, 2001, Fitzpatrick, 2004).

Therefore, even 200 million starlings are a very small number of new birds to be added to the North American avian fauna, compared to the enormous loss of perhaps as many as 5 billion passenger pigeons and a huge number of Carolina parakeets, among other bird species, due to anthropogenic activities. The extermination of these species in North America was accomplished with stunning speed, in only a few decades, mainly between the middle of the 19th century and the very beginning of the 20th century, leaving behind empty ecological niches. Since once very abundant North American bird species like the passenger pigeon and the Carolina parakeet have been extinct for about a century, and are not coming back, it could legitimately be argued that successful exotic species such as starlings, house sparrows, and rock doves have helped in partially filling an ecological void left by all the bird extinctions of previous centuries. These introduced species are capable of surviving in and around human settlements – something relatively few species are capable of doing successfully for very long. If we try to get rid of the few bird species that can survive well in our cities and human modified landscapes, we may be left with no birds at all around human communities, with the exception of some domesticated ones.

Furthermore, as also discussed above, the impact of starlings on native North American birds may not be all that negative, after all, and, as McKeating (1990) mentioned: "Starlings also do immense good, feeding on weed seeds and insects and grubs on lawns and in city parks." In addition to feeding on a lot of small invertebrates, some of which are potentially harmful to grass and other plants, the abundant starlings also represent a very important food source for native birds of prey such as hawks (Langton, 2006). And even in a book which focuses on the threats posed by invasive species, within a chapter titled *Our feathered foes*, Williams (2011) mentions that: "Starlings often ride on the back of farm animals in order to eat their mites." Well, presumably that is good for the starlings, the farm animals, and the farmers who own these animals – in other words, a positive outcome for everyone but the mites.

Given the fact that starlings have now been an integral part of North American ecosystems for about 120 years, and given everything else we now know about them, perhaps it is time for us to start appreciating these interesting and beautiful birds and stop the persecution campaigns against them. After all this time, even the name "European" starling is more of a relic, indicating ancient origins, since these birds have been North American residents for many generations, and it is highly unlikely that the vast majority of the 200 million or so starlings currently here have ever seen Europe.

The same can be said about rock doves and house sparrows – two other very common introduced birds which have been conspicuous members of the avian North American fauna for a long time.

Rock doves or rock pigeons, known to most of us simply as pigeons – since this is the only pigeon species most people will ever see, are native to parts of Eurasia and North Africa. They have been introduced to every other continent, except Antarctica (Sandilands, 2007). Even if someone may have thought of actually introducing these pigeons to Antarctica, they would not have survived there, obviously, despite being quite resourceful in more suitable environments, and thriving particularly around human settlements. Rock doves were even introduced, and managed to survive, in some of the most remote parts of the world, such as Easter Island (Chester, 2008).

FIGURE 27 *City pigeons on rooftop in the Old Town of Luzern, Switzerland*
PHOTO BY RADU GUIAȘU

Species such as starlings and pigeons, which have been introduced to various places around the world long ago, also allow us to perhaps reflect further on the concept of "nativeness", which is now so prominent in ecology. So, what exactly does the term "native" mean in a biological context? Presumably, in the field of invasion biology, the term applies mostly to species in general, and not so much to individuals of these species (even though, actions against "non-native" species are, of course, taken against the individuals unfortunate enough to belong to such species). For instance, the familiar city pigeons found in many urban areas around the world are the descendants of rock doves (*Columba livia*) which may have originated on the shores of the Mediterranean Sea (although the exact geographic origins of species are often difficult to determine with certainty). So, city pigeons, as a species, are generally considered non-native everywhere except the cliffs of the particular Mediterranean coastal area, and perhaps some other parts of the Old World, where they may have first evolved. However, all the individual pigeons I see every day, flying around various neighbourhoods of Toronto, Canada, are almost certainly born in this city, and they are likely descended from many previous generations of pigeons which have only known North America as their home. European settlers first brought this pigeon species to North America in the 17th century – perhaps as early as 1606 (McKeating, 1990; Hughes, 2001), and the descendants of these introduced pigeons, as well as the offspring of some of the first European human colonists, have been here ever since. In fact, the North American roots of these pigeons, for example, can probably be traced much farther back in time than those of the vast majority of the people living on this continent now. Similarly, the pigeons of Paris, or London, or Amsterdam, or New York, or any number of other areas, were born in those respective cities, and their ancestors had established a presence there for many generations. So, in that sense, all the individual North American city pigeons we routinely label as "introduced" today, at the species level, are, in fact, very much native to North America as individuals, since they were all obviously born here, on this continent. This analysis can be extrapolated to almost all the individuals of almost all introduced species in most parts of the world. Thus, paradoxically perhaps, when we are fighting against so-called non-native species, we are trying to eliminate or damage individuals of these species which are almost always native to (as in: born in) the respective regions we (also) consider as our home. Therefore, some uncomfortable, but nevertheless important, questions have to be asked, yet again: how many generations have to pass before we consider that members of a particular species have the right to exist, and be left alone, in an environment perhaps outside their initial "native" range? How do we decide which species we should encourage and promote, which species we should merely tolerate, and which ones we should try to control and eliminate? Do we have universal, objective criteria to help us make these difficult decisions? Are well-worked out ecological and ethical concepts and conclusive scientific data really our guides in making such decisions, or are we influenced mainly by our own short-term self-interest, transitory whims, and the ephemeral ideas and beliefs which happen to be in fashion during a given era?

City pigeons generally receive plenty of bad press, and are often called all kinds of derogatory names, including, for example, "rats with wings", which is an absurd label both from a phylogenetic and an ecological standpoint, as well as, sadly, the height of wit, for some. However, like many other ridiculous statements, this one is often repeated, mainly by people who know virtually nothing about either pigeons or biology. Nobody likes to find pigeon droppings on their car, obviously, but this is a relatively minor nuisance, and besides, are cars a more natural, healthy, or environmentally friendly presence in our cities than pigeons? Furthermore, the so-called native birds are known to leave droppings on human property as well, from time to time. And, again, these frequently disparaged contemporary urban pigeons are, in a small way, a bit of a replacement, perhaps, for the once much more abundant native passenger pigeons we have eliminated from the world forever.

FIGURE 28 *The sign and the clear message*
PHOTO BY RADU GUIAȘU

Despite the fact that these city pigeons were introduced into North America about 400 years ago, and have been apparently found in Ontario since the early 17th century, there is little information available about the historical status and biology of *C. livia* in this province. As a result, it is obviously difficult to discuss possible long-term trends regarding the local populations of this species, due to the absence of historical data for significant periods of time. According to Sandilands (2007 and 2010), the lack of such data is related to the rather bizarre fact that early North American ornithologists simply refused to acknowledge the existence of this introduced bird species, and therefore did not record its presence on checklists of avian species until the 1950s. Incredibly, it seems that this pigeon species was not included in the well-known Christmas Bird Counts in Ontario until the 1970s. Nevertheless, despite the gaps in the data and the limited historical perspective, it appears that *C. livia* populations have remained relatively stable in Ontario, overall, during the last 40 to 45 years or so (Sandilands, 2007 and 2010).

Although they may have been ignored for a long time by early local ornithologists, rock doves were nevertheless persecuted in parts of Ontario (Sandilands, 2007). Even today, the anti-pigeon attitude remains. For example, in my neighbourhood there is a small resident group of urban pigeons. As far as I could tell, they nest under a small railway bridge, and the population never seemed to exceed about 10 to 15 adult individuals. The numbers have remained steady for many years, and there seems to be no danger that these pigeons will take over the neighbourhood, so to speak. Nevertheless, on both sides of the bridge, there are formal-looking municipal signs with the same message: "Please do not feed the pigeons." I had no intention of feeding the pigeons, and I never saw anybody else doing this either, but the sign struck me as unnecessary. Why pick on these few pigeons? Life for them must be hard enough anyway, with noisy freight trains often passing over the bridge they nest under, not to mention the harsh winters, and probably nearly constant threats from predators such as hawks, raccoons, and domestic cats. As far as I could tell, the pigeon nests were periodically destroyed, possibly to discourage them from establishing themselves there. Furthermore, I don't recall ever seeing a sign informing members of the public that they should not feed the native birds. If anything, signs like that may just remind people that the otherwise rarely seen pigeons do exist in the area, and they may require feeding...

Introduced Species Vulnerable in Their Native Ranges

Sometimes species vanish or are close to extinction in their native ranges, and only survive well and spread in the areas where they were introduced. In such cases species introductions can obviously maintain biodiversity by preventing global extinctions and keeping open the possibility of eventual reintroductions of various plants or animals to their native areas.

On the Glendon College campus, in Toronto, where I work, there are a few big dawn redwood trees (*Metasequoia glyptostroboides*), a species originally from China. It was thought that this tree had become extinct about 2 million years ago, and, until the early 1940s, only fossil specimens were known to exist. However, in 1941, one huge living dawn redwood was unexpectedly found in a remote area of China's Sichuan Province. In 1946, some branches and cones were collected from the tree and sent to botanist Hsen Hsu Hu who identified the very rare species. *M. glyptostroboides* is now recognized as a so-called "living fossil" (a rather peculiar term, since fossils, by definition, are the ancient preserved remains of dead organisms), which has been present in almost unchanged form on our planet for roughly 100 million years or so. In order to help save the dawn redwood, which is critically endangered in the wild in China, seeds from this tree were collected and given to botanical gardens and arboretums in Europe and North America. This initiative was the result of collaboration between Chinese botanists and Elmer D. Merrill, the director of Harvard University's Arnold Arboretum (Pick and Sloan, 2004). Starting in the

FIGURE 29
Dawn redwood (Metasequoia glyptostroboides) *on the Glendon campus of York University*
PHOTO BY DORIS WONG

late 1940s, very young *M. glyptostroboides* trees grown from germinated seeds were already seen far away from their native range, in European and American gardens (Johnson, 2010). The dawn redwoods planted on the Glendon campus in the 1950s were among the first trees of their species in Ontario (Anonymous, Glendon Magazine, 2004). These trees have now reached an impressive size, and tower over the nearby student residence buildings.

The property where Glendon College is now located used to be the country estate of Canadian financier Edward Rogers Wood and his wife Agnes. The couple moved into the Glendon manor in 1925. Mr. and Mrs. Wood had a common interest in botany, and planted many different types of trees and shrubs – including quite a few exotic ones – on their estate. Edward Wood died in 1941 and his wife passed away in 1950. Agnes Wood wanted to preserve the woodlots and gardens on the estate as a botanical garden, and, according to her wishes, the property was transferred to the University of Toronto's Botany Department after her death. Unfortunately, it appears that the University of Toronto found the maintenance of the newly acquired property to be expensive, and the location was never quite transformed into a full scale facility for botanical research, although the greenhouses were used by some faculty members from the University of Toronto's Department of Botany and the Faculty of Forestry. After a few years, the estate was transferred once again, and became the first campus of York University in 1961 (Horn, 2009). Subsequently, the much larger Keele campus of York University was built in the north-western part of Toronto, and, on this newer campus, a few dawn redwoods – much smaller and younger than the ones at Glendon – can also be found.

Père David's deer (*Elaphurus davidianus*) was native to China, but became extinct in the wild there in 1900. The species was named after Père Armand David, a French missionary and naturalist, who described some of the specimens from a captive herd kept by the Chinese emperor. Père David's discovery, made in 1865, introduced the species to the Western scientific establishment. A few *E. davidianus* individuals were sent to Europe as gifts, which led to the formation of a herd held in captivity in England. This turned out to be fortunate for the long-term survival of the deer, since, as mentioned, the species disappeared from its native range shortly after. The descendants of the Père David deer which had been sent to Europe were given to the Chinese government in the mid- to late 1980s. As a result, captive herds, and even a few deliberately and accidentally released individuals, of this deer species are now once again present, and doing well, in China. Thus, sending a few deer away, far from their native range, turned out to be crucial for the survival of the species, and its eventual reintroduction to China (Boitani and Bartoli, 1983; MacKinnon, 2008).

The tubenose goby (*Proterorhinus marmoratus*) – named after its characteristic tubular nostril extensions – probably arrived in the Great Lakes region of North America from Europe in the late 1980s, likely in the ballast water of ships, at approximately the same time as the round goby and the zebra mussel. Unlike the latter two species, however, the tubenose goby did not spread much in North America, where it remains restricted to a fairly small area which covers parts of Lake St. Clair, Lake Erie, and Lake Huron. As a result of its rather limited distribution, the impact of this benthic fish species in North America is considered to be minor. The tubenose goby (whose scientific species name is also sometimes listed as *Proterorhinus semilunaris*) is threatened by habitat destruction in its native range, which includes the Black Sea and the Caspian Sea (Winterton, 2005; Holm et al., 2009).

Even successful and widespread introduced, or invasive, species can become vulnerable in their native range. This appears to be happening with the European rabbit in Spain – a case study which will be discussed in more detail in a subsequent chapter – and with the house sparrow in parts of Europe and the bullfrog in southern Ontario.

House sparrows (*Passer domesticus*) were introduced in North America in the mid-1800s, as previously mentioned, and to Ontario in particular in 1870. The species became abundant especially in areas where people lived and worked (cities and agricultural regions). However, house sparrow populations have experienced significant declines in numbers in both Ontario and the rest of Canada since the start of systematic surveys in 1968. In fact, the decline of this species may have started in Canada even earlier than that – perhaps more than 70 years ago. House sparrow numbers are dwindling in the bird's native European range as well. This is happening in both rural and urban areas. The reasons for the steep decline of the house sparrow in both Europe and North America are not entirely clear, but it is probable that more efficient agricultural practices – leading to less spilled grain, for example – fewer farms, and more modern buildings which do not allow as many good nesting sites may be contributing factors (Hole et al., 2002; Ingram, 2006; Lang, 2007; Brichetti et al., 2008).

House sparrow numbers dropped by 60% in the UK between 1970 and 2000 – and these numbers were likely already lower in 1970 than earlier in the 20th century. The once ubiquitous and abundant house sparrow, also known as the English sparrow, is now a bird of conservation concern in Britain. Studies and observations of local *P. domesticus* populations revealed that many sparrows, particularly the young birds, were starving, possibly due to fewer available insects. Eating enough insects is apparently essential for the proper development of the chicks. In addition, a growing population of sparrowhawks (*Accipiter nisus*) in Britain led to much increased predation on the sparrows, which took an additional toll, especially in cities. Domestic cats can also kill a lot of sparrows, and many other small birds, particularly in urban areas (Ingram, 2006; Todd, 2012).

Bullfrogs (*Rana catesbeiana*) are native to eastern North America but have also been introduced to many other parts of the world, including western North America, Europe, Asia, and South America, as well as several islands in the Caribbean region and the Pacific Ocean. These large amphibians are now considered invasive in many areas where they had been introduced – both on purpose and accidentally – and there have been anti-bullfrog control and eradication attempts at a number of locations (Arnold and Ovenden, 2002; Lever, 2003; Anthony, 2013). Although bullfrogs are doing well in some regions in their introduced range, the current situation is not as good for this species in Ontario – a part of its native or "natural" range. According to Shirose and Edwards (2005), bullfrogs are now "increasingly rare" in southern Ontario – an area where they used to be common. As a result of the reduction in the size of the local bullfrog populations, this species has been protected in Ontario since 1981 due to provisions in the Game and Fish Act of this province.

However, while bullfrogs are protected in Ontario, because of their native and vulnerable status there, they are currently being exterminated in British Columbia, in western Canada. So, bullfrogs are left alone as long as they stay "home" in eastern Canada, but are actively fought against, through expensive control programs, on the West Coast of the same country. *R. catesbeiana* can eat smaller frogs and is therefore considered a threat to other frog species (Anthony, 2013). However, interestingly, this threat is only taken seriously outside the perceived native range of the bullfrog. For example, the decline of the northern leopard frog (*Rana pipiens*) in parts of Colorado is blamed on *R. catesbeiana* (Lever, 2003), but not much attention seems to be paid to the fact that these same two frog species appear to coexist without problems in Ontario. Despite the fact that the ranges of the northern leopard frog and the bullfrog overlap in Ontario, *R. pipiens* is not at risk in this province (MacCulloch, 2002). Since the two species can survive together in eastern North America, perhaps they will be able to do so in the western regions of the continent as well, after a period of adjustment.

Muskrats and the Aral Sea Environmental Disaster

Even initially very successful non-native species can become highly vulnerable, due to habitat destruction in the area where they were introduced. For example, the muskrat (*Ondatra zibethicus*) is native to North America, where it has an impressively wide distribution which covers most of Canada (just south of the Arctic tundra) and the continental United States (except for northern Alaska and parts of Florida, Texas, and California). This rodent species lives mainly along the shores of lakes, ponds, rivers, and marshes, and tends to prefer habitats that include both wetland plant species, such as cattails, as well as some open water (Eder, 2002; Kays and Wilson, 2009).

The muskrat was introduced into the Aral Sea region, in Central Asia, in the fall of 1943 and the spring of 1944. This species seemed to adjust quickly to its new surroundings, and the muskrat populations grew rapidly and spread out over large areas, particularly in the deltas of the Amu Darya and Syr Darya rivers and surrounding regions (Zonn et al., 2009). The Amu Darya and the Syr Darya are the two main rivers which supplied water to the Aral Sea. This hypersaline lake loses water quickly due to evaporation, and this water was replenished for thousands of years by these two rivers (Lomolino et al., 2010). However, because of the disastrous decisions taken by officials in the former Soviet Union, where the Aral Sea was located, the water of the Amu Darya and the Syr Darya was diverted for irrigation purposes, in order to allow the growth of valuable crops such as cotton in an arid area which otherwise seemed quite ill-suited for agriculture on a wide scale. As a result, the rivers no longer brought water to the Aral Sea, and, starting in 1960, the sea started to shrink (De Villiers, 2003). The water loss was dramatic, and the consequences for the health of the local human populations and ecosystems were severe. In 1960, the Aral Sea had a surface area of 67,499 square kilometres, a volume of 1,089 cubic kilometres, and an average depth of 16.1 metres. Due to the loss of vital water input from its two main rivers, by 2009, the Aral Sea was reduced to a small and much saltier fragment of its former self – a fragment with a surface area of only 8,522 square kilometres (12.6% of the 1960 area), a volume of 83 cubic kilometres (7.6% of the 1960 volume), and an average depth of 9.7 metres (Micklin, 2014). Not surprisingly, the local fishing and muskrat fur industries collapsed. The destruction of the Aral Sea led to the loss of wetlands and the coastal vegetation muskrats, and other wildlife, depended on (De Villiers, 2003; Zonn et al., 2009). Starting in 1946, only two or three years after the introduction of muskrats in the Aral Sea region, the muskrat fur industry was already thriving, and until 1960 about one million muskrat pelts were collected and processed per year. However, because of the drastic changes in the local environmental conditions, by 1968, only 9,000 muskrat pelts could be obtained, and in recent years the muskrats and quite a few other species have been extirpated or almost extirpated from this region (Zonn et al., 2009).

This tragic example of widespread environmental degradation shows that even resourceful introduced species, such as the muskrat, are not invincible after all. If ecosystems are truly destroyed by human actions, then both native and non-native species will disappear.

An additional example of this is provided by the famous zebra mussel (a case study discussed in a previous chapter of this book). Although these mussels are resourceful colonists, they disappear from polluted waters contaminated with heavy metals or excessive amounts of fertilizers used in agriculture (Barr, 1996). The conclusion is that one way of getting rid of non-native species is to thoroughly wreck the environment, but this is not a solution anyone should wish for, of course.

Meat Ants, Spinal Arthritis, and Non-native Cane Toads

The mighty invasive cane toad (*Bufo marinus*) is not quite as invulnerable in Australia as previously believed. Despite its undeniable success in colonizing a sizeable part of that continent, the range expansion came at a price. It turns out that young cane toads are much more vulnerable to predation by native

Australian meat ants (*Iridomyrmex reburrus*) than several species of frogs which are native to Australia. The small, juvenile invasive toads are more at risk from predatory ants in Australia than native frogs because of a number of features that make these toads poorly adapted to the threat posed by the ants. Unlike the species of native frogs tested, the juvenile cane toads are active mostly during the day and tend to prefer open, exposed habitats, just like the ants. Therefore, the probability that meat ants will encounter these small toads is quite high. It has been hypothesized that small young toads are more active during the day in order to escape possible cannibalism from much larger adult conspecifics, which are usually nocturnal. In addition, unlike native Australian frogs, young cane toads did not detect, and escape from, dangerous approaching ants until it was often too late. The small toads tended to become immobile when faced with immediate threats from these predators. While this passive defensive behaviour may be effective against vertebrate predators, which are generally repelled by the toad's poison, the lack of an active escape behaviour makes the toads particularly vulnerable to the potentially deadly attacks of meat ants, and possibly other invertebrate predators. The mortality rate for the small toads attacked by the ants was very high (69%) at the time of the attack, and even many toads that managed to escape the initial ant attack died within 24 hours afterwards, presumably because of the injuries inflicted by the ants. Therefore, overall, it seems that young cane toads are ill-equipped to deal with the threat posed by native Australian meat ants (Ward-Fear et al., 2009 and 2010).

In general, many native Australian invertebrate predators seem to tolerate the toad poison much better than the potential vertebrate predators, such as lizards and some snakes. For instance, Australian crayfish, crabs, water beetles, and water bugs are able to eat cane toads without being negatively affected (Shine, 2010; Cabrera-Guzmán et al., 2012). The Australian water beetle *Cybister godeffroyi* can eat large numbers of cane toad eggs and tadpoles, and would rather prey on cane toad tadpoles than on the tadpoles of native Australian frogs (Cabrera-Guzmán et al., 2012).

Other introduced species can also be vulnerable to new predators encountered in their non-native range. For example, although mute swans (*Cygnus olor*) defend their nests vigorously, they are not effective in protecting their cygnets from predation by snapping turtles (*Chelydra serpentina*) in North America (Lumsden and Drever, 2002).

Although this topic has not received sufficient attention, perhaps, members of invasive species can also develop serious health problems due, in part, to the invasion process itself and the unfamiliar conditions encountered in a novel environment. For example, some of the same features – such as longer legs – which make certain cane toads particularly good at advancing rapidly and steadily across tropical northern Australia also predispose these toads to debilitating spinal arthritis. This condition affects about 10% of the larger adult cane toads in Australia and is more prevalent among toads present in newly colonized regions of the continent – in other words, in toads found at or near the so-called "invasion front". These toads also appear to be more likely to be negatively affected by the common soil bacterium *Ochrobactrum anthropi*, which is sometimes associated with spinal infections. Damage to the toad's spine can also occur in the absence of bacterial infections. Cane toads near the invasion front may tend to be more vulnerable to spinal arthritis and related infections because of the physical stress associated with sustained rapid movements over long distances. While the evolution of longer legs may be helpful for quick dispersal into previously unoccupied territory, these longer legs also cause more powerful movements which, in turn, put more pressure on the spine. Furthermore, the overexertion associated with rapid colonization, as well as moving into possibly suboptimal new habitats, may weaken the immune systems of the toads and therefore make them more vulnerable to bacterial infections (Brown et al., 2007; Skelly, 2007).

The Collapse of Populations of Invasive Species in the Absence of Control Programs

Sometimes, even populations of well-known and seemingly well-established invasive species can collapse, at various locations, without any human intervention.

The yellow crazy ant (*Anoplolepis gracilipes*) is a widely distributed invasive species, whose native range, as previously mentioned, is unknown. Cooling and Hoffmann (2015) reported the extinction of some populations of this ant species at several sites in the Arnhem Land region of northern Australia, where *A. gracilipes* had been first observed in 1982. These population collapses of an invasive species were not caused by control or eradication programs, or other forms of human intervention.

Similarly, populations of the Argentine ant (*Linepithema humile*) collapsed at 40% of the sites examined in New Zealand, where this non-native species was first seen in 1990. Although the Argentine ant is considered to be one of the 100 worst invasive species in the world, and is known to form abundant, aggressive, and widespread colonies, the *L. humile* populations studied in New Zealand had a mean survival time of only 14.1 years. As in the case of the yellow crazy ant populations of Australia, the New Zealand populations of the Argentine ant disappeared from various locations without human management activities. As a result, the authors of this study concluded that "the long-term ecological or evolutionary effects of Argentine ants in New Zealand may not be as dire as first feared" (Cooling et al., 2012).

Such population collapses can occur in invasive freshwater species as well. The analysis of long-term data, collected since the early 1960s in 44 Swedish lakes, revealed that 41% of the populations of the invasive signal crayfish (*Pacifastacus leniusculus*) had collapsed. This crayfish species, which is native to parts of western North America, was deliberately introduced into Sweden starting in 1960 to replace the declining native crayfish *Astacus astacus*. The introduced *P. leniusculus* is an economically important species in both Sweden and Finland. The population collapses in Swedish lakes were most commonly observed 16 to 20 years after the signal crayfishes were first introduced there. In this case, population collapse was defined as a severe decline in the invasive crayfish populations. The signal crayfish did not disappear completely from these lakes, but only remained there in very low numbers, and their populations did not recover (Sandström et al., 2014).

In all of these examples, the exact reasons for the collapse of the populations of invasive species are not known, although we do know that people were not involved in deliberately reducing or eliminating these particular populations. Some of the possible causes for the population collapses or declines in these, and other, invasive species include the effects of pathogens and parasites, the depletion of local resources, changing environmental conditions, and low genetic diversity associated with inbreeding (Cooling et al., 2012; Sandström et al., 2014; Cooling and Hoffmann, 2015).

CHAPTER 7

The Endless War on Invasive Species – Control and Eradication Programs

> All warfare is based on deception.
>
> SUN-TZU

The most unsavoury and disturbing aspect of invasion biology is that it often advocates, and results in, the euphemistically called "control" programs. It is one thing for a few academics to debate esoteric issues at scientific meetings, or in the pages of specialized journals, but it is quite another matter for people to go out there, in the field, deliberately targeting and killing many animals and plants of a variety of species, often simply, or largely, because these species are believed to have originated elsewhere. This is frequently more of an ideological, rather than a well-grounded scientific, reason, and the potential for excesses and terrible mistakes is always there. It is the sort of misguided ideology that separates humans from nature even further, and sets up *Homo sapiens* as a type of global hitman, or enforcer, busy killing a variety of species that may dare to spread and do well beyond the arbitrary boundaries we impose on them. But that's O.K., because it is all really in the name of conservation... If we have to kill large numbers of animals and plants of more and more species to safeguard some people's idea of what the world's biodiversity should look like, then so be it... A small price to pay for making sure that each species is sitting pretty exactly where we would like it to be. Or at least, that is what a relatively few invasion biologists, and their followers, would have us believe.

I became a biologist mainly because I always liked many animals and plants, and I was always interested in learning more about the natural world. A sense of wonder and fascination inspired my career choice. I assume this would also apply to many other fellow biologists, educators, and researchers. So, it is somewhat puzzling and disheartening to see how some invasion ecologists have come to think that we have to have an on-going, worldwide, never ending, all out war against all kinds of "enemy" species, for reasons that are sometimes murky at best, and very difficult to understand, from a logical standpoint, in quite a few cases. Unfortunately, in our history as a species, ideology frequently trumps logic for depressingly long stretches of time. And besides, this war should keep many invasion biologists and wildlife managers busy and gainfully employed for generations to come...

 | DOI 10.1163/9789047426134_009

Charles Elton and the War on Invasive Species

Charles Elton himself set the hostile tone of the discussion about introduced species in the startling opening paragraph of the first chapter of his 1958 book:

> It is not just nuclear bombs and wars that threaten us, though these rank very high on the list at the moment: there are other sorts of explosions, and this book is about ecological explosions. An ecological explosion means the enormous increase in numbers of some kind of living organism – it may be an infectious virus like influenza, or a bacterium like bubonic plague, or a fungus like that of the potato disease, a green plant like the prickly pear, or an animal like the grey squirrel.

And just in case some readers may have thought he was perhaps joking, or at least exaggerating, Elton started his next sentence by emphasizing his previous use of violent imagery: "I use the word 'explosion' deliberately, for it means the bursting out from control of forces that were previously held in restraint by other forces." There is a lot that could be debated about this paragraph, but, in my opinion, placing the grey squirrel on the same list of concerns as major bubonic plague epidemics and nuclear bombs and wars seems rather bizarre and considerably over-the-top. Just how exactly does the grey squirrel threaten *us*? When was the last time a human being was killed by a grey squirrel?

Furthermore, the title of Elton's book, as we know, is *The ecology of invasions by animals and plants*, and yet here he was, starting the first chapter of his book by mentioning nuclear weapons and major diseases caused by viruses, bacteria, and fungi, to prepare us for the upcoming discussion of animal and plant invasions. The relevance of such comparisons is highly questionable and suspect. By and large, most successful introduced animals and plants do not behave like major epidemics caused by viruses, bacteria, or fungi, so such analogies are not only unhelpful, but also quite misleading. After all, the Black Death bubonic plague epidemic, caused by the bacterium *Yersinia pestis*, killed approximately one third of the human inhabitants of Europe between 1346 and 1350, and the influenza pandemic of 1918 may have killed between 30 and 40 million people (Ponting, 2007), although more recent sources give the much higher total death toll of 100 million people throughout the world (Zimmer and Emlen, 2013). How many introduced animal and plant species are expected to do *that*? Furthermore, when it comes to ancient disastrous pandemics – such as the waves of smallpox and plague infections which ravaged the Old World in medieval times and undoubtedly influenced the course of human history – it is very difficult to know exactly where they originated and how

FIGURE 30 *Two of Elton's perceived threats: the gray squirrel and the bomb*
ILLUSTRATION BY DORIS WONG

they spread, since people obviously did not have modern medical diagnosis techniques or any concept of viruses and bacteria in those days. Therefore, the ancient descriptions of such catastrophic diseases, when available, are bound to be quite vague and filled with mythical references. It is often very difficult, even today, to track the exact source and precise starting point of major diseases of viral and bacterial origins, and, given the speed with which influenza pandemics, for example, can spread around the world, and the fact that such major epidemics can originate in a wide variety of places at almost any given time, the application of the very notion of "native" range to such cases seems inappropriate.

Quite frankly, comparing disastrous past influenza or plague pandemics to the range expansions of species such as the rusty crayfish, or the introductions of dandelions to North America and gray squirrels to Europe, is obviously far-fetched, and seems more like a desperate attempt to raise the profile of invasion biology by terrifying the public with ill-founded apocalyptic rhetoric than a genuine attempt to inform and educate. If we go down that road, then pretty soon we might as well start comparing introduced green iguanas to fire-breathing dragons and making all sorts of other such supremely silly analogies. In order to have rational scientific debates about the real effects of most introduced plants and animals, there has to be a separate category for viruses, bacteria, and fungi capable of generating major epidemics in humans. I would argue that such pathogens fall mainly under the jurisdiction of medical doctors and researchers and should not be lumped together with the animal and plant species generally studied by conservationists and ecologists working within the field of invasion biology.

Charles Elton was certainly not the only one to relate the fight against pandemics to the struggle against exotic plants and animals. In 2001, an International Conference on Eradication of Island Invasives was held in New Zealand, and the proceedings of this conference were published in book form one year later by the IUCN (The World Conservation Union) and, more specifically, the IUCN Species Survival Commission. This was a massive, 414-page volume containing numerous articles with one main common theme – the eradication of invasive species (Veitch and Clout, 2002). In the keynote address, published as the second paper in this book, Daniel Simberloff (2002) discussed the financial benefits of the elimination of smallpox from the natural environment and then moved on to hypothesize about the similar possible benefits of the eradication of species such as purple loosestrife. Simberloff's point appeared to be that eradication campaigns may cost a lot of money but such costs may still be lower in the long run than the cumulative annual expenses associated with keeping species such as purple loosestrife in check. He also mentioned that in the United States, $45 million are spent each year on regular control programs (falling short of eradication, of course) against this wetland plant species. The same figure was cited by Pimentel et al. (2005) for the annual "control costs and forage losses" associated with purple loosestrife.

While we can all agree that the eradication of the smallpox virus from the natural world was indeed a major achievement, I would argue that this was a medical triumph, which has nothing to do with invasion biologists and the eradication campaigns against non-native animals and plants. Therefore, using the smallpox eradication example to "rally the troops", and encourage invasion biologists to try even harder to wipe out widely distributed species such as purple loosestrife in North America or European rabbits in Australia seems misguided. Can anyone imagine what it would take to even attempt to eradicate a widely distributed and resilient plant such as *Lythrum salicaria* from North America? Furthermore, given all the information presented in the detailed purple loosestrife case study, in a previous section of this book, it is not at all clear that spending so much money each year to control this species is really necessary. Perhaps it would be much easier to save money by reducing or eliminating unnecessary control programs (in other words, programs unsupported by conclusive scientific evidence), rather than asking for even more expensive eradication campaigns, because they might one day save us from regular control programs... We don't have to take it for granted that the annual $45 million expense to control purple loosestrife in the US, and the millions of dollars regularly spent for the control of other non-native species, are always really needed, and, in fact, closer scrutiny of the funds allocated for these purposes may be required.

In an article about the estimated current and potential costs of "nonindigenous" species in Canada, Colautti et al. (2006) presented a list of only three major expenses related to purple loosestrife. All of these so-called "characterized costs" (given in Canadian dollars) were related to control programs – Alberta control costs ($20,000 per year); Saskatchewan eradication project ($100,000 per year); and the Ontario biological control program ($90,000 per year) – and each of the estimates was based on information received through "personal communication". It should be fairly clear that it is illogical to blame purple loosestrife for the cost of the control and eradication programs we choose to initiate and implement against it. In other words, if the only costs associated with this plant in Canada are those of control programs, we can obviously eliminate the expense by stopping the programs. Using the information given by Colautti et al. (2006), if there are no more control programs against purple loosestrife, then, this plant will no longer cost us anything, so the "problem" of our own making would be solved.

The time has come to separate ideology from science before we decide which conservation initiatives genuinely deserve funding and attention. What would happen if we don't control purple loosestrife at all in North American wetlands? My guess, based on my review of the relevant literature, is that nothing terrible would occur, overall, and a lot of money could be saved for more positive and productive conservation projects, such as planting more trees or reducing soil contamination and freshwater pollution, for example.

The most expensive items shown by Colautti et al. (2006) on their list of characterized (as opposed to projected) costs were, by far, a couple of diseases (the severe acute respiratory syndrome, or SARS, which affects humans, and bovine spongiform encephalitis, also known as BSE). Again, as discussed before, I think it is misleading and inappropriate to put the health care costs related to dealing with SARS in Canada on the same list where we can also find control programs against purple loosestrife and other expenses related to non-native plants and animals. These types of overly inclusive lists may serve more to confuse, rather than truly educate, members of the public. Diseases such as SARS, or influenza, or Ebola, can indeed travel quickly across borders with infected persons, but dealing with the effects of such illnesses is a job for medical professionals, not invasion biologists, and funding for health care is generally separate from funding for anti-invasive species programs.

Simberloff's 2002 keynote address was meant to encourage invasion biologists to contemplate even bigger, costlier, and more comprehensive eradication campaigns against invasive species, not only on islands – even though the conference was supposed to be about "island invasives" – but also on continents. His message is clearly revealed in the following quote from the article:

> If the political will and economic support could be mustered, could nutria be completely eradicated in North America? Rabbits in Australia? What about invasive plants – under what circumstances could the witchweed approach be replicated? If smallpox and citrus canker can be eradicated, are insects on continents really out of the question?

Despite Simberloff's (2002) apparent faith in the increased role of eradication campaigns (or attempted eradication campaigns) in the endless war against invasive species, such efforts would appear to be not only hugely expensive but also very unlikely to succeed, especially when dealing with species introduced on continents and even large islands. The massive destruction and "collateral damage" that would likely result from such large scale eradication attempts would also be difficult to justify, and probably impossible to fully fix. And after all that effort and turmoil, the targeted species are likely to survive and possibly adapt, in any case. This is not defeatism or "unwarranted fatalism", as Simberloff mentioned in his 2002 article, but realism and common sense. Perhaps we just don't have enough good reasons to be at war with purple loosestrife, and some of us simply realize that this attractive plant does not pose anywhere near the type of threat smallpox once did.

However, as Larson (2008) pointed out, invasive species are frequently described in the scientific literature as "diseases", which attack previously "healthy" ecosystems and rob them of their perceived original vitality, diversity, and balance. These charges are often inappropriate and sometimes demonstrably false, but linking or equating exotic plants and animals with disease, even metaphorically, is seen by some as helpful in generating the negative feelings required for the support of the control and eradication campaigns against the introduced species. After all, a disease is something you are trying to overcome by destroying the pathogens that caused it, and removing them from the infected body. Therefore, if you can plant the idea in people's minds that introduced plants and animals are very much like nasty germs, presumably people will want to destroy them by any means necessary.

Interestingly, even in the previously discussed and much cited study by Wilcove et al. (1998), alien species and disease (caused by either native or exotic pathogens) are kept as distinct categories on the list of threats to imperiled species in the United States. As we know, this controversial study identifies alien species as a major threat, but at least Wilcove et al., unlike Elton, managed to separate introduced species from disease-causing pathogens.

For the most part, we pretty much eliminated most of the large carnivores that could once threaten us in any significant way. Sharks, for instance, were always considered a threat to us, and they have gained a particularly fearsome reputation ever since Steven Spielberg's blockbuster movie *Jaws* was released in 1974. And indeed, in the interest of full disclosure, I must admit that I certainly would not want to fall in the water near a swimming great white shark. However, it is worth keeping in mind that in the year 2000, for example, there were an estimated 10 human deaths worldwide which resulted from shark attacks, whereas in that same year we killed approximately 100 million sharks. No, this is not a typographical error – the number was indeed about 100 million. Even worse, more than half of these sharks were caught as bycatch – in other words, they were caught accidentally by fishermen targeting other species – and, as a result, the dead sharks were simply discarded (Primack, 2010). This trend continues every year, with minor variations. So, even though we may not always be aware of this, we are clearly a much bigger threat to sharks, and other animal species out there, than they are to us. Through overhunting, overfishing, and other forms of overexploitation, and through the relentless physical destruction of natural habitats, we have managed to kill countless individuals of many species. Now, we have found a new reason to exterminate many successful animals and plants. And, ironically, we have named this new reason "conservation".

Something tells me that if dandelions were rare and grew with great difficulty, only under extremely narrow and restrictive environmental conditions, some people would be willing to pay thousands of dollars to acquire even just one of these plants. The race would be on to save this endangered flower, and there would be annual conservation events, where "Save the dandelion!" t-shirts would be best sellers. Studies emphasizing the ecological benefits of having dandelions around, as well as all the threats posed to dandelions by other mean and inconsiderate species – particularly foreign ones – would probably be published regularly in top conservation biology journals. Protesters dressed in dandelion costumes would likely start appearing at special events, such as those associated with the yearly celebration of Earth Day. And so on.

I often wondered why we seem to start appreciating certain species only when they become rare and vulnerable, or are about to become extinct? Is it compassion, general concern for the world's biodiversity, or just remorse for what we have done to the environment so far? Or is it just a sense of pity for a helpless species that cannot challenge our hegemony on this planet in any meaningful way? A way to feel good about ourselves by trying to keep around a few specimens (not more than a token few, though) of some weakened species, whose best days may well be behind them, from an ecological and evolutionary standpoint. And, at the same time, this "compassion" for the weak, or cuddly, or inoffensive species (at least, inoffensive to us and our interests) serves as a convenient justification for acting ruthlessly to get rid of a variety of successful species we do not want to have around, because we may tend to envy, or fear, or feel threatened by, their success. According to this philosophy, only we, and the species we approve of, have a right to thrive on the Earth. Rescuing a few of the weak and vulnerable species and trying to eradicate some of the successful ones may both be ways of showing our power, dominance and control over the rest of the living world. Two sides of the same coin. A largely delusional method of asserting our own importance as humans, by deciding which other species should live and which ones should die, in each jurisdiction of the world where we can exercise our controlling powers. The illusion of control over nature may give some people a sense of security and satisfaction, and confirm the ancient self-serving belief that humans rightfully belong at the very top of some sort of ill-defined hierarchy of species – an idea that was seriously challenged by the development of the modern life sciences. Perhaps I am overanalyzing this, but, seen from this perspective, invasion biology, with its emphasis on human control over where and how other species should be allowed to live, may not be as much of a new field of biology, as a return to a system of beliefs which places humans at the top of the living world, with absolute executive and moral powers over all other organisms.

Brendon Larson (2005) called into question the use and the long-term usefulness of militaristic metaphors in invasion biology. The author stated that such metaphors "harbor inaccuracies that contribute to public misunderstanding of invasive species and even to misperception by conservationists themselves" and "invoke militaristic ways of thinking that are inconsistent with a sustainable relation between humans and the natural world." In a subsequent thoughtful and original analysis, Larson (2008) discussed several problems related to the ways in which invasive species are portrayed by invasion biologists, and the apparent lack of concern and awareness by many in this field about how such negative portrayals may create biases and a pre-determined emphasis on control programs and a violent struggle against these species. For example, invasion biology in particular, and ecology in general, tend to overemphasize the concept of competition between species, without paying equal attention to the co-operative and potentially mutually beneficial interspecific interactions. As a result, when a new species is discovered in an ecosystem, the first reaction of invasion biologists is usually one of panic and discomfort, even before we are in a position to determine if the species is having any negative effects. The first question always seems to be: how is this new species harming native species? This type of question, asked before any proof is available, obviously creates a bias from the beginning, as discussed elsewhere in the book as well. Ecologists rarely, if ever, seem to ask: how can this new species perhaps provide useful services to, and engage in mutually beneficial interactions with, some of the other species in the ecosystem? If we focused much more on this latter question, the results of our studies, and the general direction of the field, would almost certainly be very different indeed.

More recently, Larson (2011) continued to show the limitations and drawbacks of the "advocacy by fear" – as he called it – which was the result of the inappropriate metaphors and militaristic rhetoric often employed by invasion biologists. Such drawbacks include the possible loss of scientific credibility for ecologists who overstate the claims against introduced species, before comprehensive supporting data can be gathered, and the needless polarization and antagonism these aggressive and intransigent approaches can create. In his discussion, Larson (2011) begins by taking issue with the terms "invasion" and "invader" themselves, which obviously set the stage for the negative attitudes towards successful exotic species. As Larson puts it: "we have difficulty conceptualizing invaders without immediately wanting to do something about them" – and that something, I might add, is obviously not of a friendly or benign nature. The use of the term "invasive species" is a loud and unequivocal call to violent and quick action against the intruders. There is no time to wait and think, or sit on the fence while solid data can be gathered. If you don't join the war effort, you could be accused not only of dithering, but also of sympathy for the enemy.

According to some (for example, Ricciardi and MacIsaac, 2008), Elton was ahead of his time – Simberloff (2011) even called him "a prophet" – but, according to others (Davis et. al., 2001), he was very much a man of his time, and his ideas on introduced species may have been shaped at least in part by having to live in Britain during World War II, under the threat of German invasion and bombing raids.

In 1939, British scientists were informed that they would be exempted from military service, so that they could continue their research, presumably in ways which would assist the war effort against Germany. At the time, Elton was the head of the Bureau of Animal Population – an organization he had founded in 1932. As discussed by Crowcroft (1991), the Bureau employed relatively few people and relied on rather small grants for its survival. Some of these grants came from sources such as the Royal Society and the New York

Zoological Society, but the Bureau also received funding and assistance from Imperial Chemical Industries, a company which made shotgun cartridges. In any event, the long-term survival of the Bureau prior to the Second World War was far from certain, mainly due to financial concerns.

Elton was determined to make sure the Bureau of Animal Population would continue to exist, and, as the war with Germany was looming, he suggested, in a memo sent to the Agricultural Research Council, that the scientists of the Bureau could assist the war effort by investigating the effects of certain "vertebrate pests", such as rabbits, rats and mice, on the food supplies of Great Britain. Shortly after the start of the war, Elton's proposal was approved, and, as a result, the Bureau received new funding, more staff, access to a greater number of volunteers, and enhanced relevance and prestige (Crowcroft, 1991). The main purpose of the new research was to find more effective ways to control mainly rats (both the Norway rat – *Rattus norvegicus*, and the black rat – *Rattus rattus*), house mice (*Mus musculus*), and European rabbits (*Oryctolagus cuniculus*) – four species which were still considered alien in Britain, despite the fact that all of them had been introduced there several centuries earlier. The house mouse, for example, was likely introduced into England in pre-Roman times and, as we have seen earlier, the European rabbit arrived probably at some time during the 12th century (Davis et al., 2001). As for the black rat (*Rattus rattus*), until recently, the consensus seemed to be that this species had been first introduced into Britain during the 11th century (Davis et al., 2001). However, more recent studies showed that the arrival of this species into England took place just after 43 A.D. – much earlier than previously believed (Rielly, 2010). The fact that, even for a well-known and ubiquitous species such as the black rat, new evidence can be uncovered pushing back its suspected introduction date by about a millennium illustrates once again just how little we know about the historical distributions and range changes of introduced species even in well studied regions of the world such as Britain. Although it seems that it is not always easy to differentiate the black rat from the brown rat (also known as the Norway rat) when analyzing bone fragments found in sediments deposited centuries ago, the brown rat (*Rattus norvegicus*) appears to have arrived in England in the 18th century (Davis et. al., 2001; Rielly, 2010).

During the war, the Bureau members investigated various types of poisons and traps designed to kill rodents, and studied the behaviours of the targeted species, in order to learn how to kill them more efficiently. This research was published in a three-volume work titled, predictably enough, *The control of rats and mice*. According to Crowcroft (1991), the third volume of this book "became the instruction manual for training the rodent operatives who had to continue the war against mice in buffer depots for many years after the war with Germany was over." So, although the Second World War was over, the "war against mice" went on, and the use of the war rhetoric continued to be applied to the conflict with some other species. Sometimes, the "war against mice" could have unintended consequences. For example, poison baits intended for mice could contaminate spilled grain which may otherwise have been used to feed either people or certain domestic animals (Crowcroft, 1991).

So, in a sense, it could be argued that all this wartime research on finding ways to get rid of introduced "pest" species was a precursor to the larger war on invasive species to come. The quest to control four particular alien species was expanded, and more and more introduced species were targeted. Initially, the stated reason for killing a few non-native species was practical – the safeguarding of Britain's food supplies during the war – but, some years later, the reasons for disliking and persecuting pretty much all introduced species became increasingly diverse, vague, and arbitrary. Elton remained the leader of the Bureau of Animal Population until his retirement in 1967. At that time, the Bureau itself ceased to exist (Crowcroft, 1991). Thus, Elton's landmark book on invasive species was obviously published while he was the head of the Bureau, and it is not unreasonable to think that the work done during the war on the control of introduced rodent species may have influenced Elton's overall approach to invasive species in general, in the years following the end of the war.

Elton's 1958 book on invasive species was based on a series of BBC radio broadcasts given one year earlier and intended for a broader audience. The three radio broadcasts were presented under the title "Balance and barrier" (Southwood and Clarke, 1999; Simberloff, 2000; Davis et al., 2001; Davis, 2009). This was not the first time when Elton used radio broadcasts to address the general public. In the early 1930s, Elton had given a series of BBC radio broadcasts for amateur naturalists. The content of these broadcasts was published in book form in 1933. Because they were aimed at a wider audience, in both the radio broadcasts and the resulting 1933 book, Elton's ideas were deliberately presented "in the simplest possible way" (according to Crowcroft, 1991). Perhaps a similar tendency to oversimplify, and use dramatic examples and language, may have appeared in Elton's broadcasts of the late 1950s on the subject of invasive species, and in the subsequent book based on these broadcasts.

Although Simberloff (2011) argues that Elton disliked invasive species intensely well before the Second World War, it is difficult not to think about the possible influence of that war on Elton's work and worldview. Radio was a very direct and effective way of reaching a significant segment of the public during the 1940s and 1950s. For example, Churchill's legendary and inspirational wartime radio speeches, particularly during 1940, were instrumental in lifting the spirits and strengthening the resolve of the citizens of Great Britain, at a time when the prospect of invasion of their country by a very powerful German army seemed not only very real, but also almost imminent (Johnson, 2009; Urquhart, 2010; Hastings, 2011). In fact, one of the chapters of a detailed recent book (Hastings, 2011) about Winston Churchill during World War II is titled *Invasion fever*. This fear of invasion by a hostile and menacing foreign army was widespread and shared by civilians and military commanders alike. Like almost all of his countrymen, Elton must surely have heard Churchill's speeches and was certainly influenced by them. A memorable passage from one of Churchill's speeches, given in 1940 (and cited in Johnson, 2009), should capture the spirit of the times: "…we shall defend our island, whatever the cost may be, we shall fight on the beaches, we shall fight on the landing grounds, we shall fight in the fields and in the streets, we shall fight in the hills. We shall never surrender." An analogy can be made, to some extent, between Churchill's defiant wartime speeches, which advocated a violent and determined struggle against foreign invaders, and Elton's radio addresses of the 1950s warning about the dangers posed by invasive species. This is not meant to suggest that Elton's words were as important or as eloquent as Churchill's words had been, but the purpose of the speeches was somewhat similar – to mobilize the population in the fight against a common enemy and potential invader (and perceived major threat).

The problem, of course, is that using the same methods as in World War II to fight squirrels and weeds may seem misplaced and excessive. Furthermore, unlike World War II, the war against invasive species has no end in sight. As long as we're around, there will always be species which will expand their ranges partly due to human activities, and there are already many animals and plants, not to mention microorganisms, that have reached new places with our assistance. If we insist on fighting this war, then the war will last for as long as our own species does.

On page 26 of Peter Crowcroft's book, *Elton's ecologists: a history of the Bureau of Animal Population*, there is a photograph of a biologist associated with this bureau, who is smoking his pipe, while at the same time gassing rabbit burrows with some sort of fumigating device. While this task is clearly gruesome, and described in the caption of the photograph as very effective, the apparently casual way in which this activity was conducted reminded me for a moment of the quirky comedy *Caddyshack*. Those familiar with the movie may recall that one of the main characters is the rather unhinged assistant greenskeeper Carl Spackler – brilliantly played by Bill Murray – who spends most of the movie engaged in a senseless and destructive war against gophers on a golf course. The fictional gophers shown in the movie "invaded" the golf course

FIGURE 31 *Bill Murray comes face to face with the gopher in the movie Caddyshack. Permission for the use of this image was obtained from Patricia Zline of The Rowman & Littlefield Publishing Group*

from a nearby construction site, or, perhaps more accurately, they may have been chased by the construction-related activities to the adjacent lawns of the golf club. When asked by his boss to get rid of all the gophers, Carl Spackler replies: "We can do that. We don't even have to have a reason." The film ends with massive explosions, set up by Spackler, which obviously cause incomparably greater devastation to the golf course than gophers ever could. Despite the repeated attempts on its life, the targeted gopher escaped relatively unharmed, no doubt to the delight of the audience.

There is nothing funny about the gassing of rabbit burrows, or the other methods used to kill animals and plants, in real life, but this movie highlights the absurdity, futility and unintended consequences of certain control programs (although this is probably not the only reason to see the film).

Inevitably, the controversies and debates generated by massive control programs, carried out in the name of conservation, have spread beyond academic circles and caught the interest of non-academic writers. For instance, two recent popular books – one a work of fiction (Boyle, 2011) and one a work of non-fiction (Stolzenburg, 2011) – have both tackled the uncomfortable topic of eradication of introduced mammals on islands, in order to preserve some of the vulnerable native species found there. Through such books, and general articles on the same topic, we are introduced to the sometimes strange alliances made during the war against invasive species, such as the cooperation between ecologists and professional hunters and poachers.

Despite the controversies, the war on invasive species appears to be expanding, and, perhaps not surprisingly, new methods are currently explored for the detection and the eventual control of such species. Coyle (2016) mentioned that in British Columbia, in western Canada, drones are now used to find invasive plants in remote areas, and sniffer dogs were enlisted in the fight against quagga mussels. These same methods are now also beginning to be employed in Ontario, where a new Invasive Species Centre was established in 2011, in Sault Ste. Marie, because of the perceived growing threat posed by successful foreign species. Apparently, the centre has received a $340,000 grant in 2015 to enlist the help of citizens in finding suspected invasive species in local areas (Coyle, 2016). And indeed, on the web site of this centre, I found information about a program associated with the Invasive Species Awareness Week (February 21–27, 2016). The advice directed towards members of the public on this subject was clear, as shown by the following quote from the web site: "When you see an invader – REPORT IT! Reporting allows for early detection & rapid response to an invasion which betters our chances of eradication" (http://www.invasivespeciescentre.ca, 2016). According to invasion biologist Hugh MacIsaac, recent Ontario legislation included "a very comprehensive program of invasive species prevention, eradication and rapid response" (Coyle, 2016).

Entire books can be dedicated solely to the description of the often gruesome and brutal control programs associated with invasion biology. I find this topic thoroughly unpleasant and often quite depressing, and, therefore, I will limit myself to only some of the numerous examples I could give of destructive campaigns carried out against species caught outside their presumed native ranges.

Case Study: European Rabbits in Australia, Britain, France, and Spain

European rabbits (*Oryctolagus cuniculus*) were actually brought to Sydney, Australia as early as 1788, and several additional introductions of domestic rabbits followed in the next few years. However, although these rabbits sometimes formed feral populations, they did not spread much beyond the areas where they had been released (Myers et al., 1994; McLeod and Saunders, 2014).

All that changed in 1859, when Thomas Austin released a few imported wild European rabbits on his estate, located near the city of Geelong (not far from Melbourne), in the south-eastern Australian state of Victoria. At the time, the local attitudes towards non-native species were quite different than they are now, and early settlers were eager to introduce animals and plants which reminded them of England (Dickenson, 2014; McLeod and Saunders, 2014). The European rabbit was evidently a beneficiary of these introduction programs, even though, as we have seen earlier in the book, this species was not actually native to Britain, and had been brought there by people centuries ago.

It should be mentioned that in the 1860s, before they became very abundant and widespread, rabbits were regarded as a valuable game species in Australia and received protection due to government legislation (McLeod and Saunders, 2014).

Mr. Austin – the farmer who introduced the rabbits on his property mainly because he wanted to hunt them – could not have anticipated that this simple act would be recorded as the beginning of one of the most famous and spectacular examples of the rapid spread of an exotic species in history. Apparently, just thirteen rabbits were released at first, but, after only six years, the rabbit population on Thomas Austin's estate grew to about 30,000 – of which 10,000 or so managed to avoid being shot. Naturally, the rabbits did not sit still on the Austin estate, waiting patiently to be killed, and by 1880 they expanded their range northward and westward into the neighbouring states of New South Wales and South Australia, respectively. In the early 1880s, the Australian authorities were certainly concerned enough about the rabbits to come up with legislation – such as the Rabbit Suppression Act of 1880 and the New South Wales Rabbit Nuisance Act of 1883 – which authorized the killing of these animals. Rabbits were poisoned, trapped, shot, and killed with sticks, nets, and pretty much any other type of weapon people could think of using, and rabbit burrows were gassed or destroyed by mechanical means. As a result, millions of rabbits were killed in both Victoria and New South Wales in the mid-1880s, but this did not seem to have much of an effect on the overall growth of both the population and the Australian range of *O. cuniculus* (Ponting, 1993; Dickenson, 2014; McLeod and Saunders, 2014).

We have seen earlier in the book that people managed to exterminate a very abundant bird such as the passenger pigeon in North America, and reduce the population of that species from about 5 billion to zero in just a few decades mainly through intensive hunting. However, killing millions of rabbits by all sorts of gruesome means evidently did not come close to eradicating the rabbit from Australia, and the exact reasons for this are complex and probably not fully understood. Clearly, unlike the passenger pigeons, the rabbits have a fairly high reproductive rate, and this helps the rabbit populations bounce back after the massive killing campaigns and other adverse conditions.

In any case, despite the attempted eradication campaigns against them, the rabbits continued to spread northward and westward across Australia. In the 1890s, they were crossing the forbidding Nullarbor desert and moving towards the large state of Western Australia. In order to stop the advance of the rabbits, a huge 1,600 kilometre-long fence, stretching between the south and north coasts of the continent, was erected between 1902 and 1907, in the hope that this would prove to be an impenetrable obstacle placed in the rabbits' path (Ponting, 1993). This impressive project was not the only one of its kind. Dickenson (2014) mentions that a network of so-called "rabbit-proof" wire fences, stretching over a total distance of more than 48,500 km, was built in Australia between 1886 and 1929. However, rabbits can, of course, dig tunnels under such obstacles, and these extraordinarily long fences are very difficult to maintain, so animals can also move through holes and gaps which may occasionally appear, or over collapsed portions of the fences. As a result, quite predictably, the rabbits managed to overcome these artificial barriers as well, by advancing under, around, and occasionally through parts of the fences built to keep them out of previously rabbit-free parts of Australia. By 1900, the rabbits were already in Western Australia, and by 1910, they had expanded their range over most of the continent, with the exception of the northernmost areas, and some regions along the eastern and western coasts (Ponting, 1993; Myers et al., 1994; Lomolino et al., 2010; Dickenson, 2014). According to a map shown in the book chapter by Myers et al. (1994), the rabbits are common and widespread mostly in the southern half of Australia and only scattered or small and isolated rabbit populations can be found in the northern half of the continent. The dedication at the beginning of the book this chapter is a part of reads: "In memoriam Charles Elton, our mentor" (the volume was published three years after Elton's death). And, as we know, the Bureau of Animal Population, led by Elton, was actively involved in the control and suppression of rabbits in Britain.

It is believed that the European rabbit managed to expand its range dramatically and quickly in Australia, at least in part because people had already modified large areas of the continent for the benefit of domestic grazing animals such as sheep and cattle. These domestic animals contributed to changes in the local vegetation, which led to the growth of the types of grasses rabbits prefer to eat. In fact, a lot of the anger directed at the rabbits, at least initially, undoubtedly came from farmers who saw the herbivorous rabbits as possible competitors for the sheep. In order to protect their livestock, people also killed many native Australian predators, and this action inadvertently helped the rabbits as well, by reducing the predation pressure on them. Therefore, the rabbits moved into already disturbed and much modified habitats, and took advantage of the environmental changes caused by humans and domestic animals (Myers et al., 1994; McLeod and Saunders, 2014).

Although it is clear that rabbits expanded their Australian range with remarkable speed especially in the few decades immediately after their introduction, Flux (1994) wrote that "The present distribution of rabbits in Australia is curiously uncertain", partly because people rarely visit the vast and arid central areas of this continent, and therefore it is hard to collect much data there. Although *O. cuniculus* has shown great resilience and perhaps a somewhat unexpected ability to survive and reproduce in the harsh Australian deserts, it appears that this species is still not all that well adapted for life in such difficult environmental conditions. As a result, European rabbits have a patchy distribution in these arid regions, and can become locally extinct during droughts. In addition, rabbits have been unable to colonize the tropical northernmost areas of Australia, because the climatic conditions found there are unsuitable for them (Myers et al., 1994; McLeod and Saunders, 2014).

By 1950, despite numerous control programs, there may have been as many as 500 million rabbits in Australia (Ponting, 1993), although this was probably a very rough estimate, or, in other words, not much more than a guess. Since the more traditional ways of killing rabbits did not seem to reduce their total numbers significantly, people thought about the use of even more drastic measures, such as germ warfare. And so, in 1950, the myxoma virus, which causes myxomatosis, was deliberately brought to Australia. The virus was imported from Brazil, where its natural host was the South American rabbit species *Sylvilagus brasiliensis*, also known as the tapeti. Exposure to the virus usually does not produce major illness among the South American rabbits. However, wild European rabbits had not come into contact with this New World virus before, and, therefore, they did not have the opportunity to develop immunity to it. The initial results were predictably disastrous for *O. cuniculus*. The mortality rate as a result of infection with the myxoma virus among the rabbits of Australia was a staggering 99.8% (Ponting, 1993; Fenner and Ross, 1994; Diamond, 1999; Dickenson, 2014). One can only imagine what the Australian countryside must have looked like at the time, with hundreds of millions of rabbits dead or dying in the fields and in their burrows. Mosquitoes were the most important vectors, which transmitted the virus from rabbit to rabbit in Australia, although fleas are also capable of spreading the virus (Fenner and Ross, 1994). In fact, rabbit fleas belonging to the species *Spilopsyllus cuniculi* were imported from Europe in 1966 and added to the list of Australian vectors for the myxoma virus (Myers et al., 1994). And just to make sure the disease would reach as many rabbits as possible, farmers set up traps with needles contaminated with the myxoma virus and placed infected fleas into rabbit burrows. Infected rabbits exhibited horrible symptoms which included the presence of tumours and skin lesions, as well as swollen heads, blindness, and deafness (Fenner and Ross, 1994; Dickenson, 2014).

However, despite the terrible initial death toll, the mortality rate caused by myxomatosis started to drop after a few years of exposure to the disease. Apparently, this was due to both an overall decrease in the virulence of the virus – also known as attenuation – as well as the increased levels of immunity among the surviving rabbits and their offspring. When faced with a powerful selective force, such as the deadly myxoma virus, the few rabbits with some natural genetic resistance were the only ones left to breed, and their offspring were more likely to survive myxomatosis (Ponting, 1993; Fenner and Ross, 1994; Myers et al., 1994; Diamond, 1999; Dickenson, 2014). The decreases in mortality rates can vary from location to location and are influenced by environmental conditions, the availability of vectors, and the virulence of particular myxoma strains. However, overall, the drop in mortality rates was dramatic and quite rapid. Ponting (1993 and 2007) reported that by 1957, only seven years after the myxoma virus was first introduced in Australia, only 25% of the rabbits found there were still killed by myxomatosis (as opposed to the previously mentioned 99.8% death rate of 1950). The same gradual reduction in mortality rates from 99.8% to 25% was also discussed by Diamond (1999). Dickenson (2014) mentioned a decline in myxoma virus-caused mortality among rabbits in Australia from 99.8% to 40% in some regions, and Myers et al. (1994) noted that resistance to myxomatosis is steadily increasing over time among the local rabbits. Although efforts were made to select and introduce more virulent strains of the myxoma virus, and there are outbreaks of myxomatosis affecting the rabbit populations of Australia from time to time, this disease is no longer considered capable of possibly eradicating rabbits from the continent. In fact, in the decades since the introduction of this disease to Australia, the rabbit population has recovered (Ponting, 1993; Dickenson, 2014).

Unfortunately, the myxoma virus was also intentionally brought to France in 1952 to control local rabbits, even though *O. cuniculus* is regarded as a native species there. By 1954, the disease killed about 90% of the rabbits found in the wild in France. The reaction to the introduction of this virus in France was mixed – some farmers wanted to control the rabbits, but other people valued them as game animals and were opposed to eradication attempts. Therefore, measures were taken in France to protect the rabbits from this disease through immunization programs. In addition, it appears that some genetically resistant rabbits were brought from Australia to boost the French rabbit populations. Myxomatosis now also affects rabbits in Spain, another part of their initial presumed native range (Fenner and Ross, 1994; Dickenson, 2014).

In 1953, the disease was apparently carried from France to Britain by wild birds. As in France, efforts were made to stop the proliferation of the epidemic. However, some farmers illegally used infected rabbits to spread myxomatosis, and, as a result, the disease moved rapidly throughout the country and killed approximately 99% of the rabbits of Britain. However, as in Australia, the rabbit populations of France and Britain started to recover in the years after the first disastrous exposure to the myxoma virus (Fenner and Ross, 1994; Dickenson, 2014).

Undaunted, some scientists and conservationists relentlessly continued to explore new methods of getting rid of rabbits. For instance, Australian researchers have been experimenting with genetically engineered myxoma viruses, which may sterilize female rabbits. The modified viruses are not lethal, but they contain an additional gene which codes for a protein from a tissue layer found around the female rabbit's eggs. The virus with this extra gene essentially turns the immune system of the female rabbit against her own eggs, leading to potential sterilization. Similar research was conducted on modified viruses which can make female mice infertile (O'Neill, 2002).

And just as *O. cuniculus* populations were recovering somewhat from the devastating initial myxomatosis epidemics, a new and extremely lethal rabbit virus of Asian origin was identified in China in 1984. This virus was responsible for the Rabbit Hemorrhagic Disease (also known as RHD) which killed millions of rabbits throughout Asia and Europe, before being taken to Australia and New Zealand as a possible additional biological weapon against the rabbits still found there. The virus which causes RHD is also spread by mosquitoes and fleas, just like the myxoma virus. And, as myxomatosis had done in the 1950s, RHD initially killed huge numbers of rabbits (perhaps 90% to 95% of some populations) in the 1990s, although, once again, the rabbit populations of Australia and New Zealand have survived this new attempt to eliminate them, and rabbit resistance to the RHD virus appears to be growing in these countries (Dickenson, 2014).

Sadly, RHD seems to have dealt a severe blow to the *O. cuniculus* populations in their native range, and these European rabbit populations remain much smaller and more vulnerable than before (Dickenson, 2014). RHD was first recorded in Spain in 1988, and the initial resulting mortality rate was about 80% in most of the local rabbit populations. An analysis of the wild rabbit population trends in Spain – based on 30 years of data, collected between 1973 and 2002 – showed that rabbit numbers declined by 70%, overall, in the entire country, during this period, and in some regions, the rabbit populations were reduced by as much as 90%. According to the authors of the study, such a precipitous decline in the *O. cuniculus* population of Spain, over three decades, should justify the listing of the rabbit as a vulnerable species in this country, under the relevant IUCN (World Conservation Union) criteria for conferring such a status on a species at risk. The formal identification of the European rabbit as a vulnerable species would also require the drafting of a Conservation Plan Program by the Spanish government. However, despite the plummeting population numbers, the rabbits were still listed as a species of "Least Concern" – a conservation status which requires no action. Apparently, this state of affairs was also influenced by sociological factors, since hunters and farmers generally oppose any increased protection for the rabbit and these groups have considerable influence over politicians and even some conservation researchers in Spain – for example, it appears that Spanish hunter associations make important contributions to local conservation research (Virgós et al., 2007).

The sharp decline of *O. cuniculus* in Spain has had a catastrophic impact on other species as well. The European rabbit is a keystone species in the Iberian Peninsula, and more than 40 local vertebrate predators rely on this rabbit species as a prey item. At least a couple of these predators, the Iberian lynx (*Lynx pardinus*) and the Iberian (or Spanish) imperial eagle (*Aquila adalberti*), are now critically endangered, because they prey almost entirely on European rabbits, and, as we have seen, these rabbits are becoming less and less common in Spain. Other predators which frequently hunt rabbits – such as eagle owls (*Bubo bubo*) – have suffered as well from the scarcity of their regular prey. Therefore, the conservation of *O. cuniculus* should be a priority if we want to save some of its endangered and vulnerable predators (Virgós et al., 2007; Lees and Bell, 2008).

The drop in rabbit numbers has triggered a domino effect which is actually threatening even a recently discovered parasitic invertebrate species. Isidoro's chewing louse (*Felicola isidoroi*), which was first formally described in 2001, is a host-specific parasite which lives only on the Iberian lynx. Thus, this louse could not survive without its lynx host, and since the Iberian lynx is critically endangered, so is Isidoro's chewing louse (Wheeler and Pennak, 2013). And this is how we get from the introduction of deadly exotic rabbit viruses to the possible upcoming extinction of a tiny parasite which lives on lynx… Undoubtedly, this is just one of many such examples of the unforeseen consequences and ramifications of control programs.

It is interesting to note the difference in the terms used to describe the effects of the European rabbit in its estimated native range versus the introduced range. Since the rabbit is both native and vulnerable in Spain, its crucial role in the Iberian ecosystems is currently emphasized. We are now reminded that this species is very important for the survival of its predators, as well as being useful for soil fertility and plant seed dispersal, and even instrumental for "increasing floral and faunal species richness" (Lees and Bell, 2008). In the meantime, in Australia, where the European rabbit is widely despised and subjected to various types of gruesome control programs, this species is described in all sorts of unflattering ways – for example, as "the chainsaw of the outback", an "environmental vandal", "killer creatures", "furry foes", and even as a "dark stain" which is "spreading over the face of Australia" (all these derogatory terms were gathered from a variety of sources and reported by Dickenson, 2014). And just in case some children may still be attached to the cute European rabbits, there was even apparently a campaign in Australia to replace the traditional Easter Bunny with the Easter Bilby (Dickenson, 2014), since the bilby is a native Australian marsupial with somewhat prominent (although not really rabbit-like) ears... When it comes to propaganda, I suppose nothing is sacred...

It may be worth noting that if *O. cuniculus* can play useful roles in European ecosystems, then presumably, it may also play some of the same positive roles in Australia, or other parts of its non-native range as well. For example, as I will discuss in the next paragraph, some Australian predators depend heavily on the rabbit these days. And if the rabbit can spread plant seeds, and be helpful in the maintenance of biodiversity in Europe, there is no particular reason why it may not be able to do that on other continents as well. Just a thought.

Ponting (1993) stated that the European rabbit had "no natural predators" in Australia. As we have seen, this is a commonly given, and almost automatic, general explanation for the success of various invasive animals. However, it appears that this explanation is not valid in this case. Myers et al. (1994) gave a list of 29 predators of the rabbit in Australia. This list included mammals (foxes, cats, and dingoes), quite a few birds of prey (eagles, falcons, kites, goshawks, harriers, buzzards, and owls), and reptiles (several species of monitor lizards and snakes). The most important predators on the list – the red fox (*Vulpes vulpes*) and the cat (*Felis catus*) – are also regarded as invasive pests in Australia, but these species obviously can contribute to keeping the rabbit population in check naturally, at least to some extent. The rabbit is an important food source for several native Australian species as well (Myers et al., 1994). McLeod and Saunders (2014) mention "the large number of Australian native birds, mammals and reptiles that currently prey on rabbits", and speculate that these native predators were probably feeding on European rabbits, and therefore exploiting this new resource, shortly after the arrival of this species in Australia. So, it appears that it did not take long for native predators to discover this new prey species. Rabbits are the main prey of several local avian predators of large and medium size. In some parts of Australia, for instance, the diet of the wedgetail eagle (*Aquila audax*) is almost entirely (97%) made up of rabbits. Therefore, it is quite likely that there are more wedgetail eagles in Australia now than before the introduction of the rabbits took place (Myers et al., 1994; McLeod and Saunders, 2014).

An abundant and widespread animal, such as the European rabbit in Australia, would be expected to have a large impact on its surroundings, and, indeed, it is clear that large numbers of rabbits can consume vast quantities of vegetation. However, although the rabbit is commonly blamed for the extinction of native Australian mammals, the evidence for that seems much harder to find. McLeod and Saunders (2014) state that "there are few studies that have provided direct evidence of competition between rabbits and Australian native mammals." Myers et al. (1994) agree and mention that "Little information on the direct impact of the rabbit on native fauna exists."

As mentioned, it appears that major human and livestock-generated environmental changes had occurred in many parts of Australia prior to the arrival of the rabbit. These changes may well have played a larger role in the decline of native mammals than the rabbit's presence. Furthermore, it is quite likely that the large-scale control programs against the rabbits – particularly the programs using poisons and traps – had a negative effect on the smaller native marsupials (Myers et al., 1994). McLeod and Saunders (2014) wrote "that there is insufficient evidence to unequivocally conclude that the extent of rabbit invasion was negatively correlated with the species diversity of functional guild competitors."

Mao's War on Sparrows

Between 1958 and 1962, Communist China, under the leadership of Chairman Mao Zedong, went through a disastrous period known as the Great Leap Forward, when forced industrialization and the drive to produce record amounts of steel led to widespread starvation and deforestation in many parts of the country. In 1958, Mao Zedong initiated an eradication campaign against various species considered as "pests". Sparrows were one of the main targeted species, because they were accused of eating the seeds of crops such as grain, and therefore reducing agricultural yields. The people of China were asked to exterminate the birds by any means necessary. Nests, eggs, and nestlings were destroyed and adult birds were shot. Millions of people were dispatched to the hills of the countryside and the tops of buildings in the cities, where they banged pots and pans and made a lot of noise for countless hours to prevent the sparrows from landing and resting. As a result, severely harassed and exhausted sparrows could no longer fly and fell out of the sky. Children – some as young as five years old – were involved in the massive country-wide anti-sparrow killing campaign as well. Some of them were asked to make posters with slogans such as "Sparrows are mankind's number one enemy" and "Eradicate sparrows", while many others were encouraged to use slings to shoot down as many birds as possible, often quite indiscriminately. Overall, huge numbers of birds were killed and the sparrow populations crashed and did not begin to recover for years (Shapiro, 2001; Dikötter, 2010; Todd, 2012).

However, the campaign against sparrows did not have the desired results. With so many people involved in such a widespread and reckless campaign, accidents were inevitable. Some people fell off buildings and others shot themselves while trying to get rid of the sparrows. There was also confusion about the exact types of birds killed – undoubtedly plenty of individuals of other bird species must have become unintended casualties of the war on sparrows. Furthermore, sparrows also eat insects, and after the birds were virtually eradicated, there was a sharp rise in insect populations. Some of the insects – for example, locusts – damaged the crops. Therefore, the misguided elimination of the sparrows actually lowered agricultural production even further, rather than enhancing it. In any case, it is quite clear that the sparrows had been used as scapegoats for the catastrophic political decisions which led to widespread famine in the first place. Eventually, in 1960, the sparrows were removed from the list of enemy pests and were replaced with bedbugs. By then, however, the lasting damage had been done (Shapiro, 2001; Dikötter, 2010; Todd, 2012).

In some ways, the war conducted by scientists from Elton's Bureau of Animal Population against four "vertebrate pests" (two species of rats, the house mouse, and the European rabbit), and the subsequent battles against a variety of blacklisted exotic species, are similar to the Four Pests Campaign started in the late 1950s in Communist China. Aside from sparrows, the other pests initially targeted in the Chinese campaign were rats, flies, and mosquitoes (Shapiro, 2001). Ironically, and no doubt coincidentally, the Four Pests Campaign started in 1958 – the same year when Elton's book on invasive species was published.

I am not suggesting that Chairman Mao was one of the original invasion biologists, and I am not linking invasion biology with the disastrous policies associated with the Great Leap Forward period in Communist China, of course. However, the parallels between Mao's pointless and destructive campaign against sparrows and the on-going war against certain so-called invasive species are uncomfortable. In both cases a "guilty until proven innocent" or "guilty because we said so" mentality took hold and was applied on a wide scale. Whether we are talking about people being enlisted to kill sparrows in Communist China decades ago, or the members of the public mobilized to destroy cane toads, purple loosestrife, or garlic mustard in modern day Western countries, we are dealing with examples of attempted control and eradication campaigns where propaganda and emotion may supersede scientific evidence and reason.

For example, a so-called "Toad day out" event was organized in northern Queensland a few years ago. The purpose was to have local people collect and kill as many cane toads as possible. The captured toads were brought to a collection site, where they were weighed and then killed through freezing or gassing methods. Apparently, prizes were given to the people who collected the heaviest toads and the largest quantity of toads (measured by the total weight of the toads captured). One of the top prizes consisted of "several movie passes and a trophy made out of a cane toad". A Queensland politician who organized this event is quoted as saying that: "The cane toad is probably the most disgusting creature and the most destructive creature. They're killing our native wildlife, they're taking over our habitat and they're hopping all through this country." A local city councilman presumably associated with this initiative commented approvingly on the involvement of children in the killing of the toads, before concluding that: "The children really got into the character of this event." Amazingly, The Royal Society for the Prevention of Cruelty to Animals "has applauded the effort", as long as, according to a spokesman, "the toads are killed humanely – in other words they're not hit with baseball bats or cricket bats or golf clubs" (Sturcke, 2009; *The Telegraph*, 2009). Apparently, gassing the toads to death is considered a humane killing method.

Shine and Doody (2011) discussed the debates and disagreements between researchers and members of the public regarding the impact of cane toads in Australia. According to these genuine cane toad experts:

> Rarely has an invasive animal been so widely reviled by the general populace, to the point where organized community groups have sprung up to fight the toad invasion. The expenditure of more than AU$20 million in attempts to eradicate cane toads from Australia has been largely unsuccessful, and efforts are now being redirected toward mitigating toad impacts rather than eliminating the animals.

In other words, aside from the cruelty associated with such misguided ideas, the anti-cane toad campaigns were an expensive failure and a waste of time.

At this point, I should mention that I grew up under a communist dictatorship in Romania and, probably as a direct result of this, I have developed a lifelong aversion to, and suspicion of, all sorts of propaganda. If one is exposed to pervasive propaganda long enough, there is more of an opportunity to develop an enhanced awareness of it, and a heightened sensitivity to the discrepancies between slogans and reality. Perhaps that may explain at least in part my reaction to some of the anti-alien species literature.

Mistaken Identity

Aldhous (2004) reported the case of a caller to an Australian radio show, who proudly described the way in which she had contributed to the war against invasive species in her country, by driving back and forth over a group of hated cane toads. Aside from the rather disturbing image such a description brings to mind, it turned out that the cane toads (*Bufo marinus*) had not reached that particular part of Australia yet. Thus, in all likelihood, the caller to the radio show had probably destroyed some native Australian marbled frogs (*Limnodynastes convexiusculus*), which apparently look somewhat similar to the introduced cane toads.

These types of mistakes are likely to happen all the time, and have more and more serious negative consequences for both native and introduced species, if invasion biologists insist on enlisting the help of the public in the control and eradication campaigns against various animals and plants. Such often rather mindless campaigns, in my view, are also one of the plausible and predictable outcomes resulting from frequent exposure to the intense propaganda against non-native species. After being influenced, and possibly overly alarmed, by such propaganda repeatedly, there is always the risk that some members of the public – particularly those with a possible prior grudge against certain animals or plants and a lot of free time – may just take matters into their own hands and set up their own unofficial programs to get rid of species they perceive as unwanted.

The reality is that many non-native species are closely related, and look very similar, to native ones. These species can also play potentially similar roles in their respective environments. To list just a few of many possible examples: in North America, the native yellow avens (*Geum aleppicum*) plant is hard to distinguish from the supposedly invasive wood avens (*Geum urbanum*) (http://ontariowildflowers.com, 2015), and the native red mulberry tree (*Morus rubra*) is difficult to tell apart from the introduced white mulberry (*Morus alba*). In fact, despite the names of these related tree species, even the fruit colour can be the same for both sometimes (Kershaw, 2001; Dickinson and Royer, 2014). Also, the leaves of the native sugar maple (*Acer saccharum*) – a symbol of Canada – are almost indistinguishable from those of the non-native Norway maple (*Acer platanoides*), so much so, in fact, that the stylized leaf of the Norway maple may have been accidentally put in the upper left corner of new Canadian $20 bills in 2012, causing a bit of a controversy in the process (Dickinson and Royer, 2014; Larson, 2015).

As briefly mentioned before, while discussing the uncertainties related to the distributions of species, Wetzel et al. (2004) showed that the native golden crayfish (*Orconectes luteus*) has been mistaken for the similar-looking, much disliked, and presumably invasive, rusty crayfish (*Orconectes rusticus*) in parts of Iowa and Minnesota, and possibly elsewhere, as well. These authors advised that the body colorations of living specimens of these two species are quite distinct and may serve as a reliable way of separating the two species. Perhaps that may be true in some cases, but, having worked with crayfishes of various kinds for a long time, and being familiar with the relevant scientific literature, I find that crayfish body colour is a notoriously variable characteristic, which frequently differs intraspecifically from one population to another.

For instance, *Cambarus bartonii* crayfish specimens I collected in the Algonquin Park – Ragged Falls area of central Ontario often tended to be rather bluish (and sometimes dark green to dark brown) with orange claw tips, while members of the same species collected in a Ganaraska River tributary, in southern Ontario, were uniformly light brown, or even reddish-brown, and larger in overall size. I also encountered some colour and size differences among the adult crayfish of various *Cambarus robustus* populations I sampled in different Ontario streams (Guiașu, 2002a).

Other studies have shown the presence of very different colour morphs among individuals of the same crayfish species, as well. Momot and Gall (1971) and Fitzpatrick, Jr. (1987) reviewed the presence of blue colour-variants in a variety of crayfish species belonging to genera such as *Cambarus*, *Orconectes*, and *Procambarus*. Among the species mentioned in these studies, there were several crayfish also found in Ontario, such as *O. immunis*, *O. virilis*, and *C. bartonii*. Momot and Gall (1971) found both blue and the much more common brown specimens of *O. virilis* co-existing, and apparently surviving equally well, within the same populations in two Michigan lakes. Crayfish of the same species with other body colours were also occasionally found there. About two decades later, Thacker et al. (1993) collected adult *Orconectes virilis* specimens at several locations in northern Michigan, and found a variety of body colours for these crayfish, including yellow, blue, and dark brown. Dunham et al. (1979)

reported an unusual mainly white colour morph, with a broad black stripe on the dorsal side of the abdominal region, in addition to the common brown morph, for the *Orconectes propinquus* crayfish they collected in Lake Simcoe, in southern Ontario. Therefore, different crayfish body colours can be encountered even among conspecifics found in the same populations, and this makes the task of identifying different crayfish species even more challenging at times.

And, of course, the colours of living crayfish (and other aquatic animals) tend to quickly fade in preserved museum specimens, so using colour for identification purposes is generally not an option when looking at such individuals.

The rusty crayfish identification card – produced and distributed, as far as I can tell, by the University of Minnesota Sea Grant Program, the Government of Ontario, and the Ontario Federation of Anglers and Hunters, as well as, possibly, other organizations – is not necessarily all that useful in helping members of the public recognize this species. For example, on the back of the card, there is a diagram showing some of the key features of this species, but, right under the title – *How to Identify Rusty Crayfish* – above the illustration, between brackets, there is the following disclaimer: "Use the following illustration as a guide; some characteristics may be absent." Having seen and collected *O. rusticus* specimens in various streams and ponds in Ontario, I can certainly confirm that some of the characteristics shown in the illustration are indeed either absent or quite different at times, depending on the particular individual crayfish one may be looking at. Furthermore, some of the features listed are either rather broad (for example, "Claws grayish-green to reddish-brown" – that is quite a range of colours, and it may apply to crayfish of several different species), or fairly obscure ("S-shaped, movable claw"). Therefore, actually using this identification card to unambiguously find rusty crayfish may prove quite challenging, particularly for the uninitiated, and especially when one has to make a quick decision in the field, where specimens may be seen in rather murky waters, or covered by mud and debris. In my experience, even the defining pair of rust-coloured spots on the carapace may not always be as clear or obvious as advertised.

Furthermore, the difference between various shades of brown – the most common colour for most crayfish species – and a rusty colour can often be rather subtle. In a recent *Ontario nature guide*, which shows a variety of species of all sorts found in this province, the only crayfish species listed is *O. virilis*, which is described as having "a brown or rusty red" carapace (Kagume, 2008). Since these types of field guides are the ones most often used by members of the public to make sense of the species they encounter in the wild, you can imagine what would happen if we send volunteers out there to target crayfish with a somewhat rusty body colouration. In all likelihood, all sorts of species would be slaughtered, including mostly "native" ones.

As I recall, members of the public who expressed an interest in my fieldwork always referred to all crayfish, regardless of species, as simply "crayfish", or sometimes "crawfish", or even "critters". One landowner expressed intense surprise, and some vague doubts, after I explained that there were several different types of crayfish in the stream adjacent to the property where he had lived for decades. The differences between species such as the large and powerful *Cambarus robustus* and the small *Orconectes propinquus* were obvious to me, but, to most people I discussed this with, a crayfish was a crayfish and interspecific differences seemed unimportant. However, when it comes to differences between much more similar-looking species, such as *O. luteus* and *O. rusticus*, even the experts are evidently having a tough time being sure which is which. Wetzel et al. (2004) stated that these two species "can be separated using mitochondrial gene sequence data", which is fine, but not necessarily all that practical in the field, or for persons without access to specialized labs and equipment.

According to Taylor (2000), the difficult problem of separating the crayfish species within the *Orconectes juvenilis* complex preoccupied experts for more than one hundred years. Predictably, the exact number and identity of the species assigned to this complex, and therefore the boundaries between these species, kept changing over time, as various researchers tried to untangle this taxonomic mystery. This example is particularly relevant for our discussion, because Taylor (2000) assigned *O. rusticus* to this complex of very similar looking and closely related species, which also included *O. juvenilis*, *O. spinosus*, *O. putnami*, and two new species (first described by Taylor) – *O. ronaldi* and *O. cristavarius*. These species are so very similar that, in the past, various combinations of what are now (and for now) regarded as different species (for example, *O. putnami* and *O. spinosus*) were considered as one single species. Even fairly recent and very comprehensive surveys of North American crayfish biodiversity – such as those by Hobbs (1989) and Taylor et al. (1996) – did not distinguish between *O. juvenilis* and *O. rusticus*, and considered *O. juvenilis* specimens to belong to *O. rusticus*, rather than a separate species. However, the updated list of North American crayfishes given in Taylor et al. (2007) included *O. juvenilis* and *O. rusticus*, as well as all the other species listed above for the *O. juvenilis* complex, as distinct entities.

This clearly has conservation implications for at least a couple of reasons. First of all, if *O. juvenilis* is indeed a distinct species, rather than a set of populations of *O. rusticus*, then we have to adjust the range of the rusty crayfish accordingly, yet again, taking one more variable into account (in addition to some of the other ones mentioned previously in this book). Secondly, the differences between these species, as described by Taylor (2000), seem remarkably subtle, even to a crayfish expert. For instance, in the key to the members of the *O. juvenilis* complex – which is based only on Form I, or reproductive form, males – the following distinguishing characteristics between *O. juvenilis* and *O. rusticus* are given: "Pleopods long, central projection at least 45% of total length of pleopod, pleopods usually reaching base of first pair of pereiopods with abdomen flexed forward" (for *O. juvenilis*) and "Pleopods shorter, central projection less than 45% of total length of pleopod, pleopods not reaching base of first pair of pereiopods with abdomen flexed forward" (for *O. rusticus*).

Aside from the fact that even crayfish biologists (also known as astacologists, even when they work with crayfish from genera other than *Astacus*) may have to occasionally review the precise meaning of terms such as "pleopods" and "pereiopods" – terms which would normally draw blank stares, at best, from most members of the general public – we also have to deal with the issue that this key is only relevant if we happen to collect Form I males. In other words, this information becomes irrelevant if we look at adult female or Form II male crayfish, or juveniles, of these species. Also, from a practical standpoint, the pleopods of these usually rather small crayfish are only a few millimeters long. Therefore, trying to measure these structures, particularly in the field, and decide whether the central projections represent 44.99% (in which case we might be looking at *O. rusticus*) or 45% (in which case we may have a specimen of *O. juvenilis*) of the total length of the pleopod could well be too much to ask of anyone. Furthermore, live crayfish generally, and quite understandably, do not like to have their pleopods measured and do not cooperate with the researchers, making the recording of such data even more difficult. Looking at the key above, one may also wonder how far the crayfish abdomen would have to be flexed forward, before a reliable assessment of the extent of the exact reach of the pleopods can be made. And, then, we also have the word "usually" inserted there, which means that sometimes the pleopods of *O. juvenilis* may not reach the base of the first pair of pereiopods, with the abdomen flexed forward, therefore making such specimens indistinguishable from *O. rusticus*, from this point of view.

I will leave it to the readers to decide whether such differences represent a sufficient basis for distinguishing two crayfish species from one another, and, more importantly, whether such differences constitute a feasible and reasonable basis for potentially separating a non-native species from a native one. In other words, why should such tiny differences matter, from a conservation standpoint?

In fact, Taylor (2000) had to utilize multivariate and univariate statistical analyses to compare a few morphological characteristics of the two species and come up with some small differences which presumably may help in distinguishing one species from another in some cases. While this exercise may have scientific value, the findings are unlikely to be of much practical help in the field, particularly for people who may not have the time, the expertise, or the opportunity to measure and statistically analyze minute morphological differences between crayfish found at various locations, before deciding whether these crayfish "belong" there or not (and should therefore be considered either native or non-native). At the very least, since the ranges of these very similar species overlap to some extent, the potential for misidentification of these crayfish in areas where they may co-exist should be considered great.

The way we distinguish such closely related crayfish from each other may influence our interpretation of the conservation challenges faced by other species as well. For example, in a COSEWIC report, the queen snake (*Regina septemvittata*) was declared a threatened species in Ontario, due to its restricted distribution and the fact that its diet seemed to consist almost entirely of newly moulted crayfish. The author of the report assumes that the queen snake may eat mostly *O. propinquus* crayfish in Ontario, and identifies the possible replacement of this species by *O. rusticus* as a potential conservation problem for the snake. However, in the same report, we are informed that queen snakes probably prey primarily on whichever crayfish species happens to be most common in their area. In addition, it seems that in Kentucky 93.2% of the diet of the queen snake consists of *O. juvenilis* crayfish – a species almost identical to *O. rusticus* (Smith, 1999; COSEWIC, 2000). Therefore, if the queen snake can eat *O. juvenilis* in Kentucky, then, in all likelihood, it can also survive in southern Ontario by feasting on the rusty crayfish. So, in this case, the designation of the rusty crayfish as an invasive species may obscure the fact that this crayfish can be a valuable food resource for the rare queen snake, rather than a problem.

Other widely distributed North American crayfish species, such as *Fallicambarus fodiens*, *Cambarus bartonii*, *Cambarus diogenes*, and possibly *Cambarus robustus* may actually represent species complexes as well (Guiașu, 2002a; Thoma et al., 2005; Walls, 2009; Ainscough et al., 2013; Dr. John E. Cooper, North Carolina State Museum of Natural Sciences – personal communication). Careful analysis of these complexes has led to the discovery of new species, with distinct ranges and features. For example, *Cambarus carinirostris* was raised from a subspecies of *C. bartonii* to the level of full species (Thoma and Jezerinac, 1999) and new species such as *Cambarus thomai* (Jezerinac, 1993) and *Cambarus polychromatus* (Thoma et al., 2005) were separated from

C. diogenes after a closer look at the *C. diogenes* complex. Other such species undoubtedly await discovery, and each time we split a species complex into several species (new and old), our understanding of the distributions of these species changes as well. In some cases, the status of species and populations previously considered invasive may have to be changed to native, and our interpretation of the interspecific interactions among these taxa may have to evolve accordingly.

Other Unintended Consequences

Even in cases where there is no possibility of mistaken identity, serious risks associated with control programs – particularly those involving the use of poisons – remain and non-target species can be victimized. In New Zealand, for example, poisons used to eradicate mice also killed the North Island saddleback (*Philesturnus carunculatus rufusater*), a rare native forest bird. The saddleback was reintroduced to an island in New Zealand's Lake Rotorua in 1992. A couple of years prior to this reintroduction, Norway rats, goats, and sheep had been eradicated from this island. In 1996, an aerial poisoning operation was conducted to eliminate the mice found on the island as well. However, although the mouse population suffered great losses as a result of this campaign, it was not eradicated, and the poison killed approximately 45% of the adult saddlebacks on the island (Davidson and Armstrong, 2002). Such control programs are often praised for destroying unwanted invaders without paying much, if any, attention to the unintended consequences. In a book about the natural history of New Zealand, De Roy and Jones (2006) wrote uncritically about these types of controversial eradication efforts conducted by the local Department of Conservation against species seen as invasive. In my review of this book, I pointed out that I found it rather unsettling to read about the "Herculean aerial poisoning operation" – "the largest rodent control programme ever undertaken anywhere in the world", on the same page where the authors refer to the "humane integrity" associated with the conservation initiatives such massive poisoning operations are a part of (Guiașu, 2007b).

In addition, it is sometimes not at all clear if such control programs are actually helpful in achieving the intended conservation goals. The kaki or black stilt (*Himantopus novaezelandiae*) is a critically endangered New Zealand bird, and extraordinary efforts were made to save this species. These efforts included poisoning and trapping mammalian and avian predators, installing electric fences around the nests of the kaki, captive breeding programs, and the incubation of the eggs in artificial conditions. The control programs against the potential predators of the kaki targeted a variety of species such as cats, ferrets, stoats, weasels, hedgehogs, rats, possums, harriers, and gulls. The southern black-backed gulls (*Larus dominicanus*), for instance, were poisoned and their eggs were destroyed. However, a study published in 2002 revealed that, due to methodological problems, it was not clear if these anti-predator control programs, conducted over a period of about 20 years, starting in 1981, were actually useful for the conservation of the kaki. In the late 1970s, the population of the highly endangered kaki had less than ten breeding pairs, and by 2001, after two decades of intensive control programs against the perceived enemies of the kaki, there were only fourteen breeding pairs of this obviously still exceedingly rare bird, and seven of these pairs were kept in captivity (Keedwell et al., 2002). Therefore, it does not seem that any significant progress was made. Presumably, a lot of birds and mammals were killed during these decades of control programs, without any clear understanding if all that killing was actually really useful for the survival of the kaki.

The conservation of very rare species is, in theory, a noble goal. However, from a practical standpoint, one has to wonder for how long such extraordinary conservation efforts can be maintained. In other words, can we afford to maintain electric fences around nest sites, breed and raise birds in captivity, incubate eggs in laboratories, kill numerous individuals of several species, change the habitat, and take other such drastic, intrusive, and highly artificial steps, essentially forever, just so that we can maintain a few token individuals of an evidently vanishing species? And how many such species can we do this for, before the conservation budgets are used up? How do we choose which rare species we should help and which ones we should simply let go? (Yes, I know that, ideally, we should save them all, but, clearly, in a world of finite resources, the more time, effort and money we spend to save one species, the fewer such assets we will have for rescuing other species which may also be in serious trouble.) If a species becomes unable to survive without our constant and perpetual assistance, is it still truly a "wild" species? We often blame introduced species for having a supposedly "unnatural" presence in various ecosystems, because we might have helped them move to new areas, but just how natural is the existence of native species which need to be surrounded by electric fences to survive in the wild, and cannot reproduce or raise their young successfully on their own? And how do we justify decades of control programs, in the absence of conclusive evidence of their usefulness?

Also, while it may be difficult and expensive enough to maintain or introduce a rare species in a small and relatively isolated area, such as a tiny island, the task becomes much harder still, if not impossible, if we want to reintroduce this species to larger regions – for example, on the mainland – where constant care and protection simply cannot be provided. Presumably, rare species are bred in captivity, and fiercely protected in small enclosures, so that they can ultimately be released into larger areas in the wild one day, but in order to achieve that goal, we may first have to "clean up" the ecosystems by destroying quite a few other animals and/or plants which may have moved there in the decades or centuries that passed since the extirpation of the rare species from the region. There is a bitter irony in having to kill previously introduced species, so that we can introduce new ones.

Clearly, many ecosystems throughout the world have been drastically altered over the millennia, and there is no going back. According to Ponting (1993), there are about 70 million sheep and 8 million cattle in New Zealand – far outnumbering the people currently living there – and many of the habitats of this country have been changed dramatically and irreversibly as a result. It is highly unlikely that birds such as the kaki can still survive on their own in large areas of New Zealand, despite all the efforts to "rearrange" nature to suit these very vulnerable species.

It should also be obvious that when deadly poisons are released in a particular region, species other than the targeted ones may likely be killed as well. And the larger the area where the poisoned baits are dropped, the more unintended casualties we can expect. Coeurdassier et al. (2014) reported that poisoning campaigns meant to control water voles (*Arvicola terrestris*) also killed birds of prey such as the endangered red kite (*Milvus milvus*) and the common buzzard (*Buteo buteo*) in parts of France during the winter of 2011. The water vole is the main prey of the red kite in this region, and, therefore, the predators ingest the poison while feeding on the contaminated voles. The red kite is also threatened by poison in Spain. This is of particular concern, since France and Spain are home to more than 20% of the world's breeding population and over half of the global wintering population of the red kite.

Sometimes, anti-exotic species control programs can kill many more native animals than introduced ones. Such an example is presented by Charles Elton himself (1958) in his influential book on invasive species. North American muskrats (*Ondatra zibethicus*) were introduced to England, Scotland, and Ireland in the 1920s for their valuable fur. In the late 1920s and early 1930s, some of these animals escaped from fur farms, or were released when the fur farms were no longer economically viable (Lever, 2009). The muskrats never became very abundant and did not spread very far in the UK (Elton, 1958; Macdonald and Barrett, 1993; Lever, 2009). Elton (1958) estimated that there were never more than a few thousand muskrats in the British Isles. Eradication campaigns against the muskrats started in 1932, and by 1935 these introduced rodents were eliminated from England, Scotland, and Ireland (Lever, 2009). These eradication campaigns were accompanied by a study of muskrats in

Britain. The study was funded by the Ministry of Agriculture and Fisheries and was directed by Charles Elton (Simberloff, 2011). While commenting on the eradication of muskrats in Scotland, Elton (1958) acknowledged that in addition to 945 muskrats, 5,783 native animals (including 2,305 water voles and 2,178 moorhens) were caught in traps and killed. Therefore, according to the numbers given by Elton, this campaign killed roughly six times more native animals than introduced ones. From this point of view, it would be hard to label such a destructive campaign an unqualified success, unless we are completely unconcerned about the extent of the collateral damage. The numbers of native animals which may have died as a result of the anti-muskrat campaigns in England and Ireland were not published (Lever, 2009).

The elimination of introduced species can trigger a chain reaction with unexpected and undesirable ecological consequences. A clear example is shown by the eradication of cats (*Felis catus*) from Macquarie Island – an oceanic island with tundra-like vegetation located in the sub-Antarctic region. Cats and rabbits (*Oryctolagus cuniculus*) had been introduced to this island in the 19th century. As the rabbits became more abundant, they became the main prey of cats. In order to protect the local plants from excessive grazing, the *Myxoma* virus was repeatedly introduced starting in 1978 to control the rabbit population. The virus was effective in reducing the number of rabbits, but, with fewer rabbits around, the cats turned their attention to some of the seabirds on the island. As a result, conservationists started the cat eradication campaign in 1985. This program was expanded in 1998, and the last cat on the island was shot in 2000. In the absence of cats, however, rabbit numbers went up quickly again, despite the previous control programs against them. A new and very expensive eradication campaign, estimated to cost more than AU$24 million, was planned to eliminate rabbits, mice, and rats from the island (Bergstrom et al., 2009). So, here we have an example of a series of eradication and control programs which eventually led to the same problem which existed before any such programs at all were undertaken. This sequence of events does not exactly inspire confidence in the ability of conservationists to manage such programs effectively or foresee the consequences of their actions with anything approaching clarity.

Zavaleta et al. (2001) offered a sobering review of some of the unexpected, and obviously unplanned, negative effects of eradication campaigns against introduced species in many parts of the world. The rather lengthy list of such examples given in the article includes several instances when the removal of exotic herbivores (rabbits in Mauritius, Asiatic water buffalo in Australia, pigs, sheep, and goats in Hawaii, sheep and cattle in the Galápagos Islands, etc.) led to a great expansion of populations of non-native plants, which had been kept in check by these herbivores.

A Tale of Two Owls

In a nutshell, this is a cautionary tale of two North American owl species belonging to the same genus. One is rare and declining in a part of its range, while the other is expanding its range and seems to be doing well. As a result of this range expansion, the two owl species are now coming in contact with one another in areas where they did not meet previously (as far as we know). Even though the impact of these two species on each other is still unclear, control programs have been undertaken against the successful owl species, in order to "help" the more vulnerable congener. And now that the outline of the story has been shown, allow me to add some of the relevant details.

The barred owl (*Strix varia*) has a fairly wide North American distribution. Historically, this species was found mainly in the eastern half of the continent, from Florida, in the south, to Ontario, in the north. *S. varia* has been expanding its range westward dramatically for at least the last 140 years or so. As a result, this owl species was first observed in the western Canadian provinces of Alberta and British Columbia in 1934 and the early 1940s, respectively. In the 1960s, the barred owl crossed the Canadian–US border and entered Washington State. Oregon was reached in 1974 and California in 1981. As far as we can tell, this impressive range expansion was entirely natural, although there is speculation that the barred owl was assisted by factors such as climate warming and some human-generated changes in the central and western North American landscapes. Since the 1960s, the barred owl settled in forest habitats where the spotted owl was already present (Hobbs and Cannings, 2007; Campbell and Kennedy, 2009; Duncan 2013).

The spotted owl (*Strix occidentalis*) is considered to be an endangered species in Canada and the United States. Some biologists divide this species into three distinct populations which may rise to the (admittedly fuzzy) rank of subspecies. The northern spotted owl population can be found in southern British Columbia – the only part of Canada where this species exists – as well as along the western coasts of Washington State, Oregon, and northern California, in the US The California spotted owl occurs mainly in some central and southern areas of that state, although the ranges of the northern and California subspecies are continuous and may overlap slightly in certain regions (the confusing nature of common names is revealed once again here, since there are actually two spotted owl subspecies in

California, although only one bears the name of the state). The Mexican spotted owl is found not only in Mexico, but also in parts of four southwestern US states (Arizona, Colorado, New Mexico, and Utah). The morphological differences between these three subspecies, or separate populations, appear to be subtle, and the Mexican and California spotted owls are apparently pretty much identical from a genetic standpoint, which raises legitimate questions about the validity of the designation of these groups as truly distinct entities. The Mexican populations of this species are considered to be stable, and are therefore of lesser conservation concern (Hobbs and Cannings, 2007; Duncan, 2013).

The division of *S. occidentalis* into three subspecies has conservation implications as well. For instance, if all spotted owls are indeed members of only one species and belong to only one distinct evolutionary lineage, with relatively minor inter-population differences, then should the rarity and decline of the northern populations be of as much concern, and would this justify the large scale killing of barred owls? After all, the Mexican populations of spotted owls are not in trouble, as far as we know, and so the possible disappearance of the species from some areas of Canada and the US, where *S. occidentalis* is now very scarce, may simply be a shift in the overall range of the species, in response to on-going changes in environmental conditions.

In the late 20th century, the spotted owl became a symbol of the preservation of the old-growth forests found in the Pacific Northwest region of North America. This relatively small and rare owl was sometimes at the center of heated debates between environmental groups and loggers (Campbell and Kennedy, 2009), and, as a result, certain environmentalists became quite attached to the species and are determined to preserve it at all costs.

The spotted owl populations of this region continue to be threatened by habitat loss, fragmentation, and alteration. Some conservationists also saw the barred owl as an additional potential menace to the survival and recovery of the northern spotted owl. Although the barred owl was partially blamed for the decline of the spotted owl, the interactions between the two species are not well understood and the ecological consequences of the arrival of the barred owl in western North America are largely unknown (Duncan, 2013). While the addition of the barred owl can be expected to change the ecological interactions in the old-growth forests of the Pacific Northwest, to some extent, it appears that these relatively newly arrived birds generally do not evict spotted owls from their territories (Hobbs and Cannings, 2007). Forsman et al. (2011) assessed the impact of various factors on the well-being of northern spotted owls and could not find a definitive link between the presence of barred owls and the decline of *S. occidentalis*. The authors of this detailed monograph concluded that: “Taken together, results of our current study and previous studies do not prove a causal effect of Barred Owls on the demography of Northern Spotted Owls.”

Interestingly, the spotted owl is not endangered in Mexico, which was the original area where the ranges of spotted and barred owls overlapped naturally and the two species coexisted for the longest period of time (Hobbs and Cannings, 2007; Duncan, 2013). One would think that, if the barred owl has such a devastating effect on spotted owls, the spotted owls should have been in trouble in Mexico first, as a result of the longer term competition with their slightly larger relatives.

Despite the absence of any clear evidence against the barred owl, control programs were initiated against this species, supposedly in order to “save” the spotted owl. So, although we are not sure if the barred owl is really a significant threat to the few remaining northern spotted owls, conservationists decided to kill quite a few barred owls anyway, just to see what would happen. In an insightful but grim article ominously titled *There will be blood*, Cornwall (2014) discusses a six year so-called experiment, recently started by the US Fish and Wildlife Service in small areas of Washington State, Oregon, and northern California. During this “experiment”, thousands of barred owls may be shot, to see if this would help the spotted owl populations recover. The killing has already begun. The cost of this eradication campaign is estimated to be about $4 million. Presumably, at the end of this six year trial period, the killing of barred owls may well continue, perhaps indefinitely, regardless of what will happen to the spotted owl populations. If the spotted owl populations somehow recover, then conservationists can claim the control program is working and should be continued. And if the spotted owl populations do not bounce back, the advocates may well seek to expand the anti-barred owl campaign. In other words, it is hard to see how this “experiment” can turn out well for the barred owls. Apparently, some experts think that about 10,000 barred owls should be killed each year in the Pacific Northwest region of the United States, to assist the hypothetical recovery of the spotted owls. Not surprisingly, this type of plan is very controversial, leading Cornwall (2014) to ask: “What’s the exit strategy? Can we kill 10,000 barred owls every year forever?” The obvious rational answer would seem to be no, in my view, but it is not clear, at least to me, that reason is the main driving force behind such programs. Wildlife biologist Eric Forsman, the lead author of the previously cited study on the population demography of northern spotted owls, apparently agrees that this initiative is unsustainable, citing the prohibitive costs of running such a control program in the long run (Cornwall, 2014). In the meantime, however, this gruesome conservation “experiment” continues.

In a comprehensive book on North American owls, published in 2013, Dr. James Duncan – an award-winning owl expert and conservationist – stated that "The culling of Barred Owls to conserve the endangered Spotted Owl is controversial because only a little is known about how they interact." I would add that this control program is also problematic because it involves the destruction of relatively large numbers of birds essentially for the "crimes" of having expanded their range and settled down successfully in a new habitat. The killing is also done on an "experimental" basis, just to see what happens, and the conservation goals pursued in this violent and destructive manner appear to be poorly defined.

The spotted owl (*Strix occidentalis*) and the barred owl (*Strix varia*) are two rather similar looking species. According to Hobbs and Cannings (2007), the barred owl is the "slightly larger and paler" of the two species, which otherwise resemble each other closely. This, of course, once again raises the prospect of mistaken identity during the ruthless anti-barred owl control programs. If these two species are considered very similar even by the experts, how can we be sure that mistakes will not be made in the field, when hunters are trying to eliminate barred owls from a distance, while possibly viewing the moving birds in dense forests, through the potentially narrow spaces between the trees and their branches, under less than ideal lighting conditions? There is always the possibility that some of the few remaining spotted owls may be inadvertently taken out, together with the targeted barred ones.

Ironically, these two owl species are so similar, and so closely related, that hybridization between them is possible, and has occurred in the wild at least a few times. The offspring – named sparred owls – resulting from this hybridization process are fertile (Duncan, 2013). Such hybrids, which were first identified in 1986 in Washington State, are apparently able and willing to mate with owls of both parental species. Perhaps not surprisingly, this makes it even tougher to separate hybrids from so-called "pure" spotted or barred owl species (Hobbs and Cannings, 2007).

As discussed in the relevant section on this topic, earlier in the book, conservation and invasion biologists generally do not like hybridization, particularly if one of the parental species is a rare native and the other species is a more common "invader". It could be argued that this is a "glass half empty" way of looking at things. Instead of possibly celebrating the fact that at least some spotted owl genes may be passed on to future generations of owls, through interbreeding between these two similar owl species, conservationists choose to regard this type of hybridization as, in the words of Hobbs and Cannings (2007), an "insidious threat" to the survival of the spotted owl. As a result, rather than showing gratitude towards the currently more successful barred owls for helping to spread and preserve certain spotted owl genes, such conservationists actually blame the barred owl for this interspecific mating process, even though, presumably, both spotted and barred owls freely choose to interbreed in the wild, from time to time. Again, the very fact that these two species can form interspecific pairs, mate successfully with each other, and produce fertile offspring illustrates that the boundaries between the spotted owl and barred owl lineages are not all that clear. This makes it even harder to distinguish between the two species, both from a practical and philosophical standpoint, and even more difficult to understand why it is so important for certain people to destroy one of these owls while at the same time going to great lengths to save the other.

Some of the control programs discussed in this chapter reveal a strange dichotomy in our attitudes towards other species. It seems that we can show extraordinary compassion for a few individuals of certain (usually rare, cuddly, or cute) species, while at the same time exhibiting a complete lack of concern and empathy for the many more individuals of other species which are targeted by control programs.

It may be worth mentioning, in passing, that the number of authors (27) of a study on the population dynamics of northern spotted owls (Forsman et al., 2011) is greater than the actual number of spotted owls still left in south-western British Columbia (there are now fewer than 25 such owls there according to Campbell and Kennedy, 2009). While I have great sympathy for the plight of spotted owls in the northern parts of their range, and I hope they can continue to exist there for as long as possible, one has to ask if the tiny numbers of spotted owls currently found in these regions still have any sort of significant ecological role to play in their ecosystems. In other words, is the northern spotted owl more of a symbol now, rather than an integral part of the old-growth forest ecosystem in the Pacific Northwest? And is it worth killing numerous individuals of other species just to preserve a symbol with a negligible real-life ecological role? Does it make sense to destroy many owls of one type just to try to "save" a few owls of another kind, particularly when we don't even know exactly how the two species interact and influence each other? How far are conservationists willing to go to protect the last few individuals of an extremely rare species? And how far should we be allowed to go in the pursuit of such seemingly quixotic goals? At what point does the initially noble desire to protect an endangered species turn into an unhealthy obsession, which can be damaging to other species? Please note that I am not arguing against the preservation of these magnificent forests, of course. Far from it. But the old-growth forests are home to many species. And all these species – not just the spotted owl – deserve respect and consideration, including the more recently arrived ones.

The persecuted barred owl (*Strix varia*) needs large areas covered with old-growth forests since this species tends to nest in large trees found deep inside the woods. Therefore, it should come as no surprise that this owl will take advantage of the availability of relatively undisturbed forest habitats. Presumably, the conservation of the spotted owl along the western coast of North America was closely connected with, and symbolic of, the preservation of the old-growth forests. However, if we preserve such valuable habitat, we have to accept that other species, such as the barred owl, may also choose to make their home there. Surely, the fight to save these complex forest ecosystems was not just so that we can have a few spotted owls living there, while artificially and brutally excluding all potential competitors, predators, or parasites, as well as any newly arrived species. For if the main goal is to keep spotted owls around at all costs, by any means necessary, it may be easier to just move them to a zoological park, or an aviary of some sort, where we can guard them around the clock, and make sure that no other birds or mammals are allowed to interfere with the species we chose to aggressively favour over other similar ones.

In fact, the barred owl has been a victim of deforestation as well in regions such as southern Ontario, where the species appears to have become considerably less common overall since woodlands and swamps were replaced by farms, especially during the first half of the 20th century (Bezener, 2000; Allair, 2007; Sandilands, 2010). Why should we begrudge the ability of this species to expand its range westward naturally and take advantage of suitable forest habitats available there? The range of the barred owl is clearly evolving over time, in response to the changing environmental conditions. Within Canada, the species is moving out of southern Ontario and finding refuge, on its own, in the old-growth forests of southern British Columbia. This should not be considered a conservation problem unless we think perfectly normal and gradual modifications in the distributions of species should be outlawed. Instead of initiating control programs against newcomers, we could choose to be happy that new species are moving in, thereby taking advantage of the available preserved forest habitat. If you set aside these habitats, such species may well come, settle down, thrive, and ignore the "This forest is exclusively for the use of a handful of spotted owls" signs.

"Frankenfish" Alert

When the previously mentioned northern snakehead (*Channa argus*) – a fish of Asian origin called "Frankenfish" shortly after its introduction into North America – was found in a Maryland pond, local state biologists and the media reacted strongly to the new threat. Chew and Laubichler (2003) describe the over the top subsequent campaign against this species, which included the printing of anti-snakehead posters with the unambiguous message: "Kill this fish." Apparently, a local biologist said, referring to this introduced fish: "If you catch it, kill it. It's not a dead or alive thing, we want it dead." In the end, herbicides and rotenone were poured into the Maryland pond where the snakeheads were discovered, resulting in the death of all the fish found there, including many fish of other species. Clearly, this radical solution to the perceived snakehead problem was much more destructive to the biodiversity of that pond than the snakeheads could ever be. Despite the dire warnings and the fact that the snakeheads lived in the pond for about two years, it turned out that thousands of fish of other species as well as other aquatic organisms somehow managed to coexist with the snakeheads during that period of time. After the control program against the snakeheads was completed, the biodiversity of the pond was close to zero. Furthermore, northern snakeheads appear to be rather similar, from an ecological standpoint, to the bowfin (*Amia calva*) – an ancient predatory fish native to North America and present in Maryland freshwater ecosystems. Presumably, local species adapted to coexistence with bowfins may also adjust to the arrival of the ecologically similar snakeheads (Chew and Laubichler, 2003).

It turns out that the introduced northern snakehead and the native bowfin look so very similar that even experienced anglers have a tough time telling them apart. A few years ago, I read a story in a southern Ontario newspaper about a local angler who thought she might have caught an invasive northern snakehead in the Welland River. The fish quickly gained regional notoriety and was discussed in television news reports. Other experienced fishermen who were shown photographs of the fish were not sure which species it belonged to either and thought the fish was unusual looking, and therefore possibly exotic. Despite the fact that experts from Fisheries and Oceans Canada advised her that the fish she had caught was, in fact, a native bowfin, the angler remained unconvinced, and eventually killed the fish, just in case it might have escaped from the backyard pond where she had placed it (Benner, 2010).

Deliberate species introductions can have unpredictable consequences and should never be attempted casually. But the same is true of control programs against targeted species. Neither should be undertaken without the benefit of conclusive scientific evidence and careful consideration. Often, perhaps the best thing we can do for many wilderness areas, and the species they shelter, is to leave them alone, as much as possible, and allow them to continue to exist and evolve on their own (Guiașu, 2008a).

CHAPTER 8

Some Conclusions and Some Questions

> You know why there are so many whitefish in the Yellowstone River? Because the Fish and Game people have never done anything to help them.
>
> RUSSELL CHATHAM

> All animals are equal – but some animals are more equal than others.
>
> GEORGE ORWELL

> I am at two with nature.
>
> WOODY ALLEN

> Never mistake activity for achievement.
>
> JOHN WOODEN (legendary US college basketball coach)

Academic life, even for a biologist occasionally pursuing field research, provides relatively few opportunities for sustained exercise, particularly during the school year. Therefore, in order to engage even in a token amount of much needed regular outdoor physical activity, I take fairly frequent walks around an urban woodlot in my neighbourhood. It is not a very big woodlot, and it is surrounded, in the immediate vicinity, by a couple of multi-lane paved roads, a shopping mall, and a public library, as well as being located in the middle of a rather densely populated residential area. Once every year or two, it seems, workers from the local municipal Parks and Recreation department come by to cut down some trees and shrubs, particularly around the edges of the woodlot. As a biologist and resident of the area, I found these activities to be quite destructive and unnecessary, since the trees and shrubs that were cut were often not all that close to any walkways, and did not appear to pose a risk to the public. Furthermore, it seemed to me that the destruction of these plants was somewhat selective, and that certain species – generally the ones that might have been perceived by some as non-native – were particularly targeted. For some time – several years, perhaps – I did not make any enquiries about this, but, in the late fall or early winter of 2012, a particularly large section of the forest was deliberately cut down (by my quick count at the time, hundreds of trees and shrubs were removed in just a few days, sometimes even quite far from the edges), and this finally prompted me to contact the local Parks and Recreation

 | DOI 10.1163/9789047426134_010

department and ask, politely, why the destruction of all these trees was considered necessary.

After a few weeks, I did receive a fairly detailed e-mail reply to my earlier message from the Forestry Supervisor of the region, who informed me that, as I had already suspected, only invasive species were targeted for removal in order to promote native trees and shrubs. While I appreciated the fact that the Forestry Supervisor took the time to respond comprehensively to my message and I respect his expertise in forestry matters, I find the very concept of keeping thoroughly urban woodlots free of introduced plant species quite absurd. First of all, as far as I could tell, not all the trees removed belonged to introduced species. As the trees were cut, a lot of saplings belonging to both introduced and native trees must have been trampled, as well (as mentioned, this was done in the winter, when the saplings and young trees had no leaves and appeared essentially like very similar-looking sticks in the ground). Secondly, and not surprisingly, a large percentage of trees in urban forests belong to non-native species, and if we were to somehow remove them all, we would lose a large percentage of our city trees and greatly reduce the diversity of the urban plant life in the process. This is especially true given the fact that many types of trees and other plants can only grow near the edges of woodlots, and could not survive in the forest interior, due to insufficient light and other unfavourable environmental factors.

This woodlot is located in close proximity to many private backyards and gardens (where all sorts of plants grow, obviously), and is also surrounded by non-native city trees, such as Norway maples, deliberately planted along the streets. Is it any wonder that some of these plants would spread their seeds to the urban forests? Among the trees removed from the woodlot a few years ago, in a previous round of deliberate mini-deforestation, was the only (as far as I could tell) Siberian olive found there. This tree was bothering no one and provided additional food and other resources for some local birds. It seemed obvious that this particular tree species had absolutely no chance of "taking over" the maple-dominated woodlot. The way this particular tree reached the woodlot seemed quite clear, and perfectly natural, as well. About one hundred meters or so from the woodlot, beside a soccer field, there are two city-planted exotic Siberian or Russian olive trees (these trees may well have been planted by the same municipal department responsible for the removal of the trees from the forest nearby). It is very likely that a bird dropped a seed from one of these trees onto the edge of the woodlot, not far away, and this is how the only young Siberian olive which sprouted there arrived at this location. In other words, the tree was growing there due to a perfectly natural short-range dispersal event. But that didn't seem to matter in the end, and quite a few other trees, of various species, were removed as well.

FIGURE 32 *The distinctive silvery leaves of the Russian olive* (Elaeagnus angustifolia) *tree*
PHOTO BY RADU GUIAȘU

FIGURE 33 *Warning sign in the Glendon Forest. Although control programs against invasive plants have been contemplated at this location as well, the Glendon Forest is not the one referred to in this section*
PHOTO BY ILEANA GUIAȘU

Russian olive trees have spread in riparian areas of the western US, particularly in the southwest. The ecological impact of this species in its introduced range is apparently poorly known. However, bird species such as the mourning dove (*Zenaida macroura*) and the yellow-breasted chat (*Icteria virens*) prefer to nest in Russian olive trees, rather than native trees, in New Mexico. In fact, even though the Russian olive was relatively rare in the riparian vegetation of the region of New Mexico where the study was conducted, more than one third of the local bird species which built nests below 8 meters chose this tree as a nesting site. It is possible that the spines found on the branches of these introduced trees may provide better protection for the nests. In addition, the fruits of the Russian olive are eaten by certain birds and mammals (Stoleson and Finch, 2001).

The Forestry Supervisor wrote in his message that there had been a significant decline in the woodlot's species diversity because of intense competition from invasive plant species such as common buckthorn (*Rhamnus cathartica*), Russian olive (*Elaeagnus angustifolia*), and Manitoba maple (*Acer negundo*). However, as mentioned, I had only seen one Russian olive tree in the entire woodlot, and that small tree had been removed prior to the latest round of deforestation. Russian olive trees, with their silvery leaves and rough bark, are very distinctive, and really stand out from the rest of the local trees. They also tend to grow in open areas, where sufficient sunlight is available. In other words, these introduced trees would be hard to miss, and I really looked for them carefully. Therefore, as far as I could tell, there were no Russian olive trees in the woodlot at the time the tree-cutting was undertaken. And, in any case, one or two Russian olive trees obviously could not have provided "intense competition" for the other trees in the forest, native or not. Manitoba maples also appeared to be uncommon, or even rare, in the woodlot, according to my observations, even before the "invasive" trees were removed, and common, or European, buckthorn shrubs only grow around the edges of the forest, where, among other things, they form a protective barrier and prevent dogs and people from entering the woodlot and trampling on saplings or pursuing wildlife.

So, in this case, the destruction of many urban trees was, at least in part, another undeniable effect of invasion biology. Perhaps as many as several hundred healthy local trees and shrubs became the latest casualties of the war on "invasive" species. And this is not an isolated incident.

Project Crayfish

A few years ago, I was asked to participate in Project Crayfish – an initiative which brought together a number of local crayfish experts and resulted in the production of an identification guide for Ontario crayfishes (this was also discussed in a study by Hamr, 2010). Approximately one year later, the organizers of that project also invited me, and the other Project Crayfish members, to participate in what was described as a large (possibly province-wide) outreach event scheduled for May 22nd, 2009 and meant to celebrate World Biodiversity Day. The invitation came from the Curator of Conservation Programs at the Toronto Zoo, and the event was organized, as far as I understood, by the Biodiversity Education and Awareness Network (or BEAN), which was chaired by an official from the Ministry of Natural Resources. All this seemed very nice, until I read the very short list containing the two activities planned for this day: 1. Garlic mustard pull; and 2. Rusty crayfish survey. Apparently, the theme for that year's World Biodiversity Day was "Invasive Species". I politely declined to participate, since it seemed to me that the best way of celebrating local biodiversity and educating the public about this was not the massive removal of plants and a "survey" of a local crayfish species suspected of being introduced. Why pick the rusty crayfish in particular, out of all the local crayfish species, unless the preliminary survey was designed to alert the public to the presence of this supposedly undesirable species, and set the stage, potentially, for future control programs?

Some may find this unfair, perhaps, but this is also very much a part of Charles Elton's legacy. I am often puzzled by people who claim that their direct involvement in the removal and destruction of various animals and plants is due to a love of nature and a sense of duty for protecting biodiversity. In fact, what many of the participants in such activities are evidently doing is destroying a part of the biodiversity which surrounds them – the very biodiversity they claim to worship – frequently without clear and well-articulated reasons, or at least without reasons sufficiently supported by evidence.

Do we know, and are we even interested in finding out, if perhaps more harm than good is being done, overall, by these sorts of unfortunate control programs and initiatives? Is anybody keeping track, in a rigorous and systematic way, of how many harmless, or even beneficial, animals and plants were misguidedly "controlled" over the years because they were thought to belong to species with unacceptable geographic origins?

The Value of Informed Dissent

When I expressed concerns about the arbitrary nature of some key concepts in invasion biology, such as the very definitions of "native" or "introduced" species, or the exact limits of a "native range", various invasion biologists replied that these were old and settled debates, and that a lot had already been written about these topics, and the field has moved past those discussions.

Well, actually, those debates are not that old – they couldn't be, since the field of invasion biology itself is very young – and they are certainly not settled. In fact, they may be intensifying. Furthermore, we still don't have a clear way of objectively separating native from introduced species in many cases, or a good grasp of the limits of native ranges and the ways in which many organisms disperse themselves around the world, and the gulf between the propaganda against species perceived as exotic outsiders and the scientific information available about these species seems to be growing. Bottom line: there are still control programs undertaken against many non-native species, for reasons which do not appear to be well supported by science. As long as that is, or will be, the case, we need to have these discussions about the boundaries, labels, and control programs imposed and justified by invasion biology. Therefore, even though some invasion biologists may regard such debates as unnecessary and inconvenient, raising these issues remains important and relevant.

Due to the dynamic nature of the living world, preserving and restoring an ecosystem can never be the same as preserving and restoring an old painting or an ancient architectural landmark. By making sure nothing ever changes, or developing a fixation with restoring species associations and habitats to the way they might have been at a particular point in time, invasion biologists essentially stand in the way of evolution, and seem strangely oblivious to the ever changing environmental conditions. We might as well put a giant glass dome over entire ecosystems, and constantly monitor factors such as humidity, temperature, and the speed of air currents, to make sure they all fall within a preset acceptable narrow range, and not allow any species to come in or go out. This is clearly not feasible and highly intrusive, and, in the final analysis, it may be more about giving some biologists and conservationists something to do, rather than saving the planet.

After reading this book, some invasion biologists may accuse me of "cherry picking", or selecting only the most favourable examples and studies to make my case. That type of accusation can be levelled at pretty much any argument built on any number of supporting sources of evidence. As I stated in the Introduction, in this book I am presenting a particular case, based on my relevant research and experiences. One can always say: "Ah, but you missed that reference", or "You didn't pay sufficient attention to that study", and so on. However, as can be seen, hundreds of references were cited in this work, and many more were read, but could not be added, due to lack of time and space. Quite a few studies, including some which presented points of view very different from my own, were analyzed and discussed in detail in these pages. In other words, these articles were read carefully and not just mentioned in passing to artificially enlarge the literature cited section. The shelves of my offices, both at home and at the university, are full of voluminous folders, with numerous articles, notes, and reports related to this project. One of the toughest challenges in writing this book was finally deciding to wrap things up. Invasion biology is now a fashionable field, and new studies or opinion pieces are published all the time, these days. It was very tempting to keep updating this book by adding another bit of fresh information here, and another new development there. But, at some point, every project must reach an end, and every author has to have the discipline to put the final period in a particular text. I am sure there will be many more intriguing articles published while this book is being prepared for publication. Some of them may support my ideas and some of them may not. However, the many studies cited here which challenge the "conventional wisdom" associated with invasion biology and alien species cannot be dismissed as the result of mere "cherry picking", although I am sure some people might try to do just that. There is a pretty substantial bowl of cherries on display... When it comes to the anti-introduced species campaigns, there are plenty of legitimate scientific and ethical questions yet to be properly answered. These concerns cannot be simply swept under the rug and ignored, if genuine progress is to be made.

Furthermore, it can be argued that the propaganda generated by invasion biology and the anti-exotic species attitudes promoted by some in this field are very much based on "cherry picking". Only the worst examples of invasive species are always brought forward, and always presented in the worst possible light, often without nuances or full context, as an indication of what can happen any time any species is introduced anywhere.

Since we are now approaching the end of the book, and it is time to try to sum up, let me add another disclaimer. In this book, I do not mean to suggest that we should always take a passive approach, and do nothing, while facing genuine, thoroughly investigated and well-documented threats from other species (native or introduced). Obviously, if there is an epidemic which threatens human health, or if the spread of a particular species has a demonstrably massive negative effect on our well-being and economic interests, then of course we should take action to protect ourselves, as long as such action is really required and can genuinely help us improve the situation. After all, no matter how open-minded, sophisticated, and philosophical we may try to be about such issues, and no matter how much we might try to pretend otherwise, we will always look at the world through a human-centered perspective and we will always try to protect our interests first and foremost. It would be impossible for us to do otherwise.

However, we can also try to live up to our flattering species name, *Homo sapiens*, at least once in a while, by finding imaginative ways of safeguarding our truly vital interests while at the same time also attempting to coexist with as many other species (native or not) as possible. Often, this may not even involve a lot of creativity and effort – just more tolerance and understanding. Clearly, out of sheer self-interest, if nothing else, when analyzing and addressing perceived threats, we should try to develop and apply rational and objective standards, as much as possible, and avoid overreactions and distractions. Given the limited resources, tightening budgets, and growing lists of conservation and environmental protection-related priorities, we have to be selective, put things in a proper and broader context, and make intelligent choices about which battles we should fight and which matters we would be wise to leave alone. We simply cannot control and monitor every species whose range expansion may have been assisted by us at some point. That would be obviously absurd, thoroughly impractical, and probably damaging to the environment, in both the short and the longer term. Invasion biology should not provide an all too convenient and all-encompassing excuse for the creation of perpetually expanding, and often arbitrary, lists of "enemy" species, on which every conservation biologist or wildlife manager, in every jurisdiction, can place all the species he or she may consider as undesirable "outsiders", for any number of reasons. That would also lead to heated and often pointless debates about whose invasive species problem is worse, and therefore deserves more attention and funding. In the increasingly tough competition for funds and headlines, the propaganda can only become more and more intense and farther removed from the facts.

In the last few years, I have become increasingly concerned about the inclusion of more and more species on the list of unwanted "invasives" (a sort of most unwanted list, so to speak). To some, this is clear evidence that invasive species are a growing problem. Perhaps that may be true. However, it is also quite possible that the rapidly increasing numbers of so-called invasive species we now consider as our enemies are a direct result of the fact that there are now more and more invasion biologists and their disciples diligently looking for invaders behind every bush and under every stone, and trying very, very hard to add new species to this infamous list. This is a particular mind set, and if you look hard enough for exotic species and reasons to dislike them, then, at some point, you may convince yourself that you found them, and the quest has not been in vain. Besides, bringing a new villainous alien species to the public's attention can be quite a feather in a conservationist's cap as well as a way to maintain the siege mentality invasion biology thrives on.

Furthermore, with the previously discussed vague and unreliable definitions and standards in this field, almost any species which is expanding its range and its numbers, at any location, is at serious risk of being declared invasive, particularly in the current anti-introduced species climate. This is a convenient label, if some biologists, or even members of the public, want to settle scores with particular species they may not like, for whatever reason. Pretty soon, if we are not careful, quite a bit wiser, and considerably more selective, we will end up declaring most successful species as our enemies. After all, how dare these species thrive within and around human settlements, even though we did nothing to consciously encourage them or commercialize them (the way we do with, say, pets, house plants, or livestock). And, as we have seen, once a species is identified as "invasive", then all the normal activities of the individuals of that species, and even the very existence of these individuals, can be used against them.

At the same time, in a bizarre twist, the costs of fighting against these exotic species (which may well be unnecessary, since wars against invasive species are often wars of choice which do nothing to "improve" the environment or the human condition, overall) are then added to the reasons to fight the species in the first place... We have to spray toxic herbicides to get rid of otherwise harmless dandelions, the circular argument goes, because the control programs against dandelions cost us a lot of money, and therefore dandelions are an expensive nuisance. Thus, in a classic example of misdirected blame, we hold the dandelions responsible for our own misguided decision to spray them to death with plant poisons. It's really not all that complicated: if you don't want to spend the money and energy required to eradicate dandelions, and other so-called "weeds", from your lawn, then just don't do it. The sky will not fall, no one will be hurt, you will save some money and effort, and fewer herbicides and pesticides will get into the soil and the groundwater. Otherwise, if you really want a perfectly obedient lawn, where nothing ever changes or looks "out of place", then simply replace all plant species (which can be "messy" and unpredictable) with some type of artificial green carpet, which can be repainted every few years to keep the colour vivid, and you'll never have to lose sleep about the changing biodiversity in your yard again. An alternative, and very effective, method of reducing local biodiversity was employed by several of the neighbours on my street, who have either largely or entirely replaced the already fairly small green spaces in front of their homes with expanded paved driveways. This way, driveways initially designed for one car can now accommodate two, or even three, vehicles (if one gets creative with the parking methods), and the street is beginning to resemble a giant parking lot, where there are fewer and fewer plant species (native or introduced), and more and more cars and motor oil stains.

Every year, wherever I look, green spaces are dwindling and paved surfaces are expanding. Perhaps that should be a greater concern than the initial geographic origin of the plant species still found in our harsh urban landscapes.

As mentioned earlier, while discussing the suitability of the common carp for local freshwater ecosystems, summer temperatures in the Toronto region can frequently top 30 degrees Celsius (often feeling much warmer with all the unpleasantly high humidity), while in the winter temperatures of –20 degrees Celsius are certainly possible (sometimes feeling even colder with the wind chill factor). Adding to these great temperature fluctuations between seasons, we have factors such as the smog days of summer and the road salt of winter – no wonder paved roads have to be frequently repaired and resurfaced around here. Under these circumstances, again, I feel we should be grateful that any plant species can cope with life in this city (and other large urban areas around the world), rather than spending a lot of time and effort checking the "passports" of our fellow urban species, and trying to weed out the "intruders". For many people living in cities these much maligned but very resilient non-native plants offer a much more vital and immediate connection to the natural world than any distant wilderness areas most of us will never visit.

Coming back, briefly, to the funds spent on propaganda and control programs against invasive species, some invasion biologists will undoubtedly say that this is money well spent (particularly since some of this money is used to fund and assist their careers and projects) and that we cannot afford not to spend it. That may be true in some cases, but do we really know this to be a fact for all introduced species that are currently being demonized? It seems to me that you can't just cavalierly declare some species guilty from the very beginning, without evidence, simply because they are new to an area, and then decide that control programs are therefore automatically justified. This type of "Let's take action now, before the facts become available!" mentality has no place in science, and is not indicative of rational thought or analysis. Surely, the evidence should inform the decisions, and that should happen before the allocation of funds for propaganda and eradication purposes.

Writer Emma Marris reported in the November 17th, 2005 issue of the reputable journal *Nature* that in the USA the federal government was spending more than $1 billion per year to control invasive species. That is obviously a huge expense, and it is perhaps not unreasonable to ask if the substantial funds used for all these anti-invasive species programs and initiatives are always spent as wisely as possible.

Control programs unsupported by solid evidence can be justifiably regarded as little more than vanity projects for certain conservation biologists and organizations. In such cases, we may well ask if we can, or should, afford such projects, and all the destruction they inevitably lead to.

In a previously discussed study, Colautti et al. (2006) recognized that a lack of relevant data, as well as subjective interpretations of concepts such as "ecosystem value", made it difficult to estimate the costs associated with non-native species in Canada. The authors tried to fill the gaps in the data sets with an empirical model, which, in their own words, was "admittedly crude", and had a number of significant limitations. In the Discussion section of their article, the authors stated that the pervasive lack of data made the task of estimating the economic impact of non-native species in Canada impossible. However, this did not prevent Colautti et al. (2006) from finding that the "characterized costs" linked to a number of non-native species in Canada added up to $187 million (in Canadian currency) per year, and that these costs only represented a paltry 1% of the "actual costs" of dealing with such species. After reading the article, I was left wondering how they could be so sure, and what the basis for their confidence in these estimates could possibly be. Too much essential information was simply unavailable. The only logical conclusion I could come up with was that we don't really know the overall net costs associated with exotic species.

I will admit that, as a biologist, I am not in a position to independently assess some of the costs cited in relation to the effects of certain invasive species. On the other hand, I will also point out the obvious fact that most biologists are not economists and most economists are not biologists, so I would certainly like to take a closer look at some of the published estimates together with independent experts who can analyze these costs carefully, item by item, and help me understand the totals and place them in the proper context. For example, how many of these costs are hypothetical, and, for comparative purposes, what are the costs incurred due to coexistence with native species? As a general comment, I would assume that living with, and within, nature would always come with some inevitable financial costs, as well as many benefits of various kinds, regardless of the origins of the species that surround us. Do we bother to quantify the value of all the ecosystem services provided by introduced species, and subtract these benefits from the perceived costs?

A More Positive Approach to Non-native Species

This brings me to another suggestion: what if we spent equal time, effort, and energy focusing on the positive effects of introduced species and the value of so-called novel ecosystems? What if we developed and funded research programs centered on analyzing the good contributions of introduced species to their new environments and the well-being of other species (including our own). We could give this new field a nicer name, as well, such as inclusive ecology. I am convinced this change of emphasis would produce an impressive literature and make many expensive control programs not only redundant, but also impossible to contemplate in the future. And, as an additional benefit, we would be spared from further exposure to the shrill and unpleasant propaganda often associated with invasion biology. That alone may be worth celebrating and striving for.

In my experience, many people outside the invasion biology field – including academic researchers working in other areas of biology and ecology or different disciplines, students, as well as members of the public – tend to be fairly receptive and open to more tolerant ideas about introduced species and their roles in new ecosystems. In April 2013, I organized a session titled *Historical, sociological and biological aspects of invasion biology* at the Annual Conference of the American Society for Environmental History, which was held in Toronto. I presented a paper titled *Charles Elton and the origins of the endless war on invasive species*, which included some of the ideas discussed in the previous chapter. The session also featured presentations by Jane Duggan – an excellent student conducting research under my supervision at Glendon College at the time, and two leading researchers in the field, Mark Davis and Brendon Larson, whose work was cited in this book. Overall, I think we presented a more balanced and open-minded view of non-native species, and our talks were well-received by the audience of environmental historians. We had an interesting panel discussion afterwards, and we received good, well-meaning questions and kind comments. However, as one of my colleagues pointed out after the session, if we had given the same presentations at an invasion biology conference, the reception would likely have been considerably less positive, to put it mildly…

As biologists, I would argue that we have a responsibility to enhance the public's appreciation and understanding of nature in general – not just the "native" components of the natural world, but the entire living world, in all of its complexity and wonder. In my view, it is ultimately wrong-headed and counterproductive to teach people that some species are bad, and don't belong, simply because they may have originated in another part of the world long ago. Many of these so-called exotic species have lived in, and contributed to, their new environments for hundreds or even thousands of years, and, for the most part, it would be absurd and potentially damaging to the environment to try to remove them now. We really don't need to encourage the human inclination to fight against nature any further. Invasion biologists do not work in a vacuum, and sometimes, the temptation to promote one's work has to be tempered by the awareness that if you give an interview calling for the control of certain species, some people may actually listen, and take matters into their own hands, even though they may know very little about the biology of the targeted species. In my experience, many members of the public who are willing to participate enthusiastically in control programs against introduced animals and plants do not know anything about these species (other than the fact that they are from elsewhere, and therefore they must be bad), and cannot even reliably tell these species apart from other similar-looking ones (some of which may be native).

In their review of the positive impacts of non-native species, which provided a refreshingly different perspective in a field which seems dominated by negativity towards these species, Schlaepfer et al. (2011) mentioned that introduced species may be particularly useful in rapidly changing ecosystems, because of their superior ability to survive in, and adjust to, difficult and unpredictable conditions. The authors also concluded, quite optimistically, that the general attitudes towards an increasing number of non-native species will gradually improve, as more and more people will recognize the useful contributions brought by these species to various ecosystems.

Controversies and Debates

Perhaps predictably, in a paper published in 2003, Daniel Simberloff rejected the accusation that xenophobia may have played any part in shaping the attitude of some invasion biologists towards non-native species. As a brief aside, I found Simberloff's 2003 article to be rather confusing, since, in the Conclusion section, for instance, he first argued that invasion biologists are only against some introduced species (the ones that may cause harm), and have nothing against the other non-native species, but then, in the final paragraph, he mentioned that it is difficult to predict in advance which introduced species will become harmful, so, when it comes to species introductions, we should switch from the "innocent until proven guilty" philosophy to the "guilty until proven innocent" mentality. Since it is also difficult to predict how long after its introduction a non-native species may become harmful (according to some), the previously discussed "guilty until proven innocent" philosophy (as Simberloff calls it) brings us back to the inevitable conclusion that invasion biologists are indeed, for the most part, suspicious of, and hostile towards, pretty much all introduced species. After all, one cannot predict exactly when, or even if, some of these species may become harmful in the future, so one has to remain vigilant in all cases. Yet another one of those classic circular arguments in invasion biology... No, of course we don't hate all introduced species – we just dislike the bad ones, but, come to think of it, all of these species may become bad one day, so let us treat them all as such... Thank you for the clarification.

Furthermore, the "guilty until proven innocent" philosophy conveniently absolves invasion biologists from the obligation of actually collecting data and building a solid case against introduced species, before advocating their control or eradication. How can you really find out which non-native species may be "bad" and which ones may not cause any trouble, if you consider all of them to be "guilty" from the very beginning? And, as we have seen throughout the book, definitions and standards in this field can be quite subjective and unreliable. As a result, any invasion biologist or local wildlife manager can pick a fight pretty much with any introduced species, at almost any time. Also, just because a particular introduced species may have caused some problems in, say, Hawaii, or Florida, this doesn't necessarily mean that the same species will have the same (or any) negative effects in Ontario, for example. Sometimes, it seems that if a non-native species gets a bad reputation in one part of the world, this reputation will follow the species everywhere, including to other regions where it may not play a harmful role.

For instance, in an article about the exotic species of Lake Champlain, which stretches from southern Québec to the US states of New York and Vermont, Marsden and Hauser (2009) list these species and place an asterisk beside the names of species which, in the authors' words, "are nuisances in Lake Champlain or elsewhere." As a result of this overly broad labelling criterion, the common carp and the rusty crayfish had an asterisk placed beside their names, even though, according to the authors, neither species is abundant or known to cause problems in Lake Champlain.

Before taking action against an exotic animal or plant species in Ontario, for instance, one should assess what role the species actually plays in local Ontario ecosystems, obviously, rather than rely solely on claims based on studies which may have been conducted in very different faraway locations and ecosystems. Frankly, to me, it seems absurd to remove a lone, small Russian olive tree from a woodlot found within a large urban area in southern Ontario, possibly because of some reports or claims that this tree species may do well in certain wet meadows, and along the banks of streams and shores of lakes, particularly in parts of the western US We are obviously talking about completely different types of ecosystems and very different and distant parts of North America – in other words, apples and oranges, so to speak – and the chances of having Russian olive trees take over the local urban Ontario woodlot, or any other woodlot in this area, seem very, very close to zero. Therefore, allowing one or two such trees to grow naturally on the edges of the woodlot would not harm anybody, and would actually increase local biodiversity and provide a few additional resources for some animal species.

The simple act of calling a species invasive, or potentially invasive, even without any supporting evidence, can be enough to justify taking action against that species. That is unacceptable. It seems to me that if you are going to ask for substantial funds to fight against, and try to eradicate, a particular species, the least you can do is gather and present comprehensive, convincing, and relevant data showing why such funding and efforts would be appropriate, and then offer the data for adequate scrutiny by experts in the field. That is not just a matter of fairness and due process, but also one of common sense and accountability.

In fact, the burden of proof should be even greater for those who advocate the necessity of radical control programs or other such extreme and potentially cruel and destructive measures. If you're going to go out there to, say, shoot or poison a large number of animals in the name of conservation, you better be able to make a compelling fact-based case which not only justifies such drastic actions, but also shows why that would be the best way to proceed. Feelings and biases are not enough. And "guilty until proven innocent" is neither a great rallying slogan nor a good starting point.

A consistent problem is the lack of pre-invasion data for comparative purposes. This is particularly true for "invasions" which took place many decades or even centuries ago – well before any comprehensive records on the distributions and abundance of "native" species were kept. In many cases, at numerous locations, we simply don't know how the native species were doing before the non-native ones arrived (Heard and Sax, 2013). Sometimes, we also lack data on the status of native species shortly after the arrival of introduced species, as we have seen during the discussion of the cane toad case study (Phillips et al., 2003). Without such vital information it is difficult to know exactly what the effects of non-native species really are, especially in the longer term.

As far as I am concerned, biologists should be free, of course, to conduct research on non-native species, and present their findings, without having to fear immediate accusations of xenophobia, if these findings happen to portray certain introduced species in a somewhat negative light.

At the same time, critics of the methods, assumptions, and unsubstantiated claims found in some invasion biology-related studies and publications should also be able to make their criticisms and express their opinions, without being suspected of promoting and abetting species invasions, or having to face accusations of disloyalty, or even lack of patriotism. "But don't you care about *our* nature?" is a type of question I've heard more than a couple of times (with minor variations), while I was publicly (and politely) expressing some reservations about the rhetoric and programs used against certain introduced species. This is the type of rhetoric I never used to hear a couple of decades ago, before invasion biology started to influence public opinion. Has the overall threat from invasive species gotten worse in local ecosystems such as the forested ravines of the Toronto region in the last twenty or thirty years, and are these ecosystems deteriorating as a result? I see no evidence of that – when left pretty much alone, these urban forests, consisting of an interesting mixture of native and introduced species, seem to be thriving. And yet, the anti-introduced species propaganda goes on and on and even appears to become gradually more and more pervasive. Besides, I never felt a sense of ownership when it comes to the species surrounding me, regardless of which part of the world I happened to be in, so expressions such as "our nature" or "our plants and animals" never really came into my mind. After all, we are talking about wild species here, and not some sort of collective farm.

Furthermore, while it may be good, in general, to foster increased public interest in nature, limiting this interest to only a few privileged local species can ultimately have negative consequences in a changing world and may lead to extreme and unwise measures undertaken solely for the preservation of the favoured poster species, at great financial cost and also at the expense of many other species, habitats, and ecological common sense. Simply put, stopping natural changes that do not affect human well-being, solely for ideological reasons, is clearly a form of blatant anthropogenic interference with potentially damaging and difficult to predict consequences. And treating some natural range expansions as "crimes" punishable by control programs just because these changes may interfere with some conservation initiatives is obviously absurd.

As I always tell the students completing research projects under my supervision, the important part is to follow the data sets, wherever they may lead us. Of course, the ways in which we collect, analyze, interpret and disseminate data also matter, and should be scrutinized. For instance, if someone conducts a comparative study on two plant species, and determines that under certain environmental conditions plant species X produces more seeds, on average, than plant species Y, that would be a scientific conclusion based on the data. However, if someone were to write that plant species X produces too many seeds, then that would be simply a subjective opinion or value judgement. Similarly, while one can potentially determine that the range of species X may be greater than the range of species Y, stating that species X occupies too much space, or has too great a distribution, would be expressing an opinion and showing a bias against species X. The implication would be that species X really shouldn't occupy such a large area, and shouldn't produce so many seeds, and perhaps we should do something about that. In my experience, there are too many such subjective opinions, disguised as facts, in the invasion biology literature.

In her review of the benefits provided by invasive species for native species, Rodriguez (2006) stated that: "Most studies of biological invasions have focused solely on the direct negative impacts of non-indigenous species on native biota." Rodriguez also concluded that, despite their potential importance, the positive impacts of non-native species were relatively ignored by invasion biologists, which makes it harder for researchers in this field to develop a more complete understanding of novel ecosystems and so-called biological invasions.

In their previously discussed survey, Young and Larson (2011) found that even among invasion biologists, the vast majority agreed that "There should be more scientific objectivity and less emotional xenophobia regarding invasive species." An even greater majority of the invasion biologists surveyed agreed that scientists working in this field should "gather, interpret and communicate information" about invasive species "as accurately and objectively as possible." These findings suggest that even many invasion biologists are aware that invasion biology is currently plagued by too much subjectivity, and presumably also too much unwarranted or insufficiently substantiated negativity regarding non-native species. On the other hand, if asked, most people would probably agree that they should try to be as objective and fair-minded as possible. In a way, this is like asking people if they would like to be good, rather than bad, at their jobs. We can easily predict in advance what the answers will be. However, just because people recognize the need to be competent or objective, in a general sense, this doesn't mean that they will necessarily be more aware of their biases, from a practical standpoint.

Some of these biases may be understandable, at least in part. For example, if you are a young graduate student, starting your work in the lab of an invasion biologist who has spent his, or her, entire career talking about how bad invasive species are, you may, quite legitimately, feel that in order for your thesis to be approved, you would have to confirm that the particular non-native species you are studying is very bad indeed, and should be fought against. As previously discussed in this book, if one sets out with that kind of attitude, then one may well find exactly what one expects. Many may feel that in order to belong to the invasion biology club, they would have to dislike and mistrust invasive species, since such a mentality is likely one of the main reasons, if not the main reason, for the very existence of this field. This may account for the sometimes overly defensive and aggressive reactions of some invasion biologists to any criticisms of their field, no matter how thoughtful, legitimate, and fact-based such criticisms may be. If you dare to question the predominant rhetoric about the overwhelmingly negative role of introduced species or the need for some control programs, then you must be the enemy, or at least part of the problem...

In a remarkably one-sided dismissal of the criticisms of what they called "invasion science", David M. Richardson and Anthony Ricciardi (2013) referred to the critics of their field as "a relatively small but vocal number of scientists and academics – naysayers in various guises" and declared that "many of the criticisms against invasion science simply do not withstand scrutiny." It should be noted that this article was not a detailed review paper, but a rather brief editorial – consisting of about three and a half pages of text, including the Acknowledgements, a graph, and a full page table. This editorial – titled *Misleading criticisms of invasion science: a field guide* – was published in the journal *Diversity and Distributions* (Editor-in-Chief: David M. Richardson). Within the article, Table 1 is labelled as "A field guide to misleading criticisms of invasion science", but this isn't really a field guide in any realistic sense, of course. The table summarizes six major criticisms of invasion biology and the so-called rebuttals to the criticisms, and provides a fairly small number of references, or sources, for each side of the argument. While this may appear to be an even-handed examination of the field, it is quite clear that the authors of the editorial are far from neutral – in fact, they even refer to the articles they criticize as being part of "a cottage industry of criticisms". The rebuttals are given in considerably more detail than the actual criticisms, and there is no way to decide, based on the limited information provided, whether the sources cited in support of the rebuttals provide comprehensive and convincing responses to each of the criticisms of the field. The impression given is that each of the criticisms listed in the table is without merit, and has been fully addressed and debunked in the relevant literature. However, as we have seen throughout this book, that is not the case, and it is certainly possible to identify legitimate problems in invasion biology. In other words, the table about the supposedly "misleading criticisms" appears to be quite misleading itself. At least some of the already rather few sources cited for the rebuttals are very brief opinion pieces which do not offer any new data – for example, the Wilson et al. (2009) reference is a one page article published in the Letters Response category of the journal *Trends in Ecology and Evolution*, and the Simberloff et al. (2011) source, which will be discussed in more detail shortly, is a very brief letter to the editor. Furthermore, plenty of additional studies and examples could have been selected, and a more detailed and nuanced description of the criticisms of invasion biology could have been included, of course. Another table could have been put together just as easily, showing studies and data which contradicted the rebuttals presented by Richardson and Ricciardi (2013).

In discussing some of his fairly intense scientific disagreements with Charles Elton during their time together at the Bureau of Animal Population, ecologist

Dennis Chitty (1996) wrote about "how valuable for scientific integrity is the role of unbridled criticism." I completely agree, especially when the criticisms are rational and well-founded, and based on suitable expertise and thorough knowledge and understanding of the relevant scientific facts.

Perhaps understandably, some invasion biologists reject outright the notion that there is any anti-introduced species bias in their studies, and claim that their work is purely scientific and objective in nature. That may well be true, at least in some cases, but I would encourage invasion biologists to ask themselves a couple of simple questions. Would you still refer to the environmental effects of a particular species as negative, and would you contemplate taking action against that species, if you suddenly found out the species is actually native and not, as you may have thought at the beginning of the study, introduced? In other words, given the exact same data on the impact of a certain species, would your interpretation of the results, conclusions, and recommendations be different if you knew the species was native after all, rather than exotic?

Would most invasion biologists even be interested in analyzing the environmental impact of a species at all, unless the species was suspected of being invasive?

Ultimately, to me, the most important question is not whether some invasion biologists may harbor deep-seated xenophobic tendencies towards non-native species, but whether the scientific evidence against these species measures up to the propaganda and can be used to justify the expense, effort, and destruction associated with control and eradication programs. It is even more difficult to measure the damage such propaganda can cause to people's general appreciation of, and respect for, nature, and to the long-term relationship between the public and ecologists.

Our approach to conservation may need to evolve and undergo a paradigm shift. We now live in a world where certain "conservationists" and managers seem to think that feeding city pigeons should be forbidden, but shooting lots of barred owls and ruddy ducks is just fine. By minimizing positive interactions with other species and emphasizing radical and destructive actions, these types of measures take us even farther away from nature. Wiping out an introduced species because "it doesn't belong" can never be equivalent to cleaning up the mess left behind by an oil spill. That is why the term "biological pollution" can be so toxic – because it makes it seem that a field full of naturally growing, self-perpetuating beautiful exotic wildflowers is ecologically similar to the site of an industrial accident. Perhaps reexamining the delicate topics of conservation ethics and logic may be in order.

Also, while some invasion biologists may be biased against invasive species, but not necessarily xenophobic (the assumption here being that a mere bias does not rise to the level of xenophobia), that doesn't mean that all members of the public who may seem overly eager to kill introduced species are entirely free of xenophobia. Certainly, the deep animosity, and sometimes outright hatred, some members of the public are showing towards gray squirrels in the UK or cane toads in Australia, for example, seem to owe very little to scientific facts, and appear quite irrational, at least to me. Such extreme negative attitudes towards these non-native species are discussed in various articles (for instance: Max, 2007; Sturcke, 2009; *The Telegraph*, 2009; and Shine and Doody, 2011, among others). That is precisely why, as biologists working in this area, I think we all have a responsibility to put facts and reason first, and not incite the public by needlessly demonizing some of our fellow species. Do we really want armies of vigilantes going into wilderness areas destroying all the species they think should not be there? Is this really a version of conservation biology that most of us can embrace? And, most importantly, would such destructive actions really lead to a better version of nature? If you think the answer to any of these questions is yes, please explain why. And please try to make your explanations scientific. In my view, the scientific case for the "improvement", "purification", and "restoration" of the natural world through the constant persecution of introduced species has not been made, and can probably never be made. Perhaps, it is time to move on to worthier and more realistic pursuits.

FIGURE 34 *Gray squirrel* (Sciurus carolinensis) *in a park beside the Royal Ontario Museum in Toronto, on the side of the Atlantic where this species is considered to be native and is therefore reasonably safe*
PHOTO BY RADU GUIAȘU

On the other hand, even within invasion biology, there may be glimmers of hope. For instance, as I was writing this book, and had already completed about half of the first draft, I became aware of a recent controversy in the field. One of the newer chapters in the ongoing debate between ecologists who questioned the rationale behind the discrimination against exotic species and more traditional invasion biologists who thought that such discrimination is entirely legitimate and justified was recorded in the summer of 2011, in the pages of the prestigious journal *Nature*. The discussion was started in an article written by 19 ecologists and titled *Don't judge species on their origins* (Davis et al., 2011). As the title indicates, this article questioned the usefulness of the rigid dividing line between native and non-native species, and took issue with some of the biases against introduced species which seem prevalent in invasion biology. As already shown throughout the book, my views on these subjects are clearly much more closely aligned with those expressed by Davis et al. (2011) in their article. The negative reactions to this article were swift, and they were published in the form of several letters, about a month later, in a subsequent issue of the journal *Nature*. The first letter was signed by Daniel Simberloff and 140 other scientists (Simberloff et al., 2011), and there were also letters from Alyokhin (2011), Lockwood et al. (2011), and Lerdau and Wickham (2011). All in all, 19 biologists signed the article in support of a more open-minded attitude towards non-native species, and 147 scientists came out essentially against this article. This may reflect the general (numerical) balance, or imbalance, if you will, of forces between the two sides in this debate. However, the majority view is not always the right one (throughout human history, a majority of the people, at one time or another, believed some pretty crazy things), and at least we now have more intense and meaningful debates in invasion biology, where a significant group of prominent biologists are openly questioning some of the key assumptions and principles, and perhaps the very foundation, of this field. Personally, in my more optimistic moments, I find this mildly encouraging. Perhaps, as a result of such debates, invasion biology may gradually morph into a more diverse, tolerant, and intellectually stimulating field, and move forward without some of its burdensome old assumptions, and maybe even without its alarmist old name. This may be wishful thinking, but, in my opinion, it is something worth wishing for.

Invasion Biology – Can It be Reformed, or is It Just an Unfortunate Temporary Detour from Mainstream Ecology?

In a popular current conservation biology textbook (Primack, 2012), an invasive species is defined as "an introduced species that increases in abundance at the expense of native species." This is fine, except that, as we have seen throughout the book, it is often very hard to determine conclusively that certain so-called invasive species really cause serious long-term problems for native ones. It seems that it is a lot easier to make that claim than to prove it. Furthermore, the definition also raises additional questions, such as, how big, widespread, and long-lasting does the increase in abundance have to be, before we consider the species truly invasive? And are we also taking into account, and quantifying, the possible benefits such abundant new species may provide for some native species and human communities?

Simberloff (2003) stated that the main reason for concern about introduced species is that such species "can threaten the existence of native species and communities and that they can cause staggering damage, reflected in economic terms, to human endeavors." Well, some introduced species may do that, but, then again, some native species are quite capable of similar effects. Furthermore, many introduced species are currently persecuted without being guilty of any of the "crimes" Simberloff (2003) described above.

For instance, as we have seen earlier, the tubenose goby did not spread much and has had a minor impact in the Great Lakes region since its arrival in North America from Europe in the late 1980s. And yet, despite its small impact and limited and stable distribution in this region, the tubenose goby is listed as an invasive species on the Ontario's Invading Species Awareness Program web site (www.invadingspecies.com/invaders/fish/tubenose-goby/,2015). On this site, under the heading *Impact of tubenose goby*, we can read that: "It's not yet known how tubenose goby may affect native species, but scientists are concerned about the possible long-term effects of the fish's choice of habitat and diet." This sounds very much like another version of the "guilty until proven innocent" approach. And it is not at all clear that there is much interest in proving introduced species "innocent" or absolving them of undeserved, or at least unproven, "guilt". Here we have a small introduced fish which is not expanding its range in any significant way and is not having a major negative impact on other species, and yet this non-native species was still declared "invasive" and put on the naughty list. So much for the assertion that only "bad" exotic species will receive the dreaded "invasive" label.

Similarly, the fathead minnow (*Pimephales promelas*) – a fish species native to North America – is included in a newly published field guide to the invasive species found in Britain, even though, according to the authors of this field guide, there were "no impacts yet recorded" in Great Britain which can be associated with this species. Furthermore, it appears that there was only one wild fathead minnow population in Britain, and this population existed for only about two years before being eradicated in 2010. This field guide also includes the round goby and the tubenose goby among the invasive species of Britain, even though these two fish species are not present in that country (Booy et al., 2015).

As an additional example, the previously mentioned water lettuce – a species whose native range is unknown – is listed as an invasive aquatic plant on the Ontario's Invading Species Awareness Program web site (www.invadingspecies.com/invaders/plants-aquatic/water-lettuce/, 2015). However, this is not a species which can normally tolerate cold temperatures, and it is not known whether this plant can survive during the winter or distribute itself beyond a handful of locations in this province. Booy et al. (2015) acknowledge that water lettuce is "unable to establish and so unlikely to cause significant impacts in Britain", but this does not stop the authors from including this aquatic plant among the invasive species found in Britain anyway.

Ultimately, what is the real point of dividing the living world into native and non-native species? If a species is causing problems for us, it seems to me that we would definitely take action against that species, regardless of its geographic origin. Would we deliberately let a "native" virus run rampant through a human population, and only take action when the virus leaves its "native" region? Is widespread damage from a "native" fungus preferable to similar destruction caused by an "introduced" fungal species? Are native beetles which destroy forests more acceptable than introduced ones? If destruction is caused by a species whose geographic origins are uncertain, would we wait until we can determine if the species is native or introduced before reacting? We can analyze the changing distributions of, and the interactions among, species, just as well, without necessarily having to stick the "native" or "exotic" labels on them. After all, in the final analysis, they are living species on the same relatively small planet. Life on Earth has a common origin, which goes back at least 3.5 billion years or so, and all species are interconnected in some ways. The perceived chasm between native and non-native species created and emphasized by some invasion biologists may not be a particularly useful basis for moving forward in our quest to understand the living world around us, with all of its challenging complexity and beauty.

And this brings us to some additional and perhaps uncomfortable questions. If the distinction between "native" and "non-native" species is not really all that useful, and this distinction is the cornerstone of invasion biology, then should invasion biology, as a separate field, have a future, or is it just an unfortunate temporary detour from mainstream ecology – a few decades of (scientific) extremism, a passing fad, which will eventually lead to an inevitable scientific dead end? Can invasion biology be substantially reformed and improved, or should it simply be gradually abandoned, as a distinct entity, as ecologists move beyond the simplistic native vs. non-native dichotomy? The worthy studies and contributions will survive and stand on their own, without having to be classified as being part of any specific "movement" or ideology.

I talked about the fact that dissenting voices are being heard with increasing frequency in invasion biology these days, and I expressed some moderate hope that perhaps new debates could re-invigorate the field and help it grow. This may indeed happen – from my perspective, it is an optimistic view – but it is only one of several possibilities. It is quite likely that a substantial number of invasion biologists and wildlife managers, who may feel they have invested a lot in the current traditional view of invasive species and their impact, may not wish to see substantial changes in their field. During my interactions with invasion biologists, I must admit that I did not detect an overwhelming desire for change, or a softening of the negative attitudes towards introduced species over time. For such biologists, the field is just fine the way it is, and adding more and more non-native species to unwanted lists, in various parts of the world, for an ever increasing number of reasons, may be the preferred way to move forward for the foreseeable future. In this way, invasion biology can grow (at least with respect to the identification of more and more enemy species to be fought against) without actually changing.

In fact, much too often, as already discussed, the typical response from invasion biologists to legitimate and thoughtful scientific criticisms of their field is simply to dismiss the opposing views outright, and attempt to crush all dissent. In an interesting recent note, Warren and Bradford (2013) came out forcefully against this type of "gang science", as they called it, whose purpose was "to quash unpopular ideas in peer-reviewed journals", and cited the previously mentioned letter by Simberloff and 140 other scientists (Simberloff et al., 2011) as a clear example of that. As Warren and Bradford aptly put it: "In the absence of new data, such huge conglomerates contribute little more than intimidation." After all, scientific debates should be, first and foremost, about competing ideas, and the data and analyses presented in various studies. As a brief aside, not having been involved in the writing of any articles with more than a few authors, I sometimes wondered what the exact contribution of, say, the 100th author to a one- or two-page-long article may be, and how the order of so many authors is determined.

Since invasion biology is based to a large extent on antagonism towards introduced species, it is not hard to figure out why many invasion biologists would add their names to a letter such as the one published by Simberloff et al. (2011). If we start to look more closely at non-native species, we may just begin to understand them better and develop a more balanced view about their place in our world. And that, in turn, may lead to a diminished justification for trying to wipe these species out, at least in the areas where they may have been introduced, and, therefore, to a weakening of the resolve of some people to fight against these species. Apparently, in the eyes of some invasion biologists, more positive views of invasive species amount to heresy and cannot be left unchallenged. In fact, Simberloff seems quite active in responding to various articles that even hint at a more tolerant approach towards non-native species, or criticize various aspects of invasion biology. In addition to the previously mentioned letter to *Nature* where he led a group of 141 authors (Simberloff et al., 2011), he was also the sole author, or one of the authors, of several other such replies (for example: Simberloff, 2005; Richardson et al., 2008; Vitule et al., 2009; Simberloff, 2009; Simberloff and Genovesi, 2013; Simberloff et al., 2013), as well as a relentless advocate of the righteousness of the global war against invasive species.

Other, even more recent, examples of such letters, responses, or comments (not authored or co-authored by Simberloff) can also be found in the relevant literature. For instance, a letter by Hulme et al. (2015) is only slightly longer than one page, and approximately half of this space is taken by the names and addresses of the 25 authors. According to the information provided, all 25 authors wrote the paper, which consists of three paragraphs of text. In this letter, Hulme et al. (2015) challenge and criticize the previously discussed article by Thomas and Palmer (2015a). As a reminder, Thomas and Palmer (2015a) found that non-native plants added to the diversity of British flora, overall, and did not drive native British plants to extinction. Hulme et al. (2015) agree that most non-native plants in Britain do not have significant ecological impacts, but argue that even introduced plants which are not widespread can play a negative role. These authors also refer to what they see as the "headline provocations" in the initial article by Thomas and Palmer (2015a). I did not detect any such "provocations" in the study by Thomas and Palmer (2015a) – the article seems carefully written, and the conclusions are logical extensions of the data presented and analyzed in the paper. However, such labels may be indicative of the intensity of some of the disagreements in invasion biology. Where some of us see sound and convincing arguments, others may see "provocations". In their reply to Hulme et al. (2015), Thomas and Palmer (2015b) stand by their initial study, and write that:

> Most non-native species, including those that are officially classed as invasive, are far too rare to impinge negatively on plant diversity, averaged across the country. As such, their presence in Britain adds to the nation's total floral diversity.

In their editorial about the supposedly "misleading criticisms of invasion science", Richardson and Ricciardi (2013) wrote: "To suggest that non-native species are not unequivocally a major concern for the conservation of biodiversity and ecosystem services is to ignore decades of peer-reviewed science." Well, let's analyze this statement a bit more closely. First of all, I thought that, according to many ecologists, only a small minority of non-native species become invasive and cause problems, so why are we not allowed to think, or suggest, that many non-native species are not a concern for the environment or human interests? And why should we be accused of being ignorant of the relevant literature, wilfully or otherwise, if we entertain such thoughts or make such suggestions and back them up with proper references and data? Also, the word "unequivocally" seems rather misplaced in there as well. Unequivocally? Really? Again, the very definition of the term "non-native" is often unclear in this field, and even when we can actually determine that a species is indeed not native, the impact of that species is frequently ambiguous or poorly understood. In many cases, such species may have some negative effects as well as some positive ones, and the long-term impacts may be unknown. So, to suggest that anyone who does not consider non-native species in general to be unequivocally bad or capable of causing major problems is ignoring "decades of peer-reviewed science" seems rather strange. As mentioned, the field of invasion biology is young, so we don't have much more than about two and a half decades worth of peer-reviewed studies on display. And a thorough review of many of these studies, such as the one presented in the pages of this book, can actually reveal that there is quite a bit of information that would contradict the assertion made by Richardson and Ricciardi (2013). Not all of these peer-reviewed articles point in the same direction or lead to the same conclusions. Perhaps it could be said that Richardson and Ricciardi (2013) are ignoring some of the peer-reviewed literature in their field when they are making such general statements. And let us not forget that some of the invasion biology papers, published in peer-reviewed journals, are mere opinion pieces. Such papers clearly add to the volume of the literature in this field, but they do not contribute any new evidence or data.

In their survey, Young and Larson (2011) found that 37% of invasion biologists agreed that "exotics are an unnatural, undesirable component of the biota and environment." However, 34% of invasion biologists disagreed with that statement. This shows there is no unanimity in the field with respect to the presence and role of non-native species. Furthermore, in the same survey, 37% of invasion biologists agreed that "the term 'invasive' should not be used to connote negative environmental impact", while 46% disagreed with that assertion. And, as discussed earlier in the book, Young and Larson (2011) also found that there was almost equal support in the field for two different definitions of invasive species – one which stated that such species have negative environmental or socio-economic impacts and another one which did not. Therefore, it is perhaps unjustified to refer to critics of invasion biology as merely a few "naysayers", who are ignoring the relevant scientific literature. Invasion biologists themselves appear to be quite divided on some important issues and definitions in their own field.

For the most part, the criticisms of invasion science labelled as "misleading" by Richardson and Ricciardi (2013) seem quite legitimate. According to some of these criticisms, as listed in Richardson and Ricciardi's editorial, "impacts of non-native species on biodiversity and ecosystems are exaggerated" – I suppose the authors really meant to write "negative impacts", rather than simply "impacts" in this case – and "positive (desirable) impacts of non-native species are understated". First of all, I don't think the critics of invasion biology whose work I am familiar with are saying that introduced species cannot have negative impacts or that these impacts are *always* exaggerated. Specific criticisms usually refer to particular species and examples. Can we really say that the negative impacts of non-native species are *never* exaggerated, or that the positive effects of these species are *never* ignored or understated in the invasion biology literature? Quite a few studies, including several cited in this book, would suggest otherwise. Does the overall scientific information on the impact of purple loosestrife in North American wetlands, for example, justify the control programs and propaganda against this species throughout the continent? How about the impact of the rusty crayfish in Ontario? Do we know enough about it to start the anti-rusty crayfish campaigns? Should we use data about the possibly negative impact of a species in one location to demonize that species everywhere? How often is positive information about invasive species included in the brochures, information sessions, guided nature tours, or lectures associated with, or influenced by, this field? How many invasion biologists or wildlife managers set out to find out just how much of a good contribution exotic species can bring to various ecosystems? What is the percentage of invasion biology studies which specifically focus on the positive attributes of non-native species? Without such detailed information, how can these criticisms of invasion biology be labelled as "misleading"?

And couldn't the inclusion of major human diseases caused by viruses or bacteria on the lists of costs associated mainly with invasive animals and plants be classified as potentially "misleading"?

Richardson and Ricciardi (2013) tend to describe the chosen criticisms of invasion biology in very general terms. For example, one of the criticisms is described as "increased species introductions raise biodiversity (e.g. by adding to regional species pools; generating new taxa through hybridization) and therefore do not merit concern." Well, introduced species can actually do that, and new varieties and species may eventually evolve as a result of interspecific hybridization, especially if we take a long-term evolutionary view, but that doesn't mean the critics of invasion biology are always suggesting that invasive species are never of any concern. That is an overgeneralization, and, therefore, an exaggeration, or even a caricature, of the position taken by certain critics of the field. Just because some of us take a longer term and more generous view of introduced species and processes such as natural hybridization between native and exotic species, that does not mean we are oblivious to potential problems associated with the introduction of species. And, as previously stated, non-native species deserve to be known and studied for more than the potential problems they may sometimes cause, and native species are perfectly capable of generating environmental and economic troubles as well.

The previously mentioned article by Thomas (2013) was one of only three studies cited by Richardson and Ricciardi (2013) as sources for the "increased species introductions raise biodiversity and therefore do not merit concern" criticism. I read Thomas' article carefully, and I couldn't find any reference to species introductions not being of any concern, in general. In fact, Thomas (2013) specifically mentioned that: "It is true that some invasive species damage ecosystems and can eradicate resident species." He also added: "There are excellent arguments for conserving the wildlife we already have, but it is less clear why our default attitude to novel biodiversity is antagonism or ambivalence." Sounds perfectly reasonable to me. Furthermore, Thomas's 2013 article is full of interesting specific examples which illustrate his point that biodiversity can indeed be increased as a result of species introductions.

Some leading invasion biologists stated that critics of invasion biology attack so-called "straw men", rather than genuine problems in this field (Simberloff et al., 2011; Richardson and Ricciardi, 2013). However, in the rush to dismiss or diminish criticisms, invasion biologists may construct "straw men" of their own, and trivialize or downplay legitimate and thoughtful concerns, while ignoring specific examples and case studies which may not fit comfortably within the larger or predominant invasion biology narrative.

In taking issue with the criticism that some control programs may be directed at non-native species which do not really cause any, or much, harm, Richardson and Ricciardi (2013) wrote:

> In reality, managers are constrained by limited resources and seek to prioritize species that are likely to become problematic. However, this effort is hampered by several facts that are generally ignored by the naysayers:

> (1) the impacts of most invasions have not been studied, and so important effects may remain undetected, (2) invaders that are apparently innocuous in one region can be disruptive in other regions, (3) subtle impacts that may be unrecognizable without careful technical study can produce enormous ecosystem changes over time, and (4) many non-native species that currently appear innocuous may become damaging many years later – when it is no longer feasible to eradicate them.

The four items in the paragraph above seem to me to fit within the previously mentioned "guilty until proven innocent" philosophy. Let us analyze these so-called "facts" mentioned by Richardson and Ricciardi (2013), in the order in which they were listed above. (1) If the impacts of most invasions have not been studied, why should we necessarily assume that the effects of these introduced species will be important and negative? All that really means is that we don't know anything about these species and their impacts. Consequently, we shouldn't even call these species "invasive", unless we do not link the term invasive with a significant negative environmental impact. (2) Yes, theoretically, "invaders" may be harmless in one region and "disruptive" somewhere else. However, that clearly does not mean we have to consider non-native species automatically as problematic, particularly if these species are not known to have had a negative impact anywhere. It is also possible that even introduced species which have acquired a bad reputation elsewhere may cause no problems in the regions we are interested in. This is a two-way street. (3) "Subtle impacts that may be unrecognizable without careful technical study" may actually remain subtle, or minimal. "Careful technical study" may mean that we are trying a bit too hard, perhaps, to find some type of negative impact at all costs. This may not necessarily be the most efficient use of the "limited resources" mentioned by Richardson and Ricciardi above. Subtle impacts may, theoretically, produce larger impacts over long periods of time, but just how long would these projected periods have to be before we decide we must act now? If the impact of an introduced plant or animal will not be significant for hundreds or even thousands of years, it is obviously not of immediate concern. Who knows what current ecosystems will look like centuries or millennia from now? Furthermore, sometimes the impacts of invasive species become less important over time. So, again, this works both ways. In some cases, the impact of an invasive species may be most significant immediately after the introduction. After a few years, this impact may decline naturally, without our interference. (4) Yes, some non-native species may become "damaging" many years after their introduction, but, again, this is a theoretical concern that could be applied to any species. It is an a priori charge of "guilty", or "potentially guilty one day", without any supporting evidence. Changes in environmental conditions may cause a native species to become much more abundant and widespread than before as well. If we really care about managing our "limited resources" wisely, and if invasion biologists are really not against all introduced species (as Simberloff, 2003; Simberloff et al., 2011; and Richardson and Ricciardi, 2013, among others, suggested), then does it make sense to eradicate exotic species that "currently appear innocuous"? Do invasion and conservation biologists consider all non-native species to be problematic or not? Let us make up our mind, and let us indicate clearly where we each stand on this. We cannot have it both ways.

Simberloff et al. (2011) stated that: "Pronouncing a newly introduced species as harmless can lead to bad decisions about its management." So, does that mean that we can never declare a newly arrived species to be "harmless", even in the absence of any evidence that would indicate otherwise? And couldn't the assumption that a currently harmless species should be controlled anyway also lead to potentially bad management decisions (resulting, for example, in the squandering of limited resources to fight against a species which should not be of concern and the damage to the environment that may be caused by such misguided control programs)?

In an article which attempted to find the middle-ground in some of the current debates about invasion biology, Shackelford et al. (2013) wrote: "Those who defend the removal of non-native species have been accused of xenophobia and those who are more ambivalent are charged with biological homogenization. Both sides have merit."

In trying to analyze some current debates in invasion biology from a more neutral perspective, Shackelford et al. (2013) mentioned that the field "has shifted its rhetoric in recent years to reflect a focus on species with the greatest impact." These authors also added:

> However, how much of this shift reflects a change in attitude toward non-native species, rather than just limited resources and political appeal, is unclear. This point sits at the center of much recent controversy. Without resource and methodological constraints, many, if not most, conservationists would still probably prefer to rid systems entirely of non-natives regardless of impact.

Furthermore, the very concept of "impact" can be subjective in this field. If a species – for example, purple loosestrife – has been declared to be a problem, in general, it is very likely that control or eradication campaigns will be attempted against that species at many locations where its impact may not be negative, overall. I doubt that conservationists or managers in charge of individual conservation areas are always conducting their own thorough independent research to determine whether the introduced species present within their jurisdictions are having mostly negative, or positive, impacts. Once a species is put on a list of undesirable "invasives", it will probably stay there for a long time, even if convincing new exculpatory evidence may emerge. As far as I can tell, these lists are not always based on the best and most up-to-date science.

In trying to portray as positive an image of invasion biology as possible, Richardson and Ricciardi (2013) refer to it as "a valuable and thriving meta-discipline", and mention that "one of the principal goals of the field – to predict which introduced species will become disruptive – is of increasing societal importance". However, aside from the fact that we don't know exactly what is meant by the term "disruptive", which could be quite subjective, even the most dedicated proponents of invasion biology seem to agree that the ability to make such predictions is not one of the current strengths of the field. In fact, the Abstract of a recent article (Kumschick et al., 2015) about the ecological impacts of alien species – a study with 19 authors, including Ricciardi and Richardson – begins with the following admission:

> Despite intensive research during the past decade on the effects of alien species, invasion science still lacks the capacity to accurately predict the impacts of those species and, therefore, to provide timely advice to managers on where limited resources should be allocated.

Despite the occasional debates, invasion biology remains a field dominated by those who frown upon perfectly natural phenomena, such as range expansions and hybridization, and seem to dislike and mistrust environmental and ecological changes not specifically planned and directed by conservationists.

Although, as we have seen, there has been a dramatic rise in the number of invasion biology-related publications since the early 1990s (and we can add this book to the list, of course), and the simple anti-alien species message is commonly spread these days, I am not convinced there has been all that much substantial intellectual and conceptual progress in this field, overall. As far as I can tell, there are periodic attempts to come up with better essential definitions, but no definitions can cover up or rescue flawed ideas and ideologies. Occasionally, some try to propose more useful methods, flow charts, classification schemes, or ways of making predictions, but it is not obvious to me how far we can travel down that road. We still don't have clear ways of determining which species will establish themselves successfully in particular ecosystems. Some "predictions" are of the trite variety – obviously, a species has a much better chance of adjusting to a new area, if it comes from an ecologically similar region and is already known to be a successful colonizer elsewhere – and therefore do not require a great deal of insight or expertise. However, even highly successful colonizers sometimes fail to become established in areas that should, theoretically, be perfect for them, or disappear inexplicably from regions they inhabited for quite some time. It is also obviously impossible to label a species as "invasive", before it actually "invades" something. So, the pre-screening of potentially problematic exotic species before they actually arrive somewhere remains a highly speculative and rather murky affair. While the volume of studies has undoubtedly grown quite a bit, and more and more species are fingered as "invaders", it sometimes appears that invasion biology is moving in circles, and the path forward remains shrouded in an intellectual fog.

To be fair, the field has produced several interesting studies over the years, or, to put it more accurately, some biologists working in this field have produced certain thought-provoking studies, but, for the most part, invasion biology still seems to exist mainly to promote and assist "the war against invasive species". If our general view of invasive species becomes more tolerant and nuanced, this may actually sap the strength of the field, instead of re-invigorating it, and weaken the resolve of all but its most ardent proponents. After all, if there are no clear distinctions between native and introduced species, what is the point of invasion biology? Scientists who believe that this discipline is too rigid and subjective may eventually move away from it, rather than trying to work towards its improvement. Eventually, an unbridgeable gap may open up between the different groups of ecologists interested in introduced species, and the field may fracture into a sort of uncompromising old style invasion biology for the true believers, the so-called "purs et durs" (the pure and the tough), on the one hand, and something completely different, which would have very little to do with any discipline that has the word "invasion" in its title, on the other hand.

One of the essential questions I think we should ask in ecology and conservation and invasion biology would be: what is the best way to preserve nature and natural areas? Other useful and related questions may be: which nature or versions of nature should we try to preserve and why? Do we always have to attempt to go back to the past, and restore ecosystems to the way they might have been once, at an arbitrary point in time? Such a task would surely be impossible and futile, in any case, particularly given our incomplete understanding even of current ecosystems, let alone the ones that existed or may have existed in the past. Do we always have to interfere with nature in order to "preserve" it? Can we really stop ecosystems from always changing through our constant meddling and fiddling? By interfering in nature, sometimes aggressively, in the name of conservation, do we really correct past mistakes, or are we simply making new ones? Wouldn't it be easier sometimes, when human health or major economic interests are not at stake in any meaningful way, to let nature take its course, so to speak? What if we just let new species assemblages form, on a regular basis, the way they always did, for billions of years, and new interactions among species develop, within novel ecosystems, which may ultimately be more suitable for a planet inhabited by more than seven billion people?

Sometimes doing nothing may be the best approach, but it can also be the most difficult one. We are an interfering and restless species, by nature. We like to get involved in everything, everywhere, and, while this tendency can sometimes lead to brilliant discoveries and achievements, it can also cause great destruction and devastation. Activism for the sake of activism may make some people feel busy and important, but can also be very damaging to the natural world, overall. Frequently, the hardest and most useful thing to do is nothing. Nature is resilient, and it often recovers well on its own, as long as we leave it alone and allow it to do so, unimpeded. Plant and animal species recolonized the Krakatau Islands within just a few decades, after the catastrophic volcanic eruption in 1883 that wiped out pretty much all forms of life in the area, and many species, including some vulnerable ones, appear to be doing quite well in the new wilderness around Chernobyl, following the devastating nuclear accident which took place there in 1986. It is ironic that some invasion biologists are quick to decry human-mediated changes to the distributions of species on Earth, while at the same time spending a lot of time, money, and energy trying hard to alter where and how species can exist in various ecosystems. Ultimately, invasion biology seems to lead us to a world of rigid borders, which can never fit a constantly changing and evolving natural world. Biologists are not, and should never become, mere border guards, watching over the distributions of species, ready to intervene aggressively at the slightest sign of change. The world has enough barbed wire fences already.

Literature Cited

Agapow, P.-M., O. R. P. Bininda-Emonds, K. A. Crandall, J. L. Gittleman, G. M. Mace, J. C. Marshall, and A. Purvis. 2004. The impact of species concept on biodiversity studies. The Quarterly Review of Biology 79: 161–179.

Ainscough, B. J., J. W. Breinholt, H. W. Robison, and K. A. Crandall. 2013. Molecular phylogenetics of the burrowing crayfish genus *Fallicambarus* (Decapoda: Cambaridae). Zoologica Scripta 42: 306–316.

Aldhous, P. 2004. The toads are coming! Nature 432: 796–798.

Allair, J. 2007. Barred owl. Pp. 296–297. In: Cadman, M. D., D. A. Sutherland, G. G. Beck, D. Lepage, and A. R. Couturier (Eds.). Atlas of the breeding birds of Ontario 2001–2005. Bird Studies Canada, Environment Canada, Ontario Field Ornithologists, Ontario Ministry of Natural Resources, and Ontario Nature, Toronto, Canada.

Alyokhin, A. 2011. Non-natives: put biodiversity at risk. Nature 475: 36.

Anadu, D. J. 1996. Aquaculture development in Nigeria. Aquaculture 22: 37–49.

Anderson, M. G. 1995. Interactions between *Lythrum salicaria* and native organisms: a critical review. Environmental Management 19: 225–231.

Anonymous. 2004. The gardens of Glendon. Glendon Magazine 2: 5. Spring 2004 issue.

Anthony, L. 2013. A frog for the killing. Canadian Geographic 133: 46–60. April 2013 issue.

Arnold, E. N. and D. W. Ovenden. 2002. Reptiles and amphibians of Europe. Second edition. Princeton University Press, Princeton and Oxford.

Attenborough, D. 1984. The living planet. William Collins Sons & Co. Ltd., London, UK and the British Broadcasting Corporation, London, UK.

Audesirk, T., G. Audesirk, and B. E. Byers. 2003. Life on Earth. Third edition. Pearson Education, Inc., Upper Saddle River, New Jersey, USA.

Badzinski, S. S. 2007. Mute swan. Pp. 64–65. In: Cadman, M. D., D. A. Sutherland, G. G. Beck, D. Lepage, and A. R. Couturier (Eds.). Atlas of the breeding birds of Ontario 2001–2005. Bird Studies Canada, Environment Canada, Ontario Field Ornithologists, Ontario Ministry of Natural Resources, and Ontario Nature, Toronto, Canada.

Baillie, J. E. M., C. Hilton-Taylor, and S. N. Stuart (Eds.). 2004. 2004 IUCN Red List of threatened species. A global species assessement. IUCN, Gland, Switzerland and Cambridge, UK.

Baker, H. G. 1974. The evolution of weeds. Annual Review of Ecology and Systematics 5: 1–24.

Barlow, C. 2000. The ghosts of evolution. Basic Books, New York, USA.

Barlow, G. W. 2000. The cichlid fishes – nature's grand experiment in evolution. Perseus Publishing, Cambridge, Massachusetts, USA.

Barr, D. W. 1996. Our friend the zebra mussel. Rotunda – The Magazine of the Royal Ontario Museum 29: 9–15. Fall/Winter 1996 issue.

Barrington-Johnson, J. 2005. The Zoo: the story of the London Zoo. Robert Hale Limited, London, UK.

Barry, S. L., R. A. Faitoute, S. Grusin, and E. Jones. 2008. Official guide to the Smithsonian National Museum of Natural History. Revised Edition. Smithsonian Books, Washington D.C., USA.

Baskin, Y. 1998. Winners and losers in a changing world – global changes may promote invasions and alter the fate of invasive species. BioScience 48: 788–792.

Beatty, J. 2009. Dobzhansky, Theodosius (1900–1975). Pp. 522–526. In: Ruse, M. and J. Travis (Eds.). Evolution: the first four billion years. The Belknap Press of Harvard University Press, Cambridge, Massachusetts and London, England.

Beckmann, C. and R. Shine. 2012. How many of Australia's ground-nesting birds are likely to be at risk from the invasive Cane Toad (*Rhinella marina*)? Emu 112: 83–89.

Beebee, T. J. C., J. Buckley, I. Evans, J. P. Foster, A. H. Gent, C. P. Gleed-Owen, G. Kelly, G. Rowe, C. Snell, J. T. Wycherley, and I. Zeisset. 2005. Neglected native or undesirable alien? Resolution of a conservation dilemma concerning the pool frog *Rana lessonae*. Biodiversity and Conservation 14: 1607–1626.

Benner, A. 2010. The one fish that didn't get away. The Niagara Falls Review. August 10, 2010.

Bergstrom, D. M., A. Lucieer, K. Kiefer, J. Wasley, L. Belbin, T. K. Pedersen, and S. L. Chown. 2009. Indirect effects of invasive species removal devastate World Heritage Island. Journal of Applied Ecology 46: 73–81.

Bertness, M. D. and T. C. Coverdale. 2013. An invasive species facilitates the recovery of salt marsh ecosystems on Cape Cod. Ecology 94: 1937–1943.

Bezener, A. 2000. Birds of Ontario. Lone Pine Publishing, Edmonton, Canada.

Bird, D. M. (Ed.). 2010. Birds of Canada. Dorling Kindersley, Toronto, Canada.

Blossey, B. 1996. *Lythrum salicaria* – Purple loosestrife. P. 81. In: Randall, J. M. and J. Marinelli (Eds.). Invasive plants – weeds of the global garden. Brooklyn Botanic Garden, Inc., Brooklyn, NY, USA.

Blunt, W. 2001. Linnaeus: the compleat naturalist. Princeton University Press, Princeton and Oxford.

Bohlen, P. J., P. M. Groffman, T. J. Fahey, M. C. Fisk, E. Suárez, D. M. Pelletier, and R. T. Fahey. 2004. Ecosystem consequences of exotic earthworm invasion of north temperate forests. Ecosystems 7: 1–12.

Boitani, L. and S. Bartoli. 1983. Simon and Schuster's guide to mammals. Simon and Schuster, New York, USA.

Booy, O., M. Wade, and H. Roy. 2015. Field guide to invasive plants and animals in Britain. Bloomsbury, London, UK.

Botkin, D. B. and E. A. Keller. 1998. Environmental science – Earth as a living planet. Second edition. John Wiley & Sons, Inc., New York, USA.

Boyle, T. C. 2011. When the killing's done. Viking, New York, USA.

Boysen, S. 2009. The smartest animals on the planet. Firefly Books Ltd., Richmond Hill, Ontario, Canada.

Brazner, J. C. and D. A. Jensen. 2000. Zebra mussel [*Dreissena polymorpha* (Pallas)] colonization of rusty crayfish [*Orconectes rusticus* (Girard)] in Green Bay, Lake Michigan. American Midland Naturalist 143: 250–256.

Brichetti, P., D. Rubolini, P. Galeotti, and M. Fasola. 2008. Recent declines in urban Italian sparrow *Passer* (*domesticus*) *italiae* populations in northern Italy. Ibis 150: 177–181.

Bright, C. 1998. Life out of bounds. W. W. Norton & Company, Inc., New York, USA.

Brown, N. 1997. Re-defining native woodland. Forestry 70: 191–198.

Brown, G. P., C. Shilton, B. L. Phillips, and R. Shine. 2007. Invasion, stress, and spinal arthritis in cane toads. Proceedings of the National Academy of Sciences of the United States of America 104: 17698–17700.

Burnie, D. and D. E. Wilson (Eds.). 2001. Animal. Dorling Kindersley, London and New York.

Bush, G. L. 1994. Sympatric speciation in animals: new wine in old bottles. Trends in Ecology and Evolution 9: 285–288.

Butler, M. J. IV and R. A. Stein. 1985. An analysis of the mechanisms governing species replacements in crayfish. Oecologia 66: 168–177.

Cabrera-Guzmán, E., M. R. Crossland, and R. Shine. 2012. Predation on the eggs and larvae of invasive cane toads (*Rhinella marina*) by native aquatic invertebrates in tropical Australia. Biological Conservation 153: 1–9.

Cadman, M. D. 2007. European starling. Pp. 450–451. In: Cadman, M. D., D. A. Sutherland, G. G. Beck, D. Lepage, and A. R. Couturier (Eds.). Atlas of the breeding birds of Ontario 2001–2005. Bird Studies Canada, Environment Canada, Ontario Field Ornithologists, Ontario Ministry of Natural Resources, and Ontario Nature, Toronto, Canada.

Cadotte, M. W. 2006. Darwin to Elton: early ecology and the problem of invasive species. Pp. 15–33. In: Cadotte, M. W., S. M. McMahon, and T. Fukami (Eds.). Conceptual ecology and invasion biology: reciprocal approaches to nature. Springer, Dordrecht, The Netherlands.

Campbell, R. W. and G. Kennedy. 2009. Birds of British Columbia. Lone Pine Publishing, Edmonton, Canada.

Capelli, G. M. 1982. Displacement of northern Wisconsin crayfish by *Orconectes rusticus* (Girard). Limnology and Oceanography 27: 741–745.

Carlton, J. T. 1996. Biological invasions and cryptogenic species. Ecology 77: 1653–1655.

Carpenter, S. R. and 20 other authors. 2007. Understanding regional change: a comparison of two lake districts. BioScience 57: 323–335.

Carreira, B. M., M. P. Dias, and R. Rebelo. 2014. How consumption and fragmentation of macrophytes by the invasive crayfish *Procambarus clarkii* shape the macrophyte communities of temporary ponds. Hydrobiologia 721: 89–98.

Carrington, D. 2014. Shooters set their sights on UK's last remaining ruddy ducks. The Guardian. August 8, 2014.

Carter, C. 2014. Ruddy duck culls to cost £2,400 per bird. The Telegraph. January 22, 2014.

Censky, E. J., K. Hodge, and J. Dudley. 1998. Over-water dispersal of lizards due to hurricanes. Nature 395: 556.

Chadde, S. W. 2011. Wetland plants of the Chicago region: a complete guide to the wetland and aquatic plants of Chicago and vicinity. Published by the author in the USA.

Chaiken, M. 2004. Starling song: an open-ended learner. In: Marler, P. and H. Slabbekoorn (Eds.). P. 83. Nature's music: the science of birdsong. Elsevier Academic Press, Amsterdam, San Diego, and London.

Chambers, P. A., J. M. Hanson, J. M. Burke, and E. E. Prepas. 1990. The impact of the crayfish *Orconectes virilis* on aquatic macrophytes. Freshwater Biology 24: 81–91.

Chatham, R. 1988. Summer, and other small things. Pp. 185–193. In: Chatham, R. (Ed.). Silent seasons. Clark City Press, Livingston, Montana, USA.

Chester, S. 2008. A wildlife guide to Chile. Princeton University Press, Princeton and Oxford.

Chew, M. K. and M. D. Laubichler. 2003. Natural enemies – metaphor or misconception? Science 301: 52–53.

Chitty, D. 1996. Do lemmings commit suicide?: beautiful hypotheses and ugly facts. Oxford University Press, Oxford and New York.

Ciaranca, M. A., C. C. Allin, and G. S. Jones. 2007. Mute swan (*Cygnus olor*). In: A. Poole and F. Gill (Eds.). The birds of North America, No. 273. The Academy of Natural Sciences, Philadelphia, PA, and The American Ornithologists' Union, Washington, D.C., USA.

Cirillo, C. 2010. Invasive plants a threat to biodiversity. The Toronto Star. July 18, 2010.

Claridge, M. F., H. A. Dawah, and M. R. Wilson (Eds.). 1997. Species: the units of biodiversity. Chapman & Hall, London, UK.

Clay, K. 2003. Parasites lost. Nature 421: 585–586.

Clutton-Brock, J. 1999. A natural history of domesticated mammals. Second edition. Cambridge University Press, Cambridge, UK.

Coates, P. 2006. American perceptions of immigrant and invasive species – strangers on the land. University of California Press, Berkeley and Los Angeles, USA.

Coeurdassier, M., R. Riols, A. Decors, A. Mionnet, F. David, T. Quintaine, D. Truchetet, R. Scheifler, and P. Giraudoux. 2014. Unintentional wildlife poisoning and proposals for sustainable management of rodents. Conservation Biology 28: 315–321.

Colautti, R. I., S. A. Bailey, C. D. A. van Overdijk, K. Amundsen, and H. J. MacIsaac. 2006. Characterised and projected costs of nonindigenous species in Canada. Biological Invasions 8: 45–59.

Colautti, R. I. and H. J. MacIsaac. 2004. A neutral terminology to define 'invasive' species. Diversity and Distributions 10: 135–141.

Colautti, R. I., J. R. Muirhead, R. N. Biswas, and H. J. MacIsaac. 2005. Realized vs apparent reduction in enemies of the European starling. Biological Invasions 7: 723–732.

Connelly, N. A., C. R. O'Neill Jr., B. A. Knuth, and T. L. Brown. 2007. Economic impacts of zebra mussels on drinking water treatment and electric power generation facilities. Environmental Management 40: 105–112.

Cooling, M., S. Hartley, D. A. Sim, and P. J. Lester. 2012. The widespread collapse of an invasive species: Argentine ants (*Linepithema humile*) in New Zealand. Biology Letters 8: 430–433.

Cooling, M. and B. D. Hoffmann. 2015. Here today, gone tomorrow: declines and local extinctions of invasive ant populations in the absence of intervention. Biological Invasions 17: 3351–3357.

Cooper, A., I. A. E. Atkinson, W. G. Lee, and T. H. Worthy. 1993. Evolution of the moa and their effect on the New Zealand flora. Trends in Ecology and Evolution 8: 433–437.

Cooper, J. E. and S. A. Armstrong. 2007. Locality records and other data for invasive crayfishes (Decapoda: Cambaridae) in North Carolina. Journal of the North Carolina Academy of Science 123: 1–13.

Corbet, G. B. 1986. Relationships and origins of the European lagomorphs. Mammal Review 16: 105–110.

Cornwall, W. 2014. There will be blood. Conservation 15: 40–46. Fall 2014 issue.

COSEWIC. 2000. COSEWIC assessment and status report on the queen snake *Regina septemvittata* in Canada. Committee on the Status of Endangered Wildlife in Canada. Ottawa, Canada.

Cox, D. D. 2002. A naturalist's guide to wetland plants: an ecology for eastern North America. Syracuse University Press, Syracuse, New York, USA.

Cox, D. K. and J. J. Beauchamp. 1982. Thermal resistance of juvenile crayfish, *Cambarus bartoni* (Fabricius): experiment and model. American Midland Naturalist 108: 187–193.

Coyle, J. 2016. Drones, dogs vs. nature's invaders. The Toronto Star. February 5, 2016.

Coyne, J. A. 2009. Why evolution is true. Viking. New York, New York, USA.

Crichton, M. 1990. Jurassic Park. Ballantine Books, New York, USA.

Crocker, D. W. and D. W. Barr. 1968. Handbook of the crayfishes of Ontario. University of Toronto Press, Toronto, Canada.

Cronon, W. 1996. Foreword to the Paperback Edition of *Uncommon ground: rethinking the human place in nature*. Cronon, W. (Ed.). W. W. Norton & Company, New York and London.

Crossland, M. R. 2000. Direct and indirect effects of the introduced toad *Bufo marinus* (Anura: Bufonidae) on populations of native anuran larvae in Australia. Ecography 23: 283–290.

Crowcroft, P. 1991. Elton's ecologists: a history of the Bureau of Animal Population. The University of Chicago Press, Chicago, USA.

Darwin, C. 1845. Journal of researches into the natural history and geology of the countries visited during the voyage of H.M.S. "Beagle" round the world, under the command of Capt. Fitz Roy, R. N. John Murray, London.

Darwin, C. 1859. On the origin of species by means of natural selection, or the preservation of favoured races in the struggle for life. John Murray, London.

Dash, M. 2000. Tulipomania. Three Rivers Press, New York, USA.

Davidson, R. S. and D. P. Armstrong. 2002. Estimating impacts of poison operations on non-target species using mark-recapture analysis and simulation modelling: an example with saddlebacks. Biological Conservation 105: 375–381.

Davis, M. A. 2006. Invasion biology 1958–2005: the pursuit of science and conservation. Pp. 35–64. In: Cadotte, M. W., S. M. McMahon, and T. Fukami (Eds.). Conceptual ecology and invasion biology: reciprocal approaches to nature. Springer, Dordrecht, The Netherlands.

Davis, M. A. 2009. Invasion biology. Oxford University Press, Oxford, UK.

Davis, M. A., M. D. Anderson, L. Bock-Brownstein, A. Staudenmaier, M. Suliteanu, A. Wareham, and J. J. Dosch. 2015. Little evidence of native and non-native species influencing one another's abundance and distribution in the herb layer of an oak woodland. Journal of Vegetation Science 26: 1005–1012. Doi: 10.1111/jvs.12302.

Davis, M. A., M. K. Chew, R. J. Hobbs, A. E. Lugo, J. J. Ewel, G. J. Vermeij, J. H. Brown, M. L. Rosenzweig, M. R. Gardener, S. P. Carroll, K. Thompson, S. T. A. Pickett, J. C. Stromberg, P. Del Tredici, K. N. Suding, J. G. Ehrenfeld, J. P. Grime, J. Mascaro, and J. C. Briggs. 2011. Don't judge species on their origins. Nature 474: 153–154.

Davis, M. A., A. Colehour, J. Daney, E. Foster, C. MacMillen, E. Merrill, J. O'Neil, M. Pearson, M. Whitney, M. D. Anderson, and J. J. Dosch. 2012. The population dynamics and ecological effects of garlic mustard, *Alliaria petiolata*, in a Minnesota oak woodland. The American Midland Naturalist 168: 364–374.

Davis, M. A., C. MacMillen, M. LeFevre-Levy, C. Dallavalle, N. Kriegel, S. Tyndel, Y. Martinez, M. D. Anderson, and J. J. Dosch. 2014. Population and plant community dynamics involving garlic mustard (*Alliaria petiolata*) in a Minnesota Oak Woodland: a four year study. Journal of the Torrey Botanical Society 141: 205–216.

Davis, M. A. and K. Thompson. 2001. Invasion terminology: should ecologists define their terms differently than others? No, not if we want to be of any help! Bulletin of the Ecological Society of America 82: 206.

Davis, M. A., K. Thompson, and J. P. Grime. 2001. Charles Elton and the dissociation of invasion ecology from the rest of ecology. Diversity and Distributions 7: 97–102.

Davis, M. A. and K. Thompson. 2002. "Newcomers" invade the field of invasion ecology: question the field's future. Bulletin of the Ecological Society of America 83: 196–197.

Dawkins, R. 1976. The selfish gene. Oxford University Press, Oxford, UK.

De Roy, T. and M. Jones. 2006. New Zealand: a natural history. Firefly Books Ltd., Richmond Hill, Ontario, Canada.

De Villiers, M. 2003. Water: the fate of our most precious resource. Second edition. McClelland & Stewart Ltd., Toronto, Canada.

Del Tredici, P. 2010. Wild urban plants of the northeast – a field guide. Comstock Publishing Associates, a division of Cornell University Press, Ithaca and London.

Diamond, J. 1999. Guns, germs, and steel: the fates of human societies. W. W. Norton & Company, New York and London.

Dickenson, V. 2014. Rabbit. Reaktion Books Ltd., London, UK.

Dickinson, R. and F. Royer. 2014. Plants of southern Ontario – trees, shrubs, wildflowers, grasses, ferns & aquatic plants. Lone Pine Publishing, Vancouver, Canada.

Dickinson, T., D. Metsger, J. Bull, and R. Dickinson. 2004. The ROM field guide to wildflowers of Ontario. Royal Ontario Museum and McClelland and Stewart Ltd., Toronto, Canada.

DiDonato, G. T. and D. M. Lodge. 1993. Species replacements among *Orconectes* crayfishes in Wisconsin lakes: the role of predation by fish. Canadian Journal of Fisheries and Aquatic Sciences 50: 1484–1488.

Diehl, J. K., N. J. Holliday, C. J. Lindgren, and R. E. Roughley. 1997. Insects associated with purple loosestrife, *Lythrum salicaria* L., in southern Manitoba. The Canadian Entomologist 129: 937–948.

Dikötter, F. 2010. Mao's great famine: the history of China's most devastating catastrophe, 1958–1962. Walker & Co., New York, USA.

DiStefano, R. J., M. E. Litvan, and P. T. Horner. 2009. The bait industry as a potential vector for alien crayfish introductions: problem recognition by fisheries agencies and a Missouri evaluation. Fisheries 34: 586–597.

Dodds, W. K. 2009. Laws, theories, and patterns in ecology. University of California Press, Berkeley, USA.

Doody, J. S., B. Green, R. Sims, D. Rhind, P. West, and D. Steer. 2006. Indirect impacts of invasive cane toads (*Bufo marinus*) on nest predation in pig-nosed turtles (*Carettochelys insculpta*). Wildlife Research 33: 349–354.

Dorn, N. J. and J. M. Wojdak. 2004. The role of omnivorous crayfish in littoral communities. Oecologia 140: 150–159.

Dresser, C. and B. Swanson. 2013. Preemptive legislation inhibits the anthropogenic spread of an aquatic invasive species, the rusty crayfish (*Orconectes rusticus*). Biological Invasions 15: 1049–1056.

Dubé, J. and J.-F. Desroches. 2007. Les écrevisses du Québec. Ministère des Ressources Naturelles et de la Faune, Longueuil, Québec.

Dubey, S. and R. Shine. 2008. Origin of the parasites of an invading species, the Autralian cane toad (*Bufo marinus*): are the lungworms Australian or American? Molecular Ecology 17: 4418–4424.

Duncan, J. R. 2013. The complete book of North American owls. Thunder Bay Press, San Diego, USA.

Dunham, D. W., S. D. Jordan, and M. Niglas. 1979. A new color morph of the crayfish *Orconectes propinquus* (Girard). The American Midland Naturalist 102: 384–387.

Easteal, S. 1981. The history of introductions of *Bufo marinus* (Amphibia: Anura); a natural experiment in evolution. Biological Journal of the Linnean Society 16: 93–113.

Eaton, E. R. and K. Kaufman. 2007. Kaufman field guide to insects of North America. Houghton Mifflin Company, New York, USA.

Eckenwalder, J. E. 1999. Vascular plants. Pp. 105–115. In: Roots, B. I., D. A. Chant, and C. E. Heidenreich (Eds.). Special places – the changing ecosystems of the Toronto region. UBC Press, Vancouver, Canada.

Eder, T. 2002. Mammals of Ontario. Lone Pine Publishing, Edmonton, Canada.

Edwards, B. A., D. A. Jackson, and K. M. Somers. 2009. Multispecies crayfish declines in lakes: implications for species distributions and richness. Journal of the North American Benthological Society 28: 719–732.

Ellis, R. 2009. Big fish. Abrams, New York, USA.

Ellison, A. M., N. J. Gotelli, E. J. Farnsworth, and G. D. Alpert. 2012. A field guide to the ants of New England. Yale University Press, New Haven and London.

Elton, C. S. 1958. The ecology of invasions by animals and plants. The University of Chicago Press (the University of Chicago Press edition of this classic book was published in 2000), Chicago and London.

Enberg, T. 2002. Frankenfish raises fears. Metro Today. July 30, 2002.

Ensing, D. 2015. Pike (*Esox lucius*) could have been an exclusive human introduction to Ireland after all: a comment on Pedreschi et al. (2014). Journal of Biogeography 42: 604–607.

Evans, A. V. and C. L. Bellamy. 2000. An inordinate fondness for beetles. University of California Press, Berkeley and Los Angeles, California, USA.

Falk, K. 2007. Brown-headed cowbird. Pp. 602–603. In: Cadman, M. D., D. A. Sutherland, G. G. Beck, D. Lepage, and A. R. Couturier (Eds.). Atlas of the breeding birds of Ontario 2001–2005. Bird Studies Canada, Environment Canada, Ontario Field Ornithologists, Ontario Ministry of Natural Resources, and Ontario Nature, Toronto, Canada.

Falk-Petersen, J., T. Bøhn, and O. T. Sandlund. 2006. On the numerous concepts in invasion biology. Biological Invasions 8: 1409–1424.

Farnham, T. J. 2007. Saving nature's legacy: origins of the idea of biological diversity. Yale University Press. New Haven and London.

Farnsworth, E. J. and D. R. Ellis. 2001. Is purple loosestrife (*Lythrum salicaria*) an invasive threat to freshwater wetlands? Conflicting evidence from several ecological metrics. Wetlands 21: 199–209.

Feinstein, J. 2011. Field guide to urban wildlife. Stackpole Books, Mechanicsburg, USA.

Fenner, F. and J. Ross. 1994. Myxomatosis. Pp. 205–239. In: Thompson, H. V. and C. M. King (Eds.). The European rabbit: the history and biology of a successful colonizer. Oxford Science Publications. Oxford University Press, Oxford, UK.

Figuerola, J. and A. J. Green. 2002. Dispersal of aquatic organisms by waterbirds: a review of past research and priorities for future studies. Freshwater Biology 47: 483–494.

Findlay, C. S. and J. Houlahan. 1997. Anthropogenic correlates of species richness in southeastern Ontario wetlands. Conservation Biology 11: 1000–1009.

Fitzpatrick, J. F. Jr. 1987. Notes on the so-called "blue color phase" in North American cambarid crawfishes (Decapoda, Astacoidea). Crustaceana 52: 316–319.

Fitzpatrick, J. W. 2004. Bird conservation. Pp. 10.1–10.116. In: Podulka, S., R. W. Rohrbaugh, Jr., and R. Bonney (Eds.). Handbook of bird biology. Second edition. Cornell Lab of Ornithology in association with Princeton University Press, Ithaca, New York, USA.

Flannery, T. and P. Schouten. 2001. A gap in nature. Atlantic Monthly Press, New York, USA.

Fletcher, T. S. 2006. Purple loosestrife *Lythrum salicaria*. Pp. 76–77. In: Boersma, P. D., S. H. Reichard, and A. N. Van Buren (Eds.). Invasive species in the Pacific Northwest. University of Washington Press, Seattle and London.

Flux, J. E. C. 1994. World distribution. Pp. 8–21. In: Thompson, H. V. and C. M. King (Eds.). The European rabbit: the history and biology of a successful colonizer. Oxford Science Publications. Oxford University Press, Oxford, UK.

Forbes, A. A., T. H. Q. Powell, L. L. Stelinski, J. J. Smith, and J. L. Feder. 2009. Sequential sympatric speciation across trophic levels. Science 323: 776–779.

Forsman, E. D. and 26 other authors. 2011. Population demography of northern spotted owls. Studies in Avian Biology No. 40. A publication of the Cooper Ornithological Society. University of California Press, Berkeley, Los Angeles, and London.

Foster, J. and L. A. Sandberg. 2004. Friends or foe? Invasive species and public green space in Toronto. Geographical Review 94: 178–198.

France, R. 1992. The North American latitudinal gradient in species richness and geographical range of freshwater crayfish and amphipods. American Naturalist 139: 342–354.

French, J. R. P. III 1993. How well can fishes prey on zebra mussels in eastern North America? Fisheries 18: 13–19.

Frost, O. W. 1988. Introduction to the 1988 edition of *Journal of a voyage with Bering, 1741–1742* by G. W. Steller. Stanford University Press, Stanford, California, USA.

Fuller, E. 2001. Extinct birds. Revised edition. Comstock Publishing Associates, a division of Cornell University Press, Ithaca, New York, USA.

Gallo, T. and S. Windhager. 2011. Accepting 'the new normal'? Not so fast… Conservation 12: 62. Spring 2011 issue.

Garbutt, N. and J. C. Prudente. 2006. Wild Borneo – the wildlife and scenery of Sabah, Sarawak, Brunei and Kalimantan. The MIT Press, Cambridge, Massachusetts, USA.

Garvey, J. E. and R. A. Stein. 1993. Evaluating how chela size influences the invasion potential of an introduced crayfish (*Orconectes rusticus*). The American Midland Naturalist 129: 172–181.

Gewin, V. 2014. Exotic invasion stabilizes endangered raptor. Frontiers in Ecology and the Environment 12: 374.

Goble, D. D. 2010. Recovery. Pp. 70–103. In: Baur, D. C. and Wm. R. Irvin (Eds.). Endangered Species Act: law, policy, and perspectives. Second edition. ABA Publishing, American Bar Association, Section of Environment, Energy, and Resources, Chicago, Illinois, USA.

Goddard, J. (Ed.). 2010. Concise history of science and invention: an illustrated time line. National Geographic Society, Washington, D.C., USA.

Gould, S. J. 2002. An evolutionary perspective on the concept of native plants. Pp. 335–346. In: Gould, S. J. I have landed. Harmony Books, New York, USA.

Grady, W. 2006. Requiem for a rare bird. Pp. 17–29. In: Grady, W. Bringing back the dodo. McClelland & Stewart Ltd., Toronto, Canada.

Grady, W. 2007. The Great Lakes. Greystone Books, Vancouver, Canada.

Greenlees, M. J., G. P. Brown, J. K. Webb, B. L. Phillips, and R. Shine. 2007. Do invasive cane toads (*Chaunus marinus*) compete with Australian frogs (*Cyclorana australis*)? Austral Ecology 32: 900–907.

Guiașu, R. C. 1991. Phylogeny and osteology of the genera of the family Acanthuridae (Perciformes, Pisces). Master's thesis. University of Toronto, Toronto, Canada.

Guiașu, R. C. 1997. Agonistic behavior patterns and the distribution and status of the *Cambarus* (Decapoda, Cambaridae) crayfish of Ontario, Canada. Ph.D. thesis. University of Toronto, Toronto, Canada.

Guiașu, R. C. 2002a. *Cambarus*. Pp. 609–634. In: Holdich, D. M. (Ed.). Biology of freshwater crayfish. Blackwell Science, Oxford, UK.

Guiașu, R. C. 2002b. Tunnel syndrome. Rotunda – The Magazine of the Royal Ontario Museum 35: 22–27. Summer/Fall 2002 issue.

Guiașu, R. C. 2007a. Conservation and diversity of the crayfishes of the genus *Fallicambarus* Hobbs, 1969 (Decapoda, Cambaridae), with an emphasis on the status of *Fallicambarus fodiens* (Cottle, 1863) in Canada. Crustaceana 80: 207–223.

Guiașu, R. C. 2007b. Book review of "New Zealand: a natural history" by De Roy T. and M. Jones. 2006. Firefly Books Ltd., Richmond Hill, Ontario, Canada. The Canadian Field-Naturalist 121: 109–110.

Guiașu, R. C. 2008a. Specious claims? The Royal Ontario Museum Magazine 40: 26–33. Spring 2008 issue.

Guiașu, R. C. 2008b. Invasive controversy. A reply to Nicholas Mandrak. The Royal Ontario Museum Magazine 41: 6. Summer 2008 issue.

Guiașu, R. C. 2009. Conservation, status, and diversity of the crayfishes of the genus *Cambarus* Erichson, 1846 (Decapoda, Cambaridae). Crustaceana 82: 721–742.

Guiașu, R. C. 2010. New plants are no threat to biodiversity. The Toronto Star. July 22, 2010.

Guiașu, R. C., D. W. Barr, and D. W. Dunham. 1996a. Distribution and status of crayfishes of the genera *Cambarus* and *Fallicambarus* (Decapoda, Cambaridae) in Ontario, Canada. Journal of Crustacean Biology 16: 373–383.

Guiașu, R. C. and D. W. Dunham. 1997a. Initiation and outcome of agonistic contests in male Form I *Cambarus robustus* Girard 1852 crayfish (Decapoda, Cambaridae). Crustaceana 70: 480–496.

Guiașu, R. C. and D. W. Dunham. 1997b. Agonistic interactions in male Form II *Cambarus robustus* Girard 1852 crayfish (Decapoda, Cambaridae) and a comparison between male Form I and Form II intra-form contests. Crustaceana 70: 720–736.

Guiașu, R. C. and D. W. Dunham. 1998. Inter-form agonistic contests in male crayfishes, *Cambarus robustus* (Decapoda, Cambaridae). Invertebrate Biology 117: 144–154.

Guiașu, R. C. and D. W. Dunham. 1999a. Aggressive interactions between the crayfishes *Cambarus bartonii bartonii* and *C. robustus* (Decapoda, Cambaridae): interspecific and intraspecific contests. Journal of Crustacean Biology 19: 131–146.

Guiașu, R. C. and D. W. Dunham. 1999b. Aggression and competition in the surface water *Cambarus* (Decapoda, Cambaridae) crayfish species of Ontario, Canada: an overview. Volume 1. Pp. 701–717. In: Schram, F. R. and J. C. von Vaupel Klein (Eds.). Crustaceans and the biodiversity crisis – Proceedings of the Fourth International Crustacean Congress, Amsterdam, The Netherlands, July 20–24, 1998. Brill Academic Publishers, Leiden, Boston, and Köln.

Guiașu, R. C. and D. W. Dunham. 1999c. Agonistic contests in male Form I *Cambarus bartonii bartonii* (Fabricius, 1798) (Decapoda, Cambaridae) crayfish and a comparison with contests of the same type in *Cambarus robustus* Girard, 1852. Crustaceana 72: 1079–1091.

Guiașu, R. C., D. W. Dunham, and D. W. Barr. 1996b. Interspecific agonistic contests between male *Cambarus bartonii bartonii* (Fabricius 1798) and *Cambarus robustus* Girard 1852 (Decapoda, Cambaridae) crayfish and the possible competition between the two species in Ontario, Canada. Freshwater Crayfish 11 (Proceedings of the International Association of Astacology Eleventh Symposium): 364–377.

Guiașu, R. C. and S. Guiașu. 2003. Entropy in ecology and ethology. Nova Science Publishers, Inc., New York, USA.

Guiașu, R. C. and S. Guiașu. 2010a. The Rich-Gini-Simpson quadratic index of biodiversity. Natural Science 2: 1130–1137.

Guiașu, R. C. and S. Guiașu. 2010b. New measures for comparing the species diversity found in two or more habitats. International Journal of Uncertainty, Fuzziness and Knowledge-Based Systems 18: 691–720.

Guiașu, R. C. and S. Guiașu. 2010c. Diversity measures and coarse-graining in data analysis with an application involving plant species on the Galápagos Islands. Systemics, Cybernetics and Informatics 8: 54–64.

Guiașu, R. C. and S. Guiașu. 2011. The weighted quadratic index of biodiversity for pairs of species: a generalization of Rao's index. Natural Science 3: 795–801.

Guiașu, R. C. and S. Guiașu. 2012. The weighted Gini-Simpson index: revitalizing an old index of biodiversity. International Journal of Ecology 2012, Article ID 478728, 10 pages, doi:10.1155/2012/478728.

Guiașu, R. C. and S. Guiașu. 2014. Weighted Gini-Simpson quadratic index of biodiversity for interdependent species. Natural Science 6: 455–466. http://dx.doi.org/10.4236/ns.2014.67044.

Guiașu, R. C., N. Saleh, E. Mozel, and D. W. Dunham. 2005. Low aggression in juvenile burrowing crayfish, *Fallicambarus fodiens* (Cottle, 1863) (Decapoda, Cambaridae). Crustaceana 78: 421–428.

Guiașu, R. C. and R. Winterbottom. 1993. Osteological evidence for the phylogeny of recent genera of surgeonfishes (Percomorpha, Acanthuridae). Copeia 1993: 300–312.

Guiașu, R. C. and R. Winterbottom. 1998. Yellow juvenile color pattern, diet switching and the phylogeny of the surgeonfish genus *Zebrasoma* (Percomorpha, Acanthuridae). Bulletin of Marine Science 63: 277–294.

Hager, H. A. and K. D. McCoy. 1998. The implications of accepting untested hypotheses: a review of the effects of purple loosestrife (*Lythrum salicaria*) in North America. Biodiversity and Conservation 7: 1069–1079.

Hager, H. A. and R. D. Vinebrooke. 2004. Positive relationships between invasive purple loosestrife (*Lythrum salicaria*) and plant species diversity and abundance in Minnesota wetlands. Canadian Journal of Botany 82: 763–773.

Hall, P. W., C. D. Jones, A. Guidotti, and B. Hubley. 2014. The ROM field guide to butterflies of Ontario. Royal Ontario Museum Science Publication, Toronto, Canada.

Hamilton, G. 2010. Super species – the creatures that will dominate the planet. Firefly Books Ltd., Richmond Hill, Ontario, Canada.

Hambler, C. 2004. Conservation. Cambridge University Press, Cambridge, UK.

Hamr, P. 1998. Conservation status of Canadian freshwater crayfishes. World Wildlife Fund and The Canadian Nature Federation, Toronto, Canada.

Hamr, P. 2010. The biology, distribution and management of the introduced rusty crayfish, *Orconectes rusticus* (Girard), in Ontario, Canada. Freshwater Crayfish 17: 85–90.

Harrison, C. and A. Greensmith. 1993. Birds of the world. Stoddart Publishing Co. Limited, Toronto, Canada.

Hart-Davis, A. (Ed.). 2007. History. DK Publishing, New York, USA.

Hastings, M. 2011. Winston's war: Churchill, 1940–1945. Vintage Books, New York, USA.

Heard, M. J. and D. F. Sax. 2013. Coexistence between native and exotic species is facilitated by asymmetries in competitive ability and susceptibility to herbivores. Ecology Letters 16: 206–213.

Heckenberger, M. J., J. C. Russell, C. Fausto, J. R. Toney, M. J. Schmidt, E. Pereira, B. Franchetto, and A. Kuikuro. 2008. Pre-Columbian urbanism, anthropogenic landscapes, and the future of the Amazon. Science 321: 1214–1217.

Hein, C. L., B. M. Roth, A. R. Ives, and M. J. Vander Zanden. 2006. Fish predation and trapping for rusty crayfish (*Orconectes rusticus*) control: a whole lake experiment. Canadian Journal of Fisheries and Aquatic Sciences 63: 383–393.

Hein, C. L., M. J. Vander Zanden, and J. J. Magnuson. 2007. Intensive trapping and increased fish predation cause massive population decline of an invasive crayfish. Freshwater Biology 52: 1134–1146.

Henderson, C. 2012. The book of barely imagined beings: a 21st century bestiary. Granta Books, London, UK.

Henderson, I. 2009. Progress of the UK ruddy duck eradication programme. British Birds 102: 680–690.

Hennig, W. 1950. Grundzüge einer Theorie der phylogenetischen Systematik. Deutsche Zentralverlag, Berlin.

Hennig, W. 1966. Phylogenetic systematics. University of Illinois Press, Urbana, Illinois, USA.

Hill, A. M. and D. M. Lodge. 1994. Diel changes in resource demand: competition and predation in species replacement among crayfishes. Ecology 75: 2118–2126.

H. H. Hobbs, Jr., 1989. An illustrated checklist of the American crayfishes (Decapoda: Astacidae, Cambaridae, and Parastacidae). Smithsonian Contributions to Zoology 480: 1–236.

H. H. Hobbs, III, J. P. Jass, and J. V. Huner. 1989. A review of global crayfish introductions with particular emphasis on two North American species (Decapoda, Cambaridae). Crustaceana 56: 299–316.

Hobbs, J. and R. Cannings. 2007. Spotted owls: shadows in an old-growth forest. Greystone Books, Vancouver, Canada.

Holdich, D. M. 1999. The introduction of alien crayfish species into Britain for commercial exploitation – An own goal? Pp. 85–97. In: von Vaupel Klein, J. C. and F. R. Schram (Eds.). The biodiversity crisis and Crustacea: Proceedings of the Fourth International Crustacean Congress, Amsterdam, The Netherlands, July 20–24, 1998. A. A. Balkema, Rotterdam, The Netherlands.

Hole, D. G., M. J. Whittingham, R. B. Bradbury, G. Q. A. Anderson, P. L. M. Lee, J. D. Wilson, and J. R. Krebs. 2002. Widespread local house-sparrow extinctions. Nature 418: 931–932.

Holland, B. S. 2009. Land snails. Pp. 537–542. In: Gillespie, R. G. and D. A. Clague (Eds.). Encyclopedia of islands. University of California Press, Berkeley and Los Angeles, USA.

Holland, R. E., T. H. Johengen, and A. M. Beeton. 1995. Trends in nutrient concentrations in Hatchery Bay, western Lake Erie, before and after *Dreissena polymorpha*. Canadian Journal of Fisheries and Aquatic Sciences 52: 1202–1209.

Hölldobler, B. and E. O. Wilson. 2009. The superorganism: the beauty, elegance, and strangeness of insect societies. W. W. Norton & Company, New York and London.

Holm, E., N. E. Mandrak, and M. E. Burridge. 2009. The ROM field guide to freshwater fishes of Ontario. Royal Ontario Museum Science Publication, Toronto, Canada.

Horn, M. 2009. York University: the way must be tried. McGill-Queen's University Press, Canada.

Houlahan, J. E. and C. S. Findlay. 2004. Effect of invasive plant species on temperate wetland plant diversity. Conservation Biology 18: 1132–1138.

Hughes, J. M. 2001. The ROM field guide to birds of Ontario. Royal Ontario Museum and McClelland and Stewart Ltd., Toronto, Canada.

Hulme, P. E. and 24 other authors. 2015. Challenging the view that invasive non-native plants are not a significant threat to the floristic diversity of Great Britain. Proceedings of the National Academy of Sciences of the United States of America 112: E2988–E2989.

Humair, F., P. J. Edwards, M. Siegrist, and C. Kueffer. 2014. Understanding misunderstandings in invasion science: why experts don't agree on common concepts and risk assessments. NeoBiota 20: 1–30. doi: 10.3897/neobiota.20.6043.

Huston, M. A. 1994. Biological diversity – the coexistence of species on changing landscapes. Cambridge University Press, Cambridge, UK.

Ingram, J. 2006. You may pine for it when it's gone – the house sparrow is in steep decline. The Toronto Star. August 19, 2006.

Invasive Species Centre. 2016. http://www.invasivespeciescentre.ca. Accessed on February 17th, 2016.

Isaac, N. J. B., J. Mallet, and G. M. Mace. 2004. Taxonomic inflation: its influence on macroecology and conservation. Trends in Ecology and Evolution 19: 464–469.

Ito, T. 2014. London Zoo and the Victorians 1828–1859. A Royal Historical Society publication. The Boydell Press, Woodbridge, Suffolk, UK.

IUCN (World Conservation Union) – Invasive Species Specialist Group (ISSG). 2014 and 2016. 100 of the World's Worst Invasive Alien Species list. *Cyprinus carpio* (fish). www.issg.org/worst100_species.html. Accessed on November 28th, 2014 and January 24th, 2016.

Jazdzewski, K. and A. Konopacka. 2002. Invasive Ponto-Caspian species in waters of the Vistula and Oder basins and the southern Baltic Sea. Pp. 384–398. In: Leppäkoski, E., S. Gollasch, and S. Olenin (Eds.). Invasive aquatic species of Europe: distribution, impacts, and management. Kluwer Academic Publishers, Dordrecht, Boston, and London.

Jenkins, V. S. 1994. The lawn – a history of an American obsession. Smithsonian Institution Press, Washington and London.

Jezerinac, R. F. 1993. A new subgenus and species of crayfish (Decapoda: Cambaridae) of the genus *Cambarus*, with an amended description of the subgenus *Lacunicambarus*. Proceedings of the Biological Society of Washington 106: 532–544.

Jezerinac, R. F., G. W. Stocker, and D. C. Tarter. 1995. The crayfishes (Decapoda: Cambaridae) of West Virginia. Ohio Biological Survey Bulletin New Series 10: 1–193.

Johnsgard, P. A. and M. Carbonell. 2010. Ruddy ducks and other stifftails: their behavior and biology. University of Oklahoma Press, Norman, Oklahoma, USA.

Johnson, H. 2010. The world of trees. University of California Press, Berkeley and Los Angeles, USA.

Johnson, P. 2009. Churchill. Penguin Books, New York, USA.

Johnston, R. F. and R. K. Selander. 1964. House sparrows: rapid evolution of races in North America. Science 144: 548–550.

Jonsson, L. 1993. Birds of Europe: with North Africa and the Middle East. Princeton University Press, Princeton, New Jersey, USA.

Jurmain, R., L. Kilgore, W. Trevathan, and H. Nelson. 2004. Essentials of physical anthropology. Wadsworth/Thomson Learning. Belmont, USA.

Kagume, K. 2008. Ontario nature guide. Lone Pine Publishing Inc., Edmonton, Canada.

Kahn, C. H. 1979. The art and thought of Heraclitus. Cambridge University Press, Cambridge, UK.

Karatayev, V. A., A. Y. Karatayev, L. E. Burlakova, and D. K. Padilla. 2013. Lakewide dominance does not predict the potential for spread of dreissenids. Journal of Great Lakes Research 39: 622–629.

Kays, R. W. and D. E. Wilson. 2009. Mammals of North America. Second Edition. Princeton University Press, Princeton and Oxford.

Keedwell, R. J., R. F. Maloney, and D. P. Murray. 2002. Predator control for protecting kaki (*Himantopus novaezelandiae*) – lessons from 20 years of management. Biological Conservation 105: 369–374.

Keller, R. P., D. M. Lodge, M. A. Lewis, and J. F. Shogren (Eds.). 2009. Bioeconomics of invasive species. Integrating ecology, economics, policy, and management. Oxford University Press, Oxford and New York.

Kendle, A. D. and J. E. Rose. 2000. The aliens have landed! What are the justifications for 'native only' policies in landscape plantings? Landscape and Urban Planning 47: 19–31.

Kerasote, T. 2005. The big flap. Audubon 107: 44–50. May–June 2005 issue.

Kershaw, L. 2001. Trees of Ontario. Lone Pine Publishing, Edmonton, Canada.

Kershaw, L. 2002. Ontario wildflowers. Lone Pine Publishing, Edmonton, Canada.

Kidd, K. 2006. The return of Toronto's big trees. The Toronto Star. May 7, 2006.

Kirch, P. V. 2002. On the road of the winds. University of California Press, Berkeley, USA.

Kiviat, E. 1996. American Goldfinch nests in purple loosestrife. The Wilson Bulletin 108: 182–186.

Koch, P. L. and A. D. Barnosky. 2006. Late Quaternary extinctions: state of the debate. Annual Review of Ecology, Evolution, and Systematics 37: 215–250.

Koenig, W. D. 2003. European starlings and their effect on native cavity-nesting birds. Conservation Biology 17: 1134–1140.

Kornás, J. 1990. Plant invasions in Central Europe: historical and ecological aspects. Pp. 19–36. In: di Castri, F., A. J. Hansen, and M. Debussche (Eds.). Biological invasions in Europe and the Mediterranean basin. Kluwer Academic Publishers, Dordrecht, The Netherlands.

Kortright Centre for Conservation. 2015. http://kortright.org/. Accessed on July 15th, 2015.

Kuhlmann, M. L. 2008. Do invading rusty crayfish interfere with reproduction in a native congener? Journal of Crustacean Biology 28: 461–465.

Kumschick, S., M. Gaertner, M. Vilà, F. Essl, J. M. Jeschke, P. Pyšek, A. Ricciardi, S. Bacher, T. M. Blackburn, J. T. A. Dick, T. Evans, P. E. Hulme, I. Kühn, A. Mrugala, J. Pergl, W. Rabitsch, D. M. Richardson, A. Sendek, and M. Winter. 2015. Ecological impacts of alien species: quantification, scope, caveats, and recommendations. BioScience 65: 55–63.

Kunz, W. 2012. Do species exist? Wiley-Blackwell, Wiley-VCH Verlag & Co. KGaA, Weinheim, Germany.

Lady Bird Johnson Wildflower Center – The University of Texas at Austin. 2013 and 2014. http://www.wildflower.org/. Accessed on August 2nd, 2013 and September 12th, 2014.

Lang, A. L. 2007. House sparrow. Pp. 628–629. In: Cadman, M. D., D. A. Sutherland, G. G. Beck, D. Lepage, and A. R. Couturier (Eds.). Atlas of the breeding birds of Ontario 2001–2005. Bird Studies Canada, Environment Canada, Ontario Field Ornithologists, Ontario Ministry of Natural Resources, and Ontario Nature, Toronto, Canada.

Lange, I. M. 2002. Ice Age mammals of North America. Mountain Press Publishing Company, Missoula, Montana, USA.

Langton, J. 2006. Foreign bird has made itself at home – native to Europe, starlings blanket North America. The Toronto Star. February 18, 2006.

Larson, B. M. H. 2005. The war of the roses: demilitarizing invasion biology. Frontiers in Ecology and the Environment 3: 495–500.

Larson, B. M. H. 2008. Entangled biological, cultural and linguistic origins of the war on invasive species. Pp. 169–195. In: Frank, R. M., R. Dirven, T. Ziemke, and E. Bernárdez (Eds.). Body, language and mind. Volume 2: Sociocultural situatedness. Mouton de Gruyter, Berlin and New York.

Larson, B. 2011. Metaphors for environmental sustainability: redefining our relationship with nature. Yale University Press, New Haven and London.

Larson, B. M. H. 2015. Seeking citizenship: the Norway maple (*Acer platanoides*) in Canada. Pp. 176–190. In: Sandberg, L. A., A. Bardekjian, and S. Butt (Eds.). Urban forests, trees, and greenspace: a political ecology perspective. Routledge, Taylor & Francis Group, London and New York.

Larson, E. R., C. A. Busack, J. D. Anderson, and J. D. Olden. 2010. Widespread distribution of the non-native northern crayfish (*Orconectes virilis*) in the Columbia River basin. Northwest Science 84: 108–111.

Lautenschlager, R. A. 1997. Biodiversity is dead. Wildlife Society Bulletin 25: 679–685.

Lavoie, C. 2010. Should we care about purple loosestrife? The history of an invasive plant in North America. Biological Invasions 12: 1967–1999.

Lavoie, C., M. Jean, F. Delisle, and G. Létourneau. 2003. Exotic plant species of the St Lawrence River wetlands: a spatial and historical analysis. Journal of Biogeography 30: 537–549.

Lees, A. C. and D. J. Bell. 2008. A conservation paradox for the 21st century: the European wild rabbit *Oryctolagus cuniculus*, an invasive alien and an endangered native species. Mammal Review 38: 304–320.

Lerdau, M. and J. D. Wickham. 2011. Non-natives: four risk factors. Nature 475: 36–37.

Letnic, M., J. K. Webb, and R. Shine. 2008. Invasive cane toads (*Bufo marinus*) cause mass mortality of freshwater crocodiles (*Crocodylus johnstoni*) in tropical Australia. Biological Conservation 141: 1773–1782.

Lever, C. 2001. The cane toad: the history and ecology of a successful colonist. Westbury Academic and Scientific Publishing, Otley, West Yorkshire, UK.

Lever, C. 2003. Naturalized reptiles and amphibians of the world. Oxford University Press, Oxford, UK (reprinted in 2006).

Lever, C. 2009. The naturalized animals of Britain and Ireland. New Holland Publishers (UK) Ltd., London, UK.

Lewis, A. M. and H. Keier. 2006. Star Wars – the new essential guide to alien species. Del Rey Books, New York, USA.

Linn, C. Jr., J. L. Feder, S. Nojima, H. R. Dambroski, S. H. Berlocher, and W. Roelofs. 2003. Fruit odor discrimination and sympatric host race formation in *Rhagoletis*. Proceedings of the National Academy of Sciences of the United States of America 100: 11490–11493.

Littlepage, D. 2006. Steller's island: adventures of a pioneer naturalist in Alaska. The Mountaineers Books, Seattle, USA.

Lockwood, J. L., M. F. Hoopes, and M. P. Marchetti. 2007. Invasion ecology. Blackwell Publishing, Oxford, UK.

Lockwood, J. L., M. F. Hoopes, and M. P. Marchetti. 2011. Non-natives: plusses of invasion ecology. Nature 475: 36.

Lodge, D. M. 1993. Biological invasions: lessons for ecology. Trends in Ecology and Evolution 8: 133–137.

Lodge, D. M., A. L. Beckel, and J. J. Magnuson. 1985. Lake-bottom tyrant. Natural History 94: 32–37.

Lomolino, M. V., B. R. Riddle, R. J. Whittaker, and J. H. Brown. 2010. Biogeography. Fourth edition. Sinauer Associates, Inc. Publishers, Sunderland, Massachusetts, USA.

Lougheed, T. 2008. Federal agency sows seeds for future plant experts. University Affairs 49 (1): 39–40. January 2008 issue.

Lowe, R. L. and R. W. Pillsbury. 1995. Shifts in benthic algal community structure and function following the appearance of zebra mussels (*Dreissena polymorpha*) in Saginaw Bay, Lake Huron. Journal of Great Lakes Research 21: 558–566.

Lu, W. and J. Lazell. 1996. The voyage of the beetle. Natural History 105: 36–39.

Ludyanskiy, M. L., D. McDonald, and D. MacNeill. 1993. Impact of the zebra mussel, a bivalve invader. BioScience 43: 533–544.

Lumsden, H. G. 2007. Trumpeter swan. Pp. 66–67. In: Cadman, M. D., D. A. Sutherland, G. G. Beck, D. Lepage, and A. R. Couturier (Eds.). Atlas of the breeding birds of Ontario 2001–2005. Bird Studies Canada, Environment Canada, Ontario Field Ornithologists, Ontario Ministry of Natural Resources, and Ontario Nature, Toronto, Canada.

Lumsden, H. G. and M. C. Drever. 2002. Overview of the trumpeter swan reintroduction program in Ontario, 1982–2000. Waterbirds 25 (Special Publication 1): 301–312.

Luque, G. M., C. Bellard, C. Bertelsmeier, E. Bonnaud, P. Genovesi, D. Simberloff, and F. Courchamp. 2014. The 100th of the world's worst invasive alien species. Biological Invasions 16: 981–985.

MacCulloch, R. D. 2002. The ROM field guide to amphibians and reptiles of Ontario. Royal Ontario Museum and McClelland and Stewart Ltd., Toronto, Canada.

Macdonald, D. W. and P. Barrett. 1993. Mammals of Europe. Princeton University Press. Princeton and Oxford.

MacKinnon, J. 2008. Order Artiodactyla. Pp. 451–481. In: Smith, A. T. and Y. Xie (Eds.). A guide to the mammals of China. Princeton University Press, Princeton and Oxford.

Macleod, A., P. Parker, and E. Rae. 2010. Explorers: great tales of adventure and endurance. Smithsonian Institution, DK Publishing, New York, USA.

Magner, L. N. 2002. A history of the life sciences. Third edition, revised and expanded. CRC Press, Taylor & Francis Group, Boca Raton, London, and New York.

Mallet, J. 2005. Hybridization as an invasion of the genome. Trends in Ecology and Evolution 20: 229–237.

Mallet, J. 2007. Hybrid speciation. Nature 446: 279–283.

Mandrak, N. E. 2008. Invasive controversy. Letter to the editor. The Royal Ontario Museum Magazine 41: 6. Summer 2008 issue.

Mann, C. C. 2011. 1493: uncovering the new world Columbus created. Alfred A. Knopf, New York, USA.

Marris, E. 2005. Shoot to kill. Nature 438: 272–273.

Marris, E. 2010. The new normal. Conservation Magazine 11: 12–17. April–June 2010.

Marsden, J. E. and M. Hauser. 2009. Exotic species in Lake Champlain. Journal of Great Lakes Research 35: 250–265.

Marshall, S. A. 2006. Insects – their natural history and diversity: with a photographic guide to insects of eastern North America. Firefly Books Ltd., Richmond Hill, Ontario, Canada.

Martin, A. C. 1987. Weeds. St. Martin's Press, New York, USA.

Martin, G. W. and L. D. Corkum. 1994. Predation of zebra mussels by crayfish. Canadian Journal of Zoology 72: 1867–1871.

Martin, P. S. 1984. Prehistoric overkill: the global model. Pp. 354–403. In: Martin, P. S. and R. G. Klein (Eds.). Quaternary extinctions – a prehistoric revolution. The University of Arizona Press, Tucson, USA.

Mather, M. E. and R. A. Stein. 1993. Direct and indirect effects of fish predation on the replacement of a native crayfish by an invading congener. Canadian Journal of Fisheries and Aquatic Sciences 50: 1279–1288.

Mathieu, E. 2009. Newest pest can cut you. The Toronto Star. May 26, 2009.

Maude, S. H. 1988. A new Ontario locality record for the crayfish *Orconectes rusticus* from West Duffin Creek, Durham Regional Municipality. Canadian Field-Naturalist 102: 66–67.

Max, D. T. 2007. The squirrel wars. The New York Times. October 7, 2007.

Mayr, E. 1982. The growth of biological thought. The Belknap Press of Harvard University Press, Cambridge, Massachusetts, USA.

Mayr, E. 2001. What evolution is. Basic Books, New York, USA.

McAllister, D. E. 2000. Biodiversity in Canadian fresh and marine waters. Pp. 81–105. In: Bocking, S. (Ed.). Biodiversity in Canada: ecology, ideas, and action. Broadview Press, Peterborough, Canada.

McAlpine, D. F., T. J. Fletcher, M. A. Osepchook, and J.-C. Savoie. 1999. A range extension for *Orconectes virilis* (Decapoda, Cambaridae) and a third crayfish species for New Brunswick, Canada. Crustaceana 72: 356–358.

J. E. McClellan III and H. Dorn. 1999. Science and technology in world history: an introduction. The Johns Hopkins University Press, Baltimore and London.

McGrath, S. 2005. Attack of the alien invaders. National Geographic 207 (3): 92–117. March 2005 issue.

McKeating, G. 1990. Birds of Toronto and vicinity. Lone Pine Publishing, Edmonton, Canada.

McLaren, M. A. 2007. Northern cardinal. Pp. 578–579. In: Cadman, M. D., D. A. Sutherland, G. G. Beck, D. Lepage, and A. R. Couturier (Eds.). Atlas of the breeding birds of Ontario 2001–2005. Bird Studies Canada, Environment Canada, Ontario Field Ornithologists, Ontario Ministry of Natural Resources, and Ontario Nature, Toronto, Canada.

McLeod, S. R. and G. Saunders. 2014. The European rabbit: Australia's worst mammalian invader. Pp. 429–451. In: Prins, H. H. T. and I. J. Gordon (Eds.). Invasion biology and ecological theory: insights from a continent in transformation. Cambridge University Press, Cambridge, UK.

McMullen, C. K. 1999. Flowering plants of the Galápagos. Comstock Publishing Associates, a division of Cornell University Press, Ithaca and London.

Mellars, P. 2004. Neanderthals and the modern human colonization of Europe. Nature 432: 461–465.

Menon, V. 2011. Invasive species: Is the threat over? The Toronto Star. July 24, 2011.

Meyer, S. M. 2006. The end of the wild. Boston Review, Somerville, USA and The MIT Press, Cambridge, USA.

Micklin, P. 2014. Efforts to revive the Aral Sea. Pp. 361–380. In: Micklin, P., N. V. Aladin, and I. Plotnikov (Eds.). The Aral Sea: the devastation and partial rehabilitation of a great lake. Springer-Verlag, Berlin and Heidelberg.

Millward, A. A. and S. Sabir. 2011. Benefits of a forested urban park: What is the value of Allan Gardens to the city of Toronto, Canada? Landscape and Urban Planning 100: 177–188.

Milton, K. 2000. Ducks out of water: nature conservation as boundary maintenance. Pp. 229–246. In: Knight, J. (Ed.). Natural enemies: people-wildlife conflicts in anthropological perspective. Routledge, Taylor & Francis Group, London and New York.

Mindell, D. P. 2006. The evolving world: evolution in everyday life. Harvard University Press, Cambridge, Massachusetts, USA.

Mitsch, W. J. and J. G. Gosselink. 2007. Wetlands. Fourth edition. John Wiley & Sons, Inc., Hoboken, New Jersey, USA.

Mitsch, W. J., J. G. Gosselink, C. J. Anderson, and L. Zhang. 2009. Wetland ecosystems. John Wiley & Sons, Inc., Hoboken, New Jersey, USA.

Momot, W. T. and J. E. Gall. 1971. Some ecological notes on the blue color phase of the crayfish, *Orconectes virilis*, in two lakes. The Ohio Journal of Science 71: 363–370.

Monastersky, R. 2014. Life – a status report. Nature 516: 158–161.

Myers, J. H. and D. R. Bazely. 2003. Ecology and control of introduced plants. Cambridge University Press, Cambridge, UK.

Myers, K., I. Parer, D. Wood, and B. D. Cooke. 1994. The rabbit in Australia. Pp. 108–157. In: Thompson, H. V. and C. M. King (Eds.). The European rabbit: the history and biology of a successful colonizer. Oxford Science Publications. Oxford University Press, Oxford, UK.

Naddafi, R. and L. G. Rudstam. 2014. Predation on invasive zebra mussel, *Dreissena polymorpha*, by pumpkinseed sunfish, rusty crayfish, and round goby. Hydrobiologia 721: 107–115.

Natural Heritage Information Centre, Ontario Ministry of Natural Resources. 2007, 2008, and 2011. http://nhic.mnr.gov.on.ca/nhic_.cfm. Accessed on August 7th, 2007, June 12th, 2008, and June 15th, 2011.

Noonan, J. P., G. Coop, S. Kudaravalli, D. Smith, J. Krause, J. Alessi, F. Chen, D. Platt, S. Pääbo, J. K. Pritchard, and E. M. Rubin. 2006. Sequencing and analysis of Neanderthal genomic DNA. Science 314: 1113–1118.

Olden, J. D., J. M. McCarthy, J. T. Maxted, W. W. Fetzer, and M. J. Vander Zanden. 2006. The rapid spread of rusty crayfish (*Orconectes rusticus*) with observations on native crayfish declines in Wisconsin (USA.) over the past 130 years. Biological Invasions 8: 1621–1628.

Olsen, T. M., D. M. Lodge, G. M. Capelli, and R. J. Houlihan. 1991. Mechanisms of impact of an introduced crayfish (*Orconectes rusticus*) on littoral congeners, snails, and macrophytes. Canadian Journal of Fisheries and Aquatic Sciences 48: 1853–1861.

O'Neill, G. 2002. The contraceptive plague. New Scientist 175: 4–5. August 10, 2002 issue.

Ono, M., M. Mitsuhata, and M. Sasaki. 1994. Use of introduced *Bombus terrestris* worker helpers for rapid development of Japanese native *B. hypocrita* colonies (Hymenoptera, Apidae). Applied Entomology and Zoology 29: 413–419.

Ontario Federation of Anglers and Hunters. 2003 and 2009. Purple loosestrife: what you should know, what you can do. Peterborough, Ontario, Canada.

Ontario's Invading Species Awareness Program. 2015. Tubenose goby. www.invadingspecies.com/invaders/fish/tubenose-goby/. Accessed on June 3rd, 2015.

Ontario's Invading Species Awareness Program. 2015. Water lettuce. www.invadingspecies.com/invaders/plants-aquatic/water-lettuce/. Accessed on December 31st, 2015.

Ontario Wildflowers. 2015. Yellow avens and wood avens. http://ontariowildflowers.com. Accessed on July 1st, 2015.

Page, L. M. and B. M. Burr. 2011. Peterson field guide to freshwater fishes of North America north of Mexico. Second edition. Houghton Mifflin Harcourt, Boston and New York, USA.

Parmesan, C. 2005. Biotic response: range and abundance changes. Pp. 41–55. In: Lovejoy, T. E. and L. Hannah (Eds.). Climate change and biodiversity. Yale University Press, New Haven and London.

Pattemore, D. E. and D. S. Wilcove. 2012. Invasive rats and recent colonist birds partially compensate for the loss of endemic New Zealand pollinators. Proceedings of the Royal Society B 279: 1597–1605.

Paxton, J. R. 1995. Habitats and adaptations. Pp. 32–41. In: Paxton, J. R. and W. N. Eschmeyer (Eds.). Encyclopedia of fishes. Academic Press Inc., San Diego, USA.

Pedreschi, D., M. Kelly-Quinn, J. Caffrey, M. O'Grady, and S. Mariani. 2014. Genetic structure of pike (*Esox lucius*) reveals a complex and previously unrecognized colonization history of Ireland. Journal of Biogeography 41: 548–560.

Pedreschi, D. and S. Mariani. 2015. Towards a balanced view of pike in Ireland: a reply to Ensing. Journal of Biogeography 42: 607–609.

Pennisi, E. 2006. Speciation standing in place. Science 311: 1372–1374.

Perrett, N. G. 2012. Hikes and outings of South-Central Ontario. Lone Pine Publishing, Edmonton, Canada.

Perry, W. L., D. M. Lodge, and G. A. Lamberti. 2000. Crayfish (*Orconectes rusticus*) impacts on zebra mussel (*Dreissena polymorpha*) recruitment, other macroinvertebrates and algal biomass in a lake-outlet stream. American Midland Naturalist 144: 308–316.

Peters, J. A., M. J. Cooper, S. M. Creque, M. S. Kornis, J. T. Maxted, W. L. Perry, F. W. Schueler, T. P. Simon, C. A. Taylor, R. F. Thoma, D. G. Uzarski, and D. M. Lodge. 2014. Historical changes and current status of crayfish diversity and distribution in the Laurentian Great Lakes. Journal of Great Lakes Research 40: 35–46.

Peters, J. A. and D. M. Lodge. 2009. Invasive species policy at the regional level: a multiple weak links problem. Fisheries 34: 373–380.

Peterson, R. T. and M. McKenny. 1968. A field guide to wildflowers of northeastern and north-central North America. Houghton Mifflin Company, Boston, USA.

Petrides, G. A. 1972. A field guide to trees and shrubs (northeastern and central North America). Second edition. Houghton Mifflin Company, Boston, USA.

Petrie, S. A. and R. W. Knapton. 1999. Rapid increase and subsequent decline of zebra and quagga mussels in Long Point Bay, Lake Erie: possible influence of waterfowl predation. Journal of Great Lakes Research 25: 772–782.

Phillips, B. L., G. P. Brown, and R. Shine. 2003. Assessing the potential impact of cane toads on Australian snakes. Conservation Biology 17: 1738–1747.

Phillips, B. L., G. P. Brown, J. K. Webb, and R. Shine. 2006. Invasion and the evolution of speed in toads. Nature 439: 803.

Phillips, B. L. and R. Shine. 2004. Adapting to an invasive species: Toxic cane toads induce morphological change in Australian snakes. Proceedings of the National Academy of Sciences of the United States of America 101: 17150–17155.

Phillips, B. L. and R. Shine. 2005. The morphology, and hence impact, of an invasive species (the cane toad, *Bufo marinus*): changes with time since colonisation. Animal Conservation 8: 407–413.

Phillips, I. D., R. D. Vinebrooke, and M. A. Turner. 2009. Ecosystem consequences of potential range expansions of *Orconectes virilis* and *Orconectes rusticus* crayfish in Canada – a review. Environmental Review 17: 235–248.

Pick, N. and M. Sloan. 2004. The rarest of the rare: stories behind the treasures at the Harvard Museum of Natural History. HarperCollins Publishers Inc., New York, New York, USA.

Pimentel, D., R. Zuniga, and D. Morrison. 2005. Update on the environmental and economic costs associated with alien-invasive species in the United States. Ecological Economics 52: 273–288.

Piper, R. 2009. Extinct animals. Greenwood Press, Westport, Connecticut, USA.

Ponting, C. 1993. A green history of the world: the environment and the collapse of great civilizations. Penguin Books, New York, USA.

Ponting, C. 2007. A new green history of the world. Penguin Books, New York, USA.

Pridham, D. 2012. The landowner's guide to controlling invasive woodland plants. Ontario Federation of Anglers and Hunters, Ontario Invasive Plant Council, Ontario's Invading Species Awareness Program, Victoria Stewardship Council, Government of Ontario.

Primack, R. B. 2008. A primer of conservation biology. Fourth edition. Sinauer Associates, Inc. Publishers, Sunderland, Massachusetts, USA.

Primack, R. B. 2010. Essentials of conservation biology. Fifth edition. Sinauer Associates, Inc. Publishers, Sunderland, Massachusetts, USA.

Primack, R. B. 2012. A primer of conservation biology. Fifth edition. Sinauer Associates Inc. Publishers, Sunderland, Massachusetts, USA.

Prokopy, R. J. and B. D. Roitberg. 1984. Foraging behavior of true fruit flies: concepts of foraging can be used to determine how tephritids search for food, mates, and egg-laying sites and to help control these pests. American Scientist 72: 41–49.

Pyšek, P., J. Sádlo, B. Mandák, and V. Jarošík. 2003. Czech alien flora and the historical pattern of its formation: what came first to Central Europe? Oecologia 135: 122–130.

Quammen, D. 2008. Planet of weeds. Pp. 161–188. In: Quammen, D. Natural acts. W. W. Norton & Company, New York and London.

Rahm, E. J., S. A. Griffith, D. B. Noltie, and R. J. DiStefano. 2005. Laboratory agonistic interactions demonstrate failure of an introduced crayfish to dominate two imperiled endemic crayfishes. Crustaceana 78: 437–456.

Randall, J. E. 2001. Surgeonfishes of Hawai'i and the world. Mutual Publishing and Bishop Museum Press, Honolulu, USA.

Rawles, K. 2004. Biological diversity and conservation policy. Pp. 199–216. In: Oksanen, M. and J. Pietarinen (Eds.). Philosophy and biodiversity. Cambridge Studies in Philosophy and Biology. Cambridge University Press, Cambridge, UK.

Ray, W. J. and L. D. Corkum. 1997. Predation of zebra mussels by round gobies, *Neogobius melanostomus*. Environmental Biology of Fishes 50: 267–273.

Renfrew, C. and P. Bahn. 1997. Archaeology. Second Edition. Thames and Hudson Ltd., London, UK.

Ricciardi, A and H. MacIsaac. 2008. The book that began invasion ecology. Nature 452: 34.

Richards, R. A. 2010. The species problem: a philosophical analysis. Cambridge University Press, Cambridge, UK.

Richardson, D. M. 2011. Introduction. In: Richardson, D. M. (Ed.). Fifty years of invasion ecology: the legacy of Charles Elton. Wiley-Blackwell, Oxford, UK.

Richardson, D. M. and P. Pyšek. 2008. Fifty years of invasion ecology – the legacy of Charles Elton. Diversity and Distributions 14: 161–168.

Richardson, D. M., P. Pyšek, M. Rejmánek, M. G. Barbour, F. D. Panetta, and C. J. West. 2000. Naturalization and invasion of alien plants: concepts and definitions. Diversity and Distributions 6: 93–107.

Richardson, D. M., P. Pyšek, D. Simberloff, M. Rejmánek, and A. D. Mader. 2008. Biological invasions – the widening debate: a response to Charles Warren. Progress in Human Geography 32: 295–298.

Richardson, D. M. and A. Ricciardi. 2013. Misleading criticisms of invasion science: a field guide. Diversity and Distributions 19: 1461–1467.

Richman, L. and K. Somers. 2005. Can we use zebra and quagga mussels for biomonitoring contaminants in the Niagara River? Water, Air, and Soil Pollution 167: 155–178.

Ridley, M. 1996. Evolution. Second edition. Blackwell Science, Oxford, UK.

Rielly, K. 2010. The black rat. Pp. 134–145. In: O'Connor, T. and N. Sykes (Eds.). Extinctions and invasions: a social history of British fauna. Windgather Press, Oxbow Books, Oxford, UK.

Rodriguez, L. F. 2006. Can invasive species facilitate native species? Evidence of how, when, and why these impacts occur. Biological Invasions 8: 927–939.

Rorer, W. E. and G. M. Capelli. 1978. Competitive interaction between two Mountain Lake crayfish species with life history notes. Virginia Journal of Science 29: 245–248.

Royer, F. and R. Dickinson. 2004. Weeds of the northern US and Canada. The University of Alberta Press, Edmonton, Canada and Lone Pine Publishing, Renton, Washington, USA.

Ruesink, J. L., I. M. Parker, M. J. Groom, and P. M. Kareiva. 1995. Reducing the risks of nonindigenous species introductions – guilty until proven innocent. BioScience 45: 465–477.

Russell, B. 1984. A history of Western philosophy. Unwin Hyman Limited, London, UK (first published in Great Britain by George Allen & Unwin in 1946).

Russell, J. J. 2004. Lady Bird: a biography of Mrs. Johnson. Taylor Trade Publishing, Lanham, Maryland, USA.

Sandilands, A. 2005. Birds of Ontario: habitat requirements, limiting factors and status. Nonpasserines: waterfowl through cranes. UBC Press, Vancouver, Canada.

Sandilands, A. 2007. Rock pigeon. Pp. 278–279. In: Cadman, M. D., D. A. Sutherland, G. G. Beck, D. Lepage, and A. R. Couturier (Eds.). Atlas of the breeding birds of Ontario 2001–2005. Bird Studies Canada, Environment Canada, Ontario Field Ornithologists, Ontario Ministry of Natural Resources, and Ontario Nature, Toronto, Canada.

Sandilands, A. 2010. Birds of Ontario: habitat requirements, limiting factors and status. Nonpasserines: shorebirds through woodpeckers. UBC Press, Vancouver, Canada.

Sandström, A., M. Andersson, A. Asp, P. Bohman, L. Edsman, F. Engdahl, P. Nyström, M. Stenberg, P. Hertonsson, T. Vrålstad, and W. Granèli. 2014. Population collapses in introduced non-indigenous crayfish. Biological Invasions 16: 1961–1977.

Sanguinetti, A. and R. B. Singer. 2014. Invasive bees promote high reproductive success in Andean orchids. Biological Conservation 175: 10–20.

Sax, D. F., S. D. Gaines, and J. H. Brown. 2002. Species invasions exceed extinctions on islands worldwide: a comparative study of plants and birds. The American Naturalist 160: 766–783.

Schilthuizen, M. 2001. Frogs, flies, and dandelions: speciation – the evolution of new species. Oxford University Press, Oxford, UK.

Schindler, D. W. 2001. The cumulative effects of climate warming and other human stresses on Canadian freshwaters in the new millennium. Canadian Journal of Fisheries and Aquatic Sciences 58: 18–29.

Schlaepfer, M. A., D. F. Sax, and J. D. Olden. 2011. The potential conservation value of non-native species. Conservation Biology 25: 428–437.

Schwartz, M. W. 1997. Defining indigenous species: an introduction. Pp. 7–17. In: Luken, J. O. and J. W. Thieret (Eds.). Assessment and management of plant invasions. Springer-Verlag, New York, USA.

Schwert, D. P. 1979. Description and significance of a fossil earthworm (Oligochaeta: Lumbricidae) cocoon from postglacial sediments in southern Ontario. Canadian Journal of Zoology 57: 1402–1405.

Scott, W. B. and E. J. Crossman. 1979. Freshwater fishes of Canada. Bulletin 184. Fisheries Research Board of Canada, Department of Fisheries and Oceans Scientific Information and Publication Branch, Ottawa, Canada.

Scrivener, L. 2007. Attack of the killer worms – foreign invaders threaten saplings and trilliums. The Toronto Star. April 29, 2007.

Scrivener, L. 2012a. Invasive plants taking over landscape. The Toronto Star. June 17, 2012.

Scrivener, L. 2012b. Blooms off the roads. The Toronto Star. June 29, 2012.

Shackelford, N., R. J. Hobbs, N. E. Heller, L. M. Hallett, and T. R. Seastedt. 2013. Finding a middle-ground: The native/non-native debate. Biological Conservation 158: 55–62.

Shapiro, A. M. 2002. The Californian urban butterfly fauna is dependent on alien plants. Diversity and Distributions 8: 31–40.

Shapiro, J. 2001. Mao's war against nature: politics and the environment in Revolutionary China. Cambridge University Press, Cambridge, UK.

Shea, R. E., H. K. Nelson, L. N. Gillette, J. G. King, and D. K. Weaver. 2002. Restoration of trumpeter swans in North America: a century of progress and challenges. Waterbirds 25 (Special Publication 1): 296–300.

Shine, R. 2010. The ecological impact of invasive cane toads (*Bufo marinus*) in Australia. The Quarterly Review of Biology 85: 253–291.

Shine, R. 2014. A review of ecological interactions between native frogs and invasive cane toads in Australia. Austral Ecology 39: 1–16.

Shine, R. and J. S. Doody. 2011. Invasive species control: understanding the conflicts between researchers and the general community. Frontiers in Ecology and the Environment 9: 400–406.

Shirose, L. and W. Edwards. 2005. Illustrated guide to frogs and toads. Pp. 321–337. In: Winterton, G. (Ed.). The comprehensive bait guide for eastern Canada, the Great Lakes region and northeastern United States: identifying, harvesting and culturing of baitfishes, crayfishes, frogs and leeches. University of Toronto Press, Toronto, Canada.

Shrader-Frechette, K. 2001. Non-indigenous species and ecological explanation. Biology and Philosophy 16: 507–519.

Silvertown, J. 2005. Demons in Eden: the paradox of plant diversity. The University of Chicago Press, Chicago and London.

Simberloff, D. 2000. Foreword to the 2000 edition of *The ecology of invasions by animals and plants* by C. S. Elton. The University of Chicago Press, Chicago and London.

Simberloff, D. 2002. Today Tiritiri Matangi, tomorrow the world! Are we aiming too low in invasives control? Keynote address. Pp.: 4–12. In: Veitch, C. R. and M. N. Clout (Eds.). Turning the tide: the eradication of invasive species. Proceedings of the International Conference on Eradication of Island Invasives. Occasional Paper of the IUCN Species Survival Commission No. 27. IUCN SSC Invasive Species Specialist Group. IUCN, Gland, Switzerland and Cambridge, UK.

Simberloff, D. 2003. Confronting introduced species: a form of xenophobia? Biological Invasions 5: 179–192.

Simberloff, D. 2005. Non-native species *do* threaten the natural environment! Journal of Agricultural and Environmental Ethics 18: 595–607.

Simberloff, D. 2009. Moving beyond strawmen and artificial dichotomies: adaptive management when an endangered species uses an invasive one. Journal of Agricultural and Environmental Ethics 22: 73–80.

Simberloff, D. 2011. Charles Elton: neither founder nor siren, but prophet. Pp. 11–24. In: Richardson, D. M. (Ed.). Fifty years of invasion ecology: the legacy of Charles Elton. Wiley-Blackwell, Oxford, UK.

Simberloff, D. and 140 other authors. 2011. Non-natives: 141 scientists object. Nature 475: 36.

Simberloff, D. and P. Genovesi. 2013. Anthropocene: action makes sense. Nature 502: 624.

Simberloff, D., V. Maris, and J.-L. Martin. 2013. Inaction ≠ caution: response to Larson, Kueffer, and the ZiF Working Group on Ecological Novelty. Trends in Ecology and Evolution 28: 257.

Simberloff, D. and M. Rejmánek (Eds.). 2011. Encyclopedia of biological invasions. University of California Press, Berkeley and Los Angeles, USA.

Simmons, I. G. 2001. An environmental history of Great Britain: from 10,000 years ago to the present. Edinburgh University Press, Edinburgh, UK.

Simon, T. P. and R. F. Thoma. 2006. Native and alien distribution of crayfish assemblages in drowned river mouth coastal wetlands of Lake Michigan. Pp. 375–382. In: Simon, T. P. and P. M. Stewart (Eds.). Coastal wetlands of the Laurentian Great Lakes: health, habitat, and indicators. AuthorHouse, Bloomington, Indiana, USA.

Skelly, D. K. 2007. The ailing invader. Proceedings of the National Academy of Sciences of the United States of America 104: 17561–17562.

Slobodkin, L. B. 2001. The good, the bad and the reified. Evolutionary Ecology Research 3: 1–13.

Smith, G. C., I. S. Henderson, and P. A. Robertson. 2005. A model of ruddy duck *Oxyura jamaicensis* eradication for the UK. Journal of Applied Ecology 42: 546–555.

Smith, K. 1999. COSEWIC status report on the queen snake *Regina septemvittata* in Canada. Committee on the Status of Endangered Wildlife in Canada. Ottawa, Canada.

Sodhi, N. S. 2011. Birds. Pp. 70–74. In: Simberloff, D. and M. Rejmánek (Eds.). Encyclopedia of biological invasions. University of California Press, Berkeley and Los Angeles, USA.

Somaweera, R. and R. Shine. 2012. The (non) impact of invasive cane toads on freshwater crocodiles at Lake Argyle in tropical Australia. Animal Conservation 15: 152–163.

Somers, K. M. and R. H. Green. 1993. Seasonal patterns in trap catches of the crayfish *Cambarus bartoni* and *Orconectes virilis* in six south-central Ontario lakes. Canadian Journal of Zoology 71: 1136–1145.

Somers, K. M. and D. P. M. Stechey. 1986. Variable trappability of crayfish associated with bait type, water temperature and lunar phase. American Midland Naturalist 116: 36–44.

Sorensen, P. W. and P. Bajer. 2011. Carp, common. Pp. 100–104. In: Simberloff, D. and M. Rejmánek (Eds.). Encyclopedia of biological invasions. University of California Press, Berkeley and Los Angeles, USA.

Southwood, R. and J. R. Clarke. 1999. Charles Sutherland Elton. 29 March 1900–1 May 1991. Biographical Memoirs of Fellows of the Royal Society 45: 130–146.

Stamos, D. N. 2007. Darwin and the nature of species. State University of New York Press, Albany, New York, USA.

Stearn, W. T. 1971. Linnaean classification. Pp. 96–101. In: Black, D. (Ed.). 1979. Carl Linnaeus. Travels. Charles Scribner's Sons, New York, USA.

Stewart, K. W. and D. A. Watkinson. 2004. The freshwater fishes of Manitoba. University of Manitoba Press, Winnipeg, Canada.

Stiling, P. 1999. Ecology: theories and applications. Third edition. Prentice Hall, Upper Saddle River, New Jersey, USA.

Stoleson, S. H. and D. M. Finch. 2001. Breeding use of and nesting success in exotic Russian olive in New Mexico. The Wilson Bulletin 113: 452–455.

Stolzenburg, W. 2011. Rat island: predators in paradise and the world's greatest wildlife rescue. Bloomsbury, New York, USA.

Storck, P. L. 2004. Journey to the Ice Age: discovering an ancient world. UBC Press, Vancouver, Canada.

Stringer, C. 2002. Modern human origins: progress and prospects. Philosophical Transactions of the Royal Society of London B 357: 563–579.

Stringer, C. and P. Andrews. 2005. The complete world of human evolution. Thames & Hudson Ltd., London, UK.

Stuart, A. J. 1991. Mammalian extinctions in the late Pleistocene of northern Eurasia and North America. Biological Reviews 66: 453–562.

Sturcke, J. 2009. Amphibious warfare: Australians offered prizes in cane toad cull. The Guardian. March 26, 2009.

Stutchbury, B. 2007. Silence of the songbirds. Walker & Company, New York, USA.

Sykes, N. and J. Curl. 2010. The rabbit. Pp. 116–126. In: O'Connor, T. and N. Sykes (Eds.). Extinctions and invasions: a social history of British fauna. Windgather Press, Oxbow Books, Oxford, UK.

Taylor, C. A. 2000. Systematic studies of the *Orconectes juvenilis* complex (Decapoda: Cambaridae), with descriptions of two new species. Journal of Crustacean Biology 20: 132–152.

Taylor, C. A., G. A. Schuster, J. E. Cooper, R. J. Distefano, A. G. Eversole, P. Hamr, H. H. Hobbs, III, H. W. Robison, C. E. Skelton, and R. F. Thoma. 2007. A reassessment of the conservation status of crayfishes of the United States and Canada after 10+ years of increased awareness. Fisheries 32: 372–389.

Taylor, C. A., M. L. Warren, Jr., J. F. Fitzpatrick, Jr., H. H. Hobbs, III, R. F. Jezerinac, W. L. Pflieger, and H. W. Robison. 1996. Conservation status of crayfishes of the United States and Canada. Fisheries 21: 25–38.

Taylor, R. M., P. Hamr, and A. Karstad. 2005. Illustrated guide to crayfishes. Pp. 223–255. In: Winterton, G. (Ed.). The comprehensive bait guide for eastern Canada, the Great Lakes region and northeastern United States: identifying, harvesting and culturing of baitfishes, crayfishes, frogs and leeches. University of Toronto Press, Toronto, Canada.

Temple, S. A. 2004. Individuals, populations, and communities: the ecology of birds. Pp. 9.1–9.134. In: Podulka, S., R. W. Rohrbaugh, Jr., and R. Bonney (Eds.). Handbook of bird biology. Second edition. Cornell Lab of Ornithology in association with Princeton University Press, Ithaca, New York, USA.

Thacker, P. D. 2004. California butterflies: at home with aliens. BioScience 54: 182–187.

Thacker, R. W., B. A. Hazlett, L. A. Esman, C. P. Stafford, and T. Keller. 1993. Color morphs of the crayfish *Orconectes virilis*. The American Midland Naturalist 129: 182–199.

The Globe and Mail (no author was listed). 2014. New York to revise plan to kill swans. March 1, 2014.

The Telegraph (no author was listed). 2009. Australians unite for mass cane toad cull. March 30, 2009.

Thoma, R. F. and R. F. Jezerinac. 1999. The taxonomic status and zoogeography of *Cambarus bartonii carinirostris* Hay, 1914 (Crustacea: Decapoda: Cambaridae). Proceedings of the Biological Society of Washington 112: 97–105.

Thoma, R. F., R. F. Jezerinac, and T. P. Simon. 2005. *Cambarus* (*Tubericambarus*) *polychromatus* (Decapoda: Cambaridae), a new species of crayfish from the United States. Proceedings of the Biological Society of Washington 118: 326–336.

Thomas, C. D. 2013. The Anthropocene could raise biological diversity. Nature 502: 7.

Thomas, C. D. 2015. Rapid acceleration of plant speciation during the Anthropocene. Trends in Ecology and Evolution 30: 448–455.

Thomas, C. D. and G. Palmer. 2015a. Non-native plants add to the British flora without negative consequences for native diversity. Proceedings of the National Academy of Sciences of the United States of America 112: 4387–4392.

Thomas, C. D. and G. Palmer. 2015b. Reply to Hulme et al.: Cover of non-native species is too low to adversely affect native plant diversity at a national scale. Proceedings of the National Academy of Sciences of the United States of America 112: E2990.

Todd, K. 2012. Sparrow. Reaktion Books Ltd., London, UK.

Torchin, M. E., K. D. Lafferty, A. P. Dobson, V. J. McKenzie, and A. M. Kuris. 2003. Introduced species and their missing parasites. Nature 421: 628–630.

Treberg, M. A. and B. C. Husband. 1999. Relationship between the abundance of *Lythrum salicaria* (purple loosestrife) and plant species richness along the Bar River, Canada. Wetlands 19: 118–125.

Trotter, M. M. and B. McCulloch. 1984. Moas, men, and middens. Pp. 708–727. In: Martin, P. S. and R. G. Klein (Eds.). Quaternary extinctions – a prehistoric revolution. The University of Arizona Press, Tucson, USA.

Tulloch, S. (Ed.). 1995. The Oxford dictionary and thesaurus. Oxford University Press, Oxford, UK.

Urquhart, B. 2010. How great was Churchill? The New York Review of Books LVII: 41–43. August 19, 2010.

Van Driesche, J. and R. Van Driesche. 2000. Nature out of place. Island Press, Washington, USA.

van Leeuwen, J. F. N., C. A. Froyd, W. O. van der Knaap, E. E. Coffey, A. Tye, and K. J. Willis. 2008. Fossil pollen as a guide to conservation in the Galápagos. Science 322: 1206.

van Leeuwen, J. F. N., H. Schäfer, W. O. van der Knaap, T. Rittentour, S. Björck, and B. Ammann. 2005. Native or introduced? Fossil pollen and spores may say. An example from the Azores Islands. NEOBIOTA 6: 27–34.

Veitch, C. R. and M. N. Clout (Eds.). 2002. Turning the tide: the eradication of invasive species. Proceedings of the International Conference on Eradication of Island Invasives. Occasional Paper of the IUCN Species Survival Commission No. 27. IUCN SSC Invasive Species Specialist Group. IUCN, Gland, Switzerland and Cambridge, UK.

Venter, O., N. N. Brodeur, L. Nemiroff, B. Belland, I. J. Dolinsek, and J. W. A. Grant. 2006. Threats to endangered species in Canada. BioScience 56: 903–910.

Vidal, J. 2012. Final 100 ruddy ducks in the UK facing extermination. The Guardian. March 8, 2012.

Virgós, E., S. Cabezas-Díaz, and J. Lozano. 2007. Is the wild rabbit (*Oryctolagus cuniculus*) a threatened species in Spain? Sociological constraints in the conservation of species. Biodiversity and Conservation 16: 3489–3504. DOI 10.1007/s10531-006-9054-5.

Vitule, J. R. S., C. A. Freire, and D. Simberloff. 2009. Introduction of non-native freshwater fish can certainly be bad. Fish and Fisheries 10: 98–108.

Waicheim, A., G. Blasetti, P. Cordero, C. Rauque, and G. Viozzi. 2014. Macroparasites of the invasive fish, *Cyprinus carpio*, in Patagonia, Argentina. Comparative Parasitology 81: 270–275.

Walls, J. G. 2009. Crawfishes of Louisiana. Louisiana State University Press, Baton Rouge, USA.

Ward-Fear, G., G. P. Brown., M. J. Greenlees, and R. Shine. 2009. Maladaptive traits in invasive species: in Australia, cane toads are more vulnerable to predatory ants than are native frogs. Functional Ecology 23: 559–568.

Ward-Fear, G., G. P. Brown, and R. Shine. 2010. Factors affecting the vulnerability of cane toads (*Bufo marinus*) to predation by ants. Biological Journal of the Linnean Society 99: 738–751.

Warren, C. R. 2007. Perspectives on the 'alien' versus 'native' species debate: a critique of concepts, language and practice. Progress in Human Geography 31: 427–446.

Warren, R. J. and M. A. Bradford. 2013. Science petitions are a façade of numbers. Nature 493: 480.

Webb, D. A. 1985. What are the criteria for presuming native status? Watsonia 15: 231–236.

Webb, J. K., G. P. Brown, T. Child, M. J. Greenlees, B. L. Phillips, and R. Shine. 2008. A native dasyurid predator (common planigale, *Planigale maculata*) rapidly learns to avoid a toxic invader. Austral Ecology 33: 821–829.

West, M. J. and A. P. King. 1990. Mozart's starling. American Scientist 78: 93–101.

Wetzel, J. E., W. J. Poly, and J. W. Fetzner, Jr. 2004. Morphological and genetic comparisons of golden crayfish, *Orconectes luteus*, and rusty crayfish, *O. rusticus*, with range corrections in Iowa and Minnesota. Journal of Crustacean Biology 24: 603–617.

Wheeler, Q. and S. Pennak. 2013. What on Earth? – 100 of our planet's most amazing new species. A Plume Book. Penguin Group (USA.) Inc., New York, USA.

Whitehead, P. and C. Keates. 1981. The British Museum (Natural History). Summerfield Press Ltd and Philip Wilson Publishers Ltd, London, UK.

Whitt, M. B., H. H. Prince, and R. R. Cox, Jr. 1999. Avian use of purple loosestrife dominated habitat relative to other vegetation types in a Lake Huron wetland complex. The Wilson Bulletin 111: 105–114.

Wichert, G. A. 1995. Effects of improved sewage effluent management and urbanization on fish associations of Toronto streams. North American Journal of Fisheries Management 15: 440–456.

Wiggins, G. B. 2009. Biological notes on an old farm – exploring common things in the kingdoms of life. Royal Ontario Museum, Toronto, Canada.

Wilcove, D. S., D. Rothstein, J. Dubow, A. Phillips, and E. Losos. 1998. Quantifying threats to imperiled species in the United States. BioScience 48: 607–615.

Wiley, E. O. 1981. Phylogenetics: the theory and practice of phylogenetic systematics. John Wiley & Sons, Inc., New York, USA.

Wilkins, J. S. 2009. Species: a history of the idea. University of California Press, Berkeley, Los Angeles, and London.

Williams, D. D. and T. R. Mason. 1999. Invertebrates. Pp. 151–159. In: Roots, B. I., D. A. Chant, and C. E. Heidenreich (Eds.). Special places – the changing ecosystems of the Toronto region. UBC Press, Vancouver, Canada.

Williams, G. 2011. 100 alien invaders: animals and plants that are changing our world. Bradt Travel Guides Ltd., UK.

Williams, G. 2013. Naturalists at sea: scientific travellers from Dampier to Darwin. Yale University Press, New Haven and London.

Williamson, M. 1996. Biological invasions. Chapman & Hall, London, UK.

Wilson, E. O. 1989. Threats to biodiversity. Scientific American 261: 108–116. September 1989 issue.

Wilson, E. O. 1992. The diversity of life. W. W. Norton & Company, New York and London.

Wilson, E. O. 2013. Letters to a young scientist. Liveright Publishing Corporation, A Division of W. W. Norton & Company, New York and London.

Wilson, J. R. U., E. E. Dormontt, P. J. Prentis, A. J. Lowe, and D. M. Richardson. 2009. Biogeographic concepts define invasion biology. Trends in Ecology and Evolution 24: 586.

Wilson, K. A., E. T. Howell, and D. A. Jackson. 2006. Replacement of zebra mussels by quagga mussels in the Canadian nearshore of Lake Ontario: the importance of substrate, round goby abundance, and upwelling frequency. Journal of Great Lakes Research 32: 11–28.

Wilson, R. A. (Ed.). 1999. Species: new interdisciplinary essays. The MIT Press, Cambridge, Massachusetts, USA.

Winterton, G. 2005. Illustrated guide to baitfish. Pp. 4–93. In: Winterton, G. (Ed.). The comprehensive bait guide for eastern Canada, the Great Lakes region and north-eastern United States: identifying, harvesting and culturing of baitfishes, crayfishes, frogs and leeches. University of Toronto Press, Toronto, Canada.

Wonham, M. 2006. Species invasions. Pp. 293–331. In: Groom, M. J., G. K. Meffe, and C. R. Carroll (Eds.). Principles of conservation biology. Third edition. Sinauer Associates, Inc. Publishers, Sunderland, USA.

Worthy, T. H. and R. N. Holdaway. 2002. The lost world of the moa. Indiana University Press, Bloomington and Indianapolis, USA.

Wouters, C. E. (Ed.). 2010. Protecting the Great Lakes from invasive and nonindigenous species. Nova Science Publishers, Inc., New York, USA.

WWF Living Planet Report 2014: Species and spaces, people and places. 2014. WWF International, Gland, Switzerland.

Young, A. M. and B. M. H. Larson. 2011. Clarifying debates in invasion biology: a survey of invasion biologists. Environmental Research 111: 893–898.

Zavaleta, E. S., R. J. Hobbs, and H. A. Mooney. 2001. Viewing invasive species removal in a whole-ecosystem context. Trends in Ecology and Evolution 16: 454–459.

Ziegler, A. C. 2002. Hawaiian natural history, ecology and evolution. University of Hawai'i Press, Honolulu, USA.

Zimmer, C. 2001. Evolution – the triumph of an idea. HarperCollins Publishers, Inc., New York, USA.

Zimmer, C. 2008. What is a species? Scientific American 298: 72–79. June 2008 issue.

Zimmer, C. and D. J. Emlen. 2013. Evolution: making sense of life. Roberts and Company Publishers, Inc., Greenwood Village, Colorado, USA.

Zonn, I. S., M. H. Glantz, A. G. Kostianoy, and A. N. Kosarev. 2009. The Aral Sea encyclopedia. Springer-Verlag, Berlin and Heidelberg.

Index